Springer-Verlag Berlin Heidelberg GmbH

The Molecular Repertoire of Adenoviruses II

Molecular Biology of Virus-Cell Interactions

Edited by W. Doerfler and P. Böhm

With 36 Figures

Springer

Walter Doerfler
Petra Böhm
Institut für Genetik
der Universität zu Köln
Weyertal 121
50931 Köln
Germany

Cover illustration: Integration of Ad12 DNA into one of the hamster chromosomes in the Ad12-transformed cell line T637 as demonstrated by the fluorescent in situ hybridization technique. The hamster cell line T637 contains in an integrated form about 20–26 copies of Ad12 DNA. The integrated Ad12 genomes were visualized by using the biotinylated PstI-D fragment of Ad12 DNA as the hybridization probe and fluorescent-labeled avidin for detection under UV light. (This illustration was contributed by Petra Wilgenbus and Walter Doerfler, Cologne.)

Cover design: Künkel+Lopka, Ilvesheim

ISSN 0070-217X
ISBN 978-3-642-79501-5 **ISBN 978-3-642-79499-5 (eBook)**
DOI 10.1007/978-3-642-79499-5

Softcover reprint of the hardcover 1st edititon 1995
Library of Congress Catalog Card Number 15-12910

Typesetting: Thomson Press (India) Ltd, Madras
SPIN: 10483983 27/3020/SPS – 5 4 3 2 1 0 – Printed on acid-free paper.

Preface

When asked whether they wanted to contribute a chapter to these three volumes on the *Molecular Repertoire of Adenoviruses* almost all of my colleagues working in the field responded favorably and supplied fine chapters on their areas of expertise.

In 1995, adenoviruses continue to provide a major and highly developed experimental system for biochemical, cell biological, genetic, virological, and epidemiological investigations in mammalian molecular biology. As a considerable body of information has become available thanks to the continued efforts of many, this virus system has become particularly useful to those who want to address details of biological mechanisms and their relation to structure and function.

We have tried to cover as wide a field as possible in current adenovirus research and to encompass the entire gamut of adenovirology and adenovirus molecular biology. In spite of the tendencies that come from many corners to seduce researchers into directing their efforts towards applied molecular biology, which of course has its place and merits, there are still sturdy groups who pursue their interests in basic molecular biology and in particular in adenovirus research, an "eternal archetype," as one of the experts chose to name the virus.

In the first of the three volumes, we present an overview of adenovirus research and go on to cover the topics of the structure and assembly of adenoviruses, viral infection, and viral gene products. In the following two volumes, we turn our attention to topics such as DNA replication, recombination and integration, post-transcriptional control, transformation and E1A, adenovirus genetics, pathogenesis, and gene therapy.

The three volumes appearing now have also been initiated to mark with some, perhaps unavoidable, delay the occasion of the 40th anniversary of the discovery of adenoviruses by W.P. Rowe and R.J. Huebner and, independently, by M.R. Hilleman and J.H. Werner in 1953/1954.

I was encouraged to undertake the task of editing the current three volumes on adenovirus molecular biology by visits to many libraries in different countries on several continents during the last decade and by the observation that the three volumes on *The Molecular Biology of Adenoviruses*, which I had edited for Springer-Verlag's series *Current Topics in Microbiology and Immunology* in 1983 and 1984, had apparently been diligently worked through by many readers. The craftsmanship of the book covers had barely resisted the intellectual assault and showed that the books had been consulted with curiosity.

There are few viral genomes that have not been developed into useful vector systems. An increasing number of research reports have been devoted to adenoviruses as potential vectors for human somatic gene therapy. Should that virus system indeed prove its worth for this important application, researchers will have to appreciate the subtleties of the biology and the molecular biology of this virus system. As someone who has not been directly involved in research on gene therapy, it appears to me that considerable efforts will have to be expended before all the problems concerning the therapeutic application of the adenovirus system or, more likely, of even better systems will be solved. The importance of this goal justifies and will require a great deal of very active research.

We wish to thank all the contributors, the editors of the *Current Topics* series, and in particular Springer-Verlag for their help in getting these three volumes to press.

Cologne

WALTER DOERFLER
PETRA BÖHM

List of Contents

List of Contents of Companion Volume 199/I

List of Contents of Companion Volume 199/III

List of Contributors

(Their addresses can be found at the beginning of their respective chapters.)

Adenovirus DNA Replication

P.C. VAN DER VLIET

1 Introduction

Studies on the replication of adenovirus DNA were initiated more than two decades ago and quickly led to a novel displacement model for DNA replication (SUSSENBACH et al. 1972). These studies were mainly performed using intact infected cells or isolated nuclei. It was only after the development of a system to study replication in vitro (CHALLBERG and KELLY 1979) that detailed information could be obtained about the protein-priming mechanism for initiation and about the replication proteins. The last decade has been characterized by the discovery of transcription factors as participants in initiation (NAGATA et al. 1982; PRUIJN et al.

Laboratory for Physiological Chemistry, University of Utrecht, 3508 TA Utrecht, The Netherlands

1986), by complete reconstitution of the system with purified recombinant proteins, and by structural information on some of the replication proteins. After the previous review in this series (SUSSENBACH and VAN DER VLIET 1983), several reviews on adenovirus replication have appeared (KELLY1984; CAMPBELL 1986; VAN DER VLIET et al. 1988; CHALLBERG KELLY 1989; STILLMAN 1989; HAY and RUSSELL 1989; VAN DER VLIET 1990, 1991; SALAS 1991; DE PAMPHILIS 1993a).

Here, I will give an overview of the latest developments regarding the replication in vitro and the role of the various replication proteins and present a model for the mechanisms of replication integrating the latest results.

Initiation of replication of the 36-kb adenovirus replicon requires origin sequences located in the inverted terminal repeats. The length of these repeats varies between serotypes and is 102 bp long in the adenovirus type 2 (Ad2)/Ad5 serotypes that will be discussed mainly in this review. The sequence (STEENBERGH et al. 1975) shows the presence of a region between positions 9 and 18 conserved in all serotypes (TOULIN et al. 1979; STILLMAN et al. 1982a) that constitutes the core origin (see Fig. 1). Two other blocks of sequences are mainly conserved and are located in the auxiliary region of the origin, which enhances replication up to 200-fold both in vivo and in vitro (ROSENFELD et al. 1987; WIDES et al. 1987; CHALLBERG and RAWLINS 1984; GUGGENHEIMERR et al. 1984; HAY 1985; LALLY et al. 1984). These regions bind the cellular transcription factors nuclear factor I (NFI) and NFIII/Oct-1 (NAGATA et al. 1982; PRUIJN et al. 1986) (Fig. 1). The total length of the origin is approximately 50 bp. Another important component is the 55-kDa terminal protein (ROBINSON et al. 1973) which is bound to the 5'-dCMP (deoxycytidine monophosphate) residue of both strands via a phosphodiester bond with a serine residue (REKOSH et al. 1977; CHALLBERG et al. 1980). This parental terminal protein serves several functions during replication (see Sect. 4).

The general picture that has emerged from the various studies can be shortly characterized as follows. Replication requires three essential viral proteins encoded by the E2 transcription unit. These are the precursor terminal protein (pTP), the DNA polymerase (pol), and the DNA-binding protein (DBP). The first two are

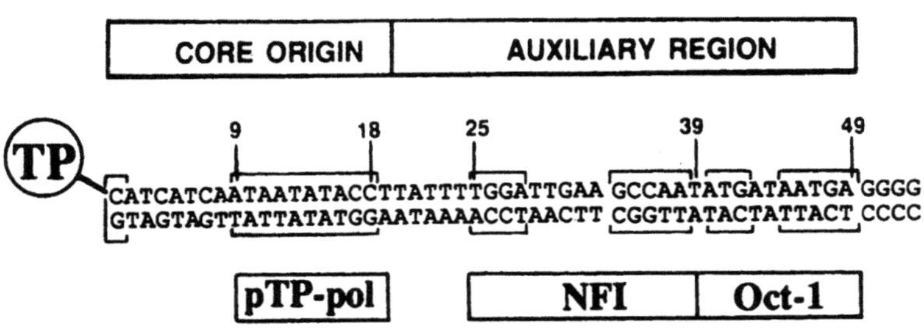

Fig. 1. The adenovirus type 2/5 origin of DNA replication. The terminal protein (*TP*) is covalently attached to the 5' dCMP (deoxycytidine monophosphate) residue through Ser-580. Regions highly conserved in various serotypes are indicated in *brackets*. Binding sites for precursor TP–DNA polymerase (*pTP-pol*) in the core origin and for nuclear factor I (*NFI*) and NFIII/Oct-1 in the auxiliary region are given

present in infected cells as a heterodimer (pTP–pol). These three proteins, together with the core origin, can sustain a low level of replication. Initiation is the rate-limiting step and can be enhanced by NFI and NFIII/Oct-1, which form a stabilized nucleoprotein structure at the origin together with the three viral proteins. Finally, a topoisomerase (NFII) is required for elongation of the total replicon. A summary of the properties of these proteins is presented in Table 1. Initiation occurs by a protein-priming mechanism in which pTP binds the first nucleotide of the nascent strand, a dCMP residue, covalently through a

Table 1. Properties of viral and cellular DNA replication proteins (adenovirus type 5)

	Molecular mass (kDa)	Function	
		Initiation	Elongation
Viral proteins			
Precursor Terminal Protein (pTP)	80	Primer bound to pol Binds core origin	
DNA polymerase (pol)	140	Binds to core origin (+pTP)Couples dCMP to pTP	Polymerizes by displacement (+DBP)
DNA-binding protein (DBP)	59 (72[a])	Reduces K_m for dCTP Enhances binding of NFI	Protects ssDNA Unwinds duplex DNA in fork Enhances processivity and rate of polymerization Changes sensitivity of pol to inhibitors
Template-bound terminal protein	55	Protects 5' ends Enhances template activity Enhances pTP–pol binding Changes origin structure	
Cellular Proteins			
Nuclear factor I (NFI)	55	Binds auxiliary origin Binds pTP–pol Positions pTP–pol to origin Stabilizes preinitiation complex	
Nuclear factor II (NFII)	30		Permits synthesis of complete DNA (topoisomerase I)
Nuclear factor III (NFIII/Oct-1)	90	Binds auxiliary origin Bends DNA Binds pTP–pol weakly	

dCMP, deoxycytidine monophosphate; dCTP, deoxycytidine triphosphate; ss, single-stranded.
[a] Apparent molecular mass in sodium dodecyl sulfate (SDS) polyacrylamide gels.

Ser–dCMP phosphodiester bond. The catalytic center for this reaction is present in the polymerase and coincides with the region required for elongation. The reaction velocity is enhanced by NFI and NFIII/Oct-1 by mechanisms to be described later. DBP also enhances initiation by changing, among others, the K_m of the reaction. Following the pTP–dCMP formation, elongation starts by a displacement mechanism. Elongation is rapid (20–30 bp/s) and processive and requires only DBP and the polymerase. Both initiation and elongation are independent of adenosine triphosphate (ATP). Unwinding is presumably caused by DBP, which has helix-destabilizing properties. Finally, the displaced strand can duplicate either by formation of a panhandle structure which regenerates an origin or by intermolecular renaturation of strands of opposite polarity originating from the use of origins at both molecular ends. A model is depicted in Fig. 2. Below, we will discuss the properties of the various replication proteins and give further details on the mechanism of replication.

2 Viral Replication Proteins

2.1 Precursor Terminal Protein–DNA Polymerase Complex

The main actors in adenovirus DNA replication are pTP and pol. The 80-kDa pTP and the 140-kDa pol are products of the E2B gene and form a stable heterodimer (Stillman et al. 1982b). They are transported together to the nucleus of infected cells employing the nuclear location signal for pTP (Zhao and Padmanabhan 1988) and attach to the nuclear matrix (Fredman and Engler 1993).

The amount of the pTP–pol complex in infected cells is very low compared to the other E2 product, DBP, despite the use of the same promoter. This has hampered investigations of these proteins considerably. Early attempts to

Fig. 2. General outline of the first round of adenovirus (*AD*) DNA replication. The parental terminal protein (*TP*) containing template DNA (*a*) forms a multiprotein–DNA complex with DNA-binding protein (*DBP*; *b*) A preinitiation complex (*c*) is assembled with the various replication proteins (not on scale). Initiation occurs by covalent coupling of a deoxycytidine monophosphate (dCMP) residue to precursor TP (*pTP*). Details are given in Fig. 5. In the presence of deoxynucleoside triphosphate (*dNTP*), elongation starts by a displacement mechanism, minimally requiring DNA polymerase (*pol*) and DBP. Here, we have drawn a molecule which started replication at both origins simulataneously (*d*). Replication can proceed from both sides, leading to partially single-stranded replication intermediates (*e*) and finally duplex daughter strands containing TP at one end and pTP at the other end (*f*). Alternatively, if replication started at either end in different molecules, single-stranded displaced strands of opposite polarities are formed (*g*). These can renature, a process enhanced by DBP, to form duplex DNA with two TP molecules (*h*). Another possibility is that intrastrand renaturation of the inverted terminal repeat occurs, leading to a panhandle structure (*i*). The regenerated double-stranded origin can be used again for protein-primed initiation, leading to a partially duplex intermediate (*j*) and finally a daughter molecule (*f*). pTP in these daughter molecules is processed later in infection by the Ad protease. It should be stressed that later in infection daughter molecules will be effectively used as templates for second and further rounds. This will lead to a considerable increase in daughter molecules containing pTP at both ends, while keeping the number of TP-containing molecules constant at the level of the input DNA

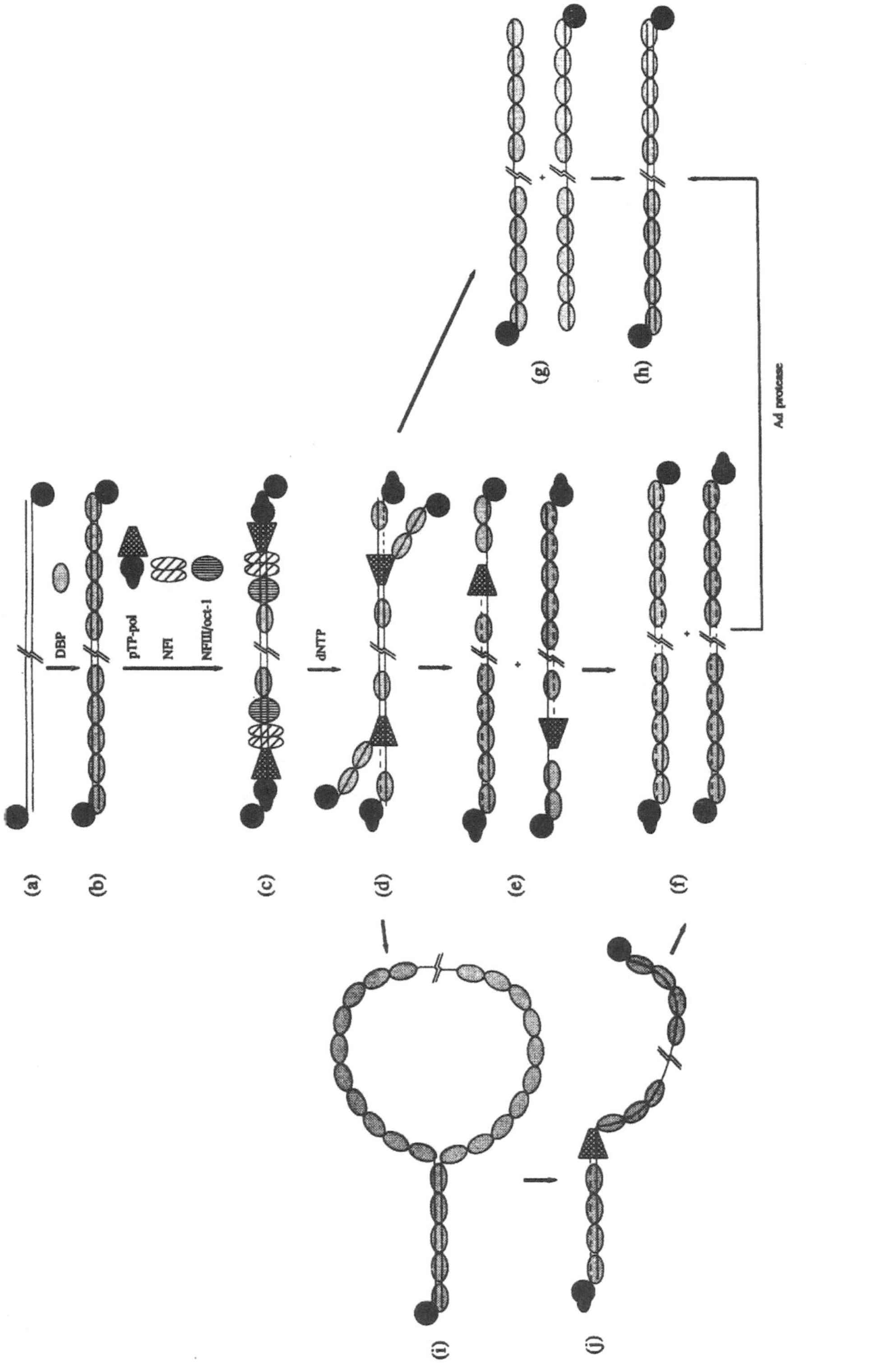
(a)
(b)
DBP
pTP-pol
NFI
NFIII/oct-1
(c)
dNTP
(d)
(e)
(f)
(g)
(h)
Ad protease
(i)
(j)

overproduce the polymerase in *Escherichia coli* were unsuccessful and gave only insoluble and inactive products. Only fusion proteins could be isolated and used for the preparation of antibodies (REKOSH et al. 1985; FRIEFELD et al. 1985). Another reason for these failures was that the mRNA for pol and pTP had not been characterized due to their low abundance, and therefore the actual open reading frame was not known.

Successful expression of functional proteins was finally achieved using various eukaryotic systems: first infection of HeLa cells with recombinant vaccinia virus (STUNNENBERG et al. 1988) or transfection of COS cells (PETTIT et al.1988; SHU et al. 1987; ZHAO and PADMANABHAN 1988) and later baculovirus (WATSON and HAY 1990; ZHAO et al. 1991; BOSHER et al. 1990). In all these systems three N-terminal amino acids were included encoded by a small exon around map unit 39 which contained the initiation codon. Apparently in earlier attempts the wrong start codon had been used for expression. Nowadays functional pTP and pol can be produced in large amounts, up to 5 mg/l infected HeLa cells using a T7-based recombinant vaccinia virus system (A. KING and P.C. VAN DER VLIET, unpublished work; NAKANO et al. 1991), and they heterodimerize spontaneously (STUNNENBERG et al. 1988; TEMPERLEY and HAY 1992).

Overexpression provided the opportunity to mutagenize pTP and pol and study functional domains. The adenovirus pol contains five out of six regions that are conserved in all α-like DNA polymerases. Moreover, it has several conserved amino acids flanking region III in common with DNA polymerases that are involved in protein priming and which form a separate subgroup (T.S.-F. WANG et al. 1989).

So far, it has not been possible to define distinctive domains involved in initiation or elongation, indicating that these functions might be closely linked. In most of the mutants constructed, several functions are lost simultaneously (CHEN and HORWITZ 1989; JOLING et al. 1991; JOUNG and ENGLER 1992; ROOVERS et al. 1991, 1993). An exception may be a region containing several Cys and His residues around amino acids 250 (Cys-His I). Mutations in this region discriminate between basal pol activity (which is retained) and replication activity (which is destroyed) (JOUNG and ENGLER 1992). Hopefully, further mutagenesis studies will reveal more functional domains.

For the pTP it has been equally difficult to identify specific regions required for the various functions of pTP such as priming activity or interaction with the pol. Single mutations scattered over the entire protein seem to eliminate most functions simultaneously, possible by changing the overall structure of pTP (FREIMUTH and GINSBERG 1986; FREDMAN et al. 1991; ROOVERS et al. 1991). The N terminus of pTP is essential for priming activity and DNA binding (PETTIT et al. 1989, R.T. HAY, personal communication). The C terminus is involved in matrix attachment (FREDMAN and ENGLER 1993) and contains the essential amino acid Ser-580 (Ad5) which binds the first dCMP residue during initiation.

The enzymatic properties of DNA polymerase have only been investigated to a limited extent, mainly employing the native pTP–pol complex isolated from Ad5-infected cells rather than the overexpressed, purified proteins. When assayed with activated DNA as template, the polymerase is sensitive to dideoxynucleoside triphosphate (ddNTP) and almost resistant to aphidicolin, which is remarkable in view of its similarity to α-like DNA polymerases. Initiation is also completely resistant to aphidicolin, but the elongation process itself is moderately sensitive to aphidicolin (SUSSENBACH and VAN DER VLIET 1983; NAGATA et al. 1983). A similar result was obtained using another inhibitor, the dATP analogue (S)-HPMPApp (MUL et al. 1989). Whereas on synthetic templates polymerization was resistant to the drug, viral DNA replication in a reconstituted system was considerably inhibited, competitive with dATP. This suggests that, in the authentic replication process, the conformation of the active site of the polymerase, or the template, is slightly different compared to synthetic templates. Presumably this is due to the presence of DBP, since DBP enhanced the sensitivity to (S)-HPMPApp on synthetic templates considerably (MUL et al. 1989). Thus, DBP modifies the sensitivity of the polymerase to at least some inhibitors. Interestingly, DBP also inhibits the 3'→5' exonuclease activity intrinsic to the polymerase and presumably involved in proofreading (LINDENBAUM et al. 1986). Another factor that influences the activity of the pol is phosphorylation of Ser-67, which may be important for initiation (RAMACHANDRA et al. 1993). Finally, pTP can inhibit the polymerase activity on synthetic templates (FIELD et al. 1984).

How does the pTP–pol complex recognize the origin and initiate replication? Direct binding studies have been performed using band shift assays and DNase I footprints (MULand VAN DER VLIET1992; TEMPERLEY and HAY1992) Both proteins can recognize the origin separately, but the pTP–pol heterodimer does so with enhanced specificity, protecting the conserved bases 8–17 from DNase I cleavage (TEMPERLEY and HAY 1992). Nevertheless, the binding specificity and affinity is not very high and at increased protein concentrations other sequences in the origin are also bound. Also single-stranded origin DNA is bound strongly (KENNY and HURWITZ 1988), indicating that the sequence recognition is not very strong. This view is supported by mutagenesis studies, indicating that, except for positions 17 and 18, single point mutations in the core origin have only a limited effect on initiation, at most fourfold (TEMPERLEY et al. 1991). A strong, site-specific recognition of DNA is not very likely in view of the role of the polymerase, which must proceed along DNA during elongation without sequence-induced roadblocks. The specificity of origin recognition as well as the binding affinity of pTP–pol is enhanced by NFI (MUL and VAN DER VLIET 1992) as well as by the parental terminal protein (TP) (PRONK and VAN DER VLIET 1993). The binding of pTP–pol to the origin induces specific changes in the origin that can be monitored by DNase I or chemical probing with $KMnO_4$ or OsO_4 (MUL and VAN DER VLIET 1992; R.T. HAY, personal communication). These changes are enhanced by NFI and DBP. Thus, origin recognition occurs apparently through the formation of a multiprotein complex consisting of at least five and possibly six polypeptides since NFIII/Oct-1 also binds pTP–pol (see Sect. 3.2.1). This is still a limited number compared to the requirements for DNA replication in prokaryotic systems or SV40. The various protein interfaces involved in protein–protein and protein–DNA recognition have not yet been well defined.

2.2 DNA-Binding Protein

Due to its abundance in infected cells, the first nonvirion viral protein to be discovered was a 72-kDa protein designated DBP (VAN DER VLIET and LEVINE 1973). DBP was found in a search for infected cell-specific proteins that could bind to the displaced single-stranded (ss) DNA originating from replication. Indeed, the protein has a high affinity for ss DNA, but later turned out to recognize double-stranded (ds) DNA and RNA as well.

Fortunately, one of a limited number of temperature-sensitive mutants that are defective for DNA replication, H5*ts*125 (ENSINGER and GINSBERG 1972), appeared to produce a temperature-sensitive DBP (VAN DER VLIET et al. 1975). This was evidence for a direct link between DNA replication and DBP and provided an opportunity to study the potential role of DBP in other aspects of the infection cycle as well. A subsequent extensive investigation of the phenotype of H5*ts*125 revealed that DBP is a multifunctional protein also involved in transcriptional control, transformation, and virus assembly (CHASE and WILLIAMS 1986; VAN DER VLIET et al. 1988). Here I will describe mainly the role of DBP in DNA replication.

2.2.1 Structure

The Ad5 DBP contains 529 amino acids (relative molecular weight, 59 049). Other serotypes encode a DBP of similar size (KRUIJER et al. 1981). Based upon sequence comparisons and also on direct proteolytic digestion (TSERNOGLOU et al. 1985), two domains can be distinguished in DBP. The N-terminal domain encompassing about one third of the molecule (1–173 in Ad5) is not well conserved among the different serotypes. It is heavily phosphorylated and contains two short sequences that constitute the nuclear location signal (MORIN et al. 1989). Mutants in this region display a modified host range by interfering with the splicing of late mRNA (ANDERSON and KLESSIG 1984). The N-terminal domain is dispensable for DNA replication.

The C-terminal domain (174–529) is well conserved and contains several stretches of highly conserved amino acids (CR 1–3) (KITCHINGMAN 1985). This domain contains the nucleic acid-binding properties and is able to stimulate DNA replication as efficiently as the intact protein. The prototype DBP mutant H5*ts*125 is located in this domain (Pro-413→Ser) as well as several other mutants affecting DNA replication, notably around positions 280–282.

Due to its abundance in infected cells, Ad5 DBP could be isolated in large quantities (0.2 mg/l infected cells) and used for studies of its three-dimensional structure. Crystals could be obtained from the C-terminal region (TSERNOGLOU et al. 1984), but not from the intact protein. Recently, the crystal structure was solved, providing for the first time information on the structure of a eukaryotic single-stranded binding protein (SSB)-type protein (TUCKER et al. 1994). DBP contains two zinc atoms in different, novel coordinations. The first one is coordinated by four cysteine residues that are all conserved, whereas the second one is coordinated by three cysteine residues and one histidine residue, again all conserved residues. Both zinc atoms have a structural role rather than being directly involved in the interaction with DNA. The need for zinc was already anticipated from the observations that Zn^{2+} is required during in vitro synthesis of DBP to obtain a functional DBP (Vos et al. 1988) and by ^{65}Zn binding to DBP (EAGLE and KLESSIG 1992).

A second, remarkable property is the presence of a 17-amino acid extension at the C terminus that hooks onto a hydrophobic pocket in a second molecule, thereby forming a protein chain. Deletion of this C-terminal "hook" destroys the cooperativity in ssDNA binding, which contributes to the overall affinity of DBP for ssDNA (KUIL et al. 1989).

2.2.2 DNA-Binding Properties

The binding of DBP to ssDNA has been studied using a number of techniques including sedimentation analysis, electron microscopy (EM), circular dichroism, and optical density measurements. The DBP binding site consists of 10–15 nucleotides, depending upon the type of DNA studied, and the complex appears to have a regular structure in which the DNA is extended and the bases are tilted (VAN AMERONGEN et al. 1987). Binding is almost completely sequence independent. The binding constant for poly rA is approximately $5 \times 10^5 M^{-1}$ at 50 mM NaCl, and the cooperativity constant ω is 20–30 (KUIL et al. 1989).

How does ssDNA bind to DBP? Cocrystals have not yet been obtained, but model building indicates that ssDNA winds around the protein chain, touching a band of positive electrostatic potential. This leads to a configuration in which the DNA has an extended structure with the bases unstacked, as predicted by circular dichroism measurements (VAN AMERONGEN et al. 1987). In this model the DNA is wound around the protein in an irregular, right-handed fashion with a pitch of 76 Å and 26 bases per turn (13 bases per protein), making the overall length of the DNA slightly shorter (14%) than dsDNA. This agrees well with previous EM measurements of the DBP–ssDNA complex (VAN DER VLIET et al. 1978). The complex is also much more rigid than protein-free DNA. Although the amino acids interacting with the DNA in this model are not exactly known, it is interesting that several residues in the proposed DNA–protein interface coincide with mutations that destroy DNA binding or areas that are proposed to be involved in DNA binding based upon cross-linking studies (CLEGHON and KLESSIG 1992).

DBP binds also to dsDNA and forms a regular multimeric protein–DNA complex (STUIVER and VAN DER VLIET 1990). In contrast to ssDNA, binding to dsDNA is hardly cooperative. DNA and protein associate readily, but DBP rapidly dissociates from dsDNA, whereas the ssDNA–DBP complex is much more stable. EM measurements show thick filament-like and beaded structures in which the length of the dsDNA is not significantly altered, again in contrast to the ssDNA–DBP complex. Although the exact structure is not known, cryoelectron micrographs suggest the presence of interwound fibers, possibly consisting of two DBP chains, around the DNA (STUIVER et al. 1992). DBP induces distinct changes in the circular dichroism (CD) spectrum of DNA indicative of structural changes.

Such changes are also apparent from the hydroxyl radical breakdown pattern, which is much more regular with DBP than without DBP. These subtle changes in the dsDNA structure may be instrumental in the enhanced binding of transcription factors (see below). No bending or twisting of DNA by DBP was observed. Thus, in conclusion DBP forms complexes with both dsDNA and ssDNA which are characterized by changes in the DNA structure which may be important for its role in DNA replication.

2.2.3 Functions of DNA-Binding Protein in DNA Replication

Based on its DNA-binding properties, a role in the elongation process seems likely and could indeed been shown, but DBP also influences initiation in several ways.

During elongation, DBP enhances the processivity and rate of polymerization considerably, in particular on ssDNA templates (LINDENBAUM et al. 1986). Moreover, it changes the sensitivity of the polymerase to inhibitors such as the dAMP analogue (S)-HPMPA (MUL et al. 1989). These properties may be explained by a direct interaction between DBP and the DNa polymerase but the evidence for this is circumstantial. Although DBP protects the DNA polymerase from thermal inactivation (LINDENBAUM et al. 1986), the possibility cannot be excluded that this is caused by the presence of trace amounts of DNA. Therefore, it is equally possible that DBP exerts its effects in an indirect way through the changes in DNA structure which it induces. A changed DNA structure may lead to different presentation of the template to the DNA polymerase and thus to a change in the kinetic properties during elongation.

Another way in which DBP may influence DNA chain elongation is by facilitating template unwinding. Although it was originally thought that DBP was not able to unwind dsDNA, recent experiments (GEORGAKI et al. 1992; ZIJDERVELD and VAN DER VLIET 1994; MONAGHAN et al. 1994) indicate that DBP is capable of helix destabilization in the replication fork. DBP can displace, in a concentration-dependent fashion and independent of ATP, at least 200 nucleotides annealed to ssDNA. Also unwinding of short, fully duplex DNA is facilitated by DBP, but unwinding of longer fragments is only possible if single-stranded protruding ends are present (ZIJDERVELD and VAN DER VLIET 1994). These results suggest that DBP, through formation of a protein chain at the displaced strand, may destabilize duplex DNA ahead of the replication fork, thus assisting in strand displacement during elongation.

A role for DBP in initiation has been debated for a long time, but has now been firmly established employing a reconstituted system and purified components. DBP stimulates initiation in at least two different ways. First, it enhances the binding of NFI (CLEAT and HAY 1989; STUIVER and VAN DER VLIET 1990; BOSHER et al. 1991). This leads to a moderate increase in V_{max} at suboptimal NFI concentrations (MUL and VAN DER VLIET 1993). This effect is specific for NFI and no increase in binding of NFIII/Oct-1 was observed despite its close proximity. Second, a much larger stimulation was caused by DBP independent of NFI. This was based on a reduction of the K_m for dCTP (MUL and VAN DER VLIET 1993). The mechanism of this reduction is not known, but might occur via enhanced unwinding of the origin by DBP, thereby improving the presentation of the dG residue in the template strand. Alternatively, DBP might induce a better fit for dCTP in the active site of the polymerase or change the active site of pTP through DNA structural changes. These two possibilities are not mutually exclusive and both will result in a higher activity of the initiation complex.

Finally, DBP also has an effect on the later stages of replication, after completion of one round of displacement synthesis. This assumption is based on the obervation that DBP enhances intermolecular renaturation considerably compared to protein-free DNA (ZIJDERVELD et al. 1993). This could lead to enhanced formation of double-stranded daughter molecules from displaced strands of opposite polarity, originating from initiation at different molecular ends (see Fig. 2). The mechanism of enhancement is likely related to the structure of the ssDNA–DBP complex and might be due to shielding of electrostatic repulsion of the two DNA strands by DBP or to the removal of secondary structures from ssDNA.

Interestingly, intramolecular renaturation between two complementary strands of a single-stranded molecule was severely inhibited by DBP. The most likely reason for this is that the rigid structure of the protein chain to which the DNA is bound increases the persistence length. Thus, molecular ends cannot be brought together as easily as in protein-free DNA. The implications for DNA replication are that panhandle formation is inhibited by DBP. This is somewhat surprising, since such panhandle formation can occur in vivo, as suggested by the repair of partially deleted inverted repeats (STOW 1981). This, as well as later results with minichromosomes (HAY et al. 1984; K. WANG and PEARSON 1985), shows that panhandle formation is an essential step for correction of damage in inverted terminal repeats. However, this may occur infrequently and with a low efficiency in the presence of DBP. The main pathway for duplex formation from displaced strands may well be renaturation of displaced strands rather than panhandle formation.

An overview of the DNA-binding properties of DBP is presented below:

1. It binds ssDNA cooperatively.
2. It binds dsDNA noncooperatively and rapidly dissociates from dsDNA.
3. It enhances intermolecular renaturation.
4. It inhibits intramolecular renaturation between complementary ends of ssDNA.
5. It unwinds partially duplex DNA in an ATP-independent fashion.

3 Cellular Replication Proteins

3.1 Nuclear Factor I

By straightforward protein purification guided by enhancement of DNA replication, the first cellular protein involved in initiation of viral replication was isolated and designated NFI (NAGATA et al. 1982) NFI was originally purified from HeLa cells as a 47–kDa protein, but later appeared to consist of a family of related proteins of

52–66 kDa originating from differential splicing (SANTORO et al. 1988; PAONESSA et al. 1988; MEISTERERNST et al. 1989). NFI binds specifically to a region in the origin between nucleotides 25 and 40. A detailed contact point analysis (DE VRIES et al. 1987) showed that NFI makes base-specific and backbone contacts in two major grooves of the DNA located one helical turn apart. Thus, the protein binds at one side of the helix, presumably as a dimer in agreement with the twofold rotational symmetry of the binding site (see Fig. 1). Indeed, subsequent protein–protein interaction studies confirmed the dimeric character (GOUNARI et al. 1990). Interestingly, the NFI consensus binding site is present in many different promoters and enhancers of cellular and viral genes. Several of these cellular binding sites could functionally substitute for the viral NFI sites (ADHYA et al. 1986). Since the NFI recognition sequence has similarities with the CCAAT box and since NFI binds to monomeric CCAAT sites, the protein was also called CTF (CCAAT binding factor; K.A. JONES et al. 1987). Binding to the CCAAT site, however, is very weak (ZORBAS et al. 1992). Cloning of NFI/CTF cDNA from human, rat, chicken, and porcine origin showed that the various translation products differed in their C-'termini, but contained a conserved N-terminal domain. This domain, stretching from amino acids 1 to approximately 220, contains the DNA-binding properties and is therefore called the NFI-binding domain (NFI-BD; MERMOD et al. 1989; GOUNARI et al. 1990). The dimerization domain is also located in this region, although it does not coincide exactly with the DNA-binding domain. The regions interacting with DNA have not been determined yet. Binding requires four conserved cysteine residues (NOVAK et al. 1992), but none of the DNA-binding motifs found in other eukaryotic transcription factors, such as helix–turn–helix, Zn-finger, b-zip, or b-HLH, is present. However, there is a putative α-helical domain located between residues 37 and 65 with basic amino acid residues spaced at intervals of seven which has been suggested to be involved in DNA recognition (MEISTERERNST et al. 1989).

When MERMOD et al. (1989) and GOUNARI et al. (1990) investigated the domains required for the stimulation of adenovirus DNA replication, they found to their surprise that NFI-BD was sufficient. This was in contrast with the requirement for a proline-rich transactivation domain for transcription enhancement. It also contrasts with the need for transcription activation domains for enhancement of papovavirus DNA replication by several other transcription factors (DE PAMPHILIS, 1993b). So far, such a need for just the DNA-binding domain, which also holds for activation by NFIII/ Oct-1 (see below) appears to be unique for adenoviruses. Possibly the virus selected conserved regions present in families of common transcription factors for optimal replication enhancement in order to enlarge its range of infection of different cell types.

3.1.1 How Does Nuclear Factor I Binding Domain Stimulate Initiation?

Upon reconstitution of replication with purified proteins, the level of stimulation appeared strongly dependent on the amount of the pTP–pol complex added to the reaction. At low pTP–pol concentrations, NFI-BD stimulated up to 60-fold, whereas at high concentration the level of stimulation dropped to only twofold (MUL et al. 1990). This suggested that NFI could increase the binding of the pTP–pol complex to the core origin by functioning as a recruitment factor for the polymerase. Indeed, a complex between pol and NFI was observed, both in the presence and the absence of DNA (CHEN et al. 1990; BOSHER et al. 1990; MUL et al. 1990). In agreement with its role in DNA replication, the interacting interface is located in NFI-BD, between amino acids 68 and 150 (CHEN et al. 1990). This leads to a model in which NFI binds pTP–pol and, by using its own recognition site, positions the pTP–pol complex at the core origin. Such a model explains the strict positional requirement for the NFI-binding site in the Ad2 origin (ADHYA et al. 1986; WIDES et al. 1987; COENJAERTS et al. 1991; BOSHER et al. 1990). The domain in pol involved in interaction with NFI has not been mapped in detail, but seems to be different from the regions involved in DNA recognition and polymerization based upon mutation of a potential Zn^{2+}-binding region in pol (CHEN et al. 1990).

By tethering pTP–pol, NFI may stabilize the formation of a preinitiation complex consisting of pTP, pol, and origin DNA. Indeed, by template challenge experiments, we were able to show that NFI reduced the dissociation of pTP–pol from the origin. Also a direct effect on the amount of pTP–pol–DNA complex was shown by band shift experiments (MUL and VAN DER VLIET 1992). Kinetic studies confirmed that NFI increases the amount of active initiation complex rather than increasing the K_m of the reaction (MUL and VAN DER VLIET 1993).

It is difficult to establish whether this is the only effect of NFI. NFI stimulates initiation up to 60-fold, whereas the level of stabilization of an initiation complex is only about threefold. Possibly NFI also acts at a later stage, for instance by facilitating changes in the DNA structure at the origin (ZORBAS et al. 1989).

Is NFI essential in vivo? Clearly mutation of the NFI-binding site leads to a strong reduction in viral replication in vivo (HAY 1985; K. WANG and PEARSON 1985). Moreover, NFI is targeted to discrete subnuclear sites which coincide with replication foci together with other replication proteins including DBP (BOSHER et al. 1992; VOELKERDING and KLESSIG 1986). Although this is indirect evidence, it is suggestive of a role of NFI also in vivo.

3.2 Nuclear Factor III

A second replication-stimulating cellular protein NFIII, was discovered when fractionated nuclear extracts from HeLa cells were assayed for additional stimulation in the presence of NFI and all three viral replication proteins (PRUIJN et al. 1986). NFIII binds to a site in the Ad5 origin next to the NFI site, approximately from bp 38 to 49, making contact with a contiguous stretch of bases (PRUIJN et al. 1988). Early studies showed that viruses containing the terminal 45 bp were fully infectious, indicating that an intact NFIII site was not important in vivo (HAY and McDOUGALL 1986). However, later studies (HATFIELD and HEARING 1993) using viral deletion mutants showed that an intact NFIII site is important in vivo for efficient growth and DNA replication.

NFIII was purified as a 90-kDa protein and, based upon several criteria, appeared identical to the ubiquitous transcription factor Oct-1 (O'Neill et al. 1988; Pruijn et al. 1989). This protein preferentially recognizes the octamer sequence ATGCAAAT present in many promoters and enhancers of cellular genes. Examples are the histone 2B genes, the immunoglobulin genes, and U-snRNA genes. In Ad4 and Ad10, the consensus octamer sequence is present, while slightly degenerated sequences are present in other serotypes. For Ad2/Ad5, the sequence is ATGATAAT, which is bound with a threefold lower affinity (Verrijzer et al. 1990a).

Oct-1 belongs to the POU protein family of transcription factors. These proteins, with the prototypes Pit-1, Oct-1, and Unc-86, are characterized by a common DNA-binding domain called the POU domain (Herr et al. 1988). POU domain transcription factors from different subclasses, including Pit-1, Oct-2, Oct-4, Oct-6, and even the distantly related zebrafish POU[c], all stimulate adenovirus replication (Verrijzer et al. 1992b). By deletion analysis (Verrijzer et al. 1990b), we showed that the POU domain suffices for stimulation of Ad DNA replication. Thus, both for NFI and NFIII/Oct-1 the DNA-binding domains alone are sufficient. One should realize, however, that the function of these DNA-binding domains is not limited to just the recognition of DNA. NFI-BD also contains the dimerization domain and at least one interface for contact with the pTP–pol complex. Similarly, the POU domain makes contacts with other proteins. It is involved in homo- and heterodimerization (Ingraham et al. 1990; Verrijzer et al. 1992c) and it interacts with the herpes simplex virus tegument protein VP16, which contains a strong *trans*-activating domain (O'Hare 1993), and with a complex of various other related proteins collectively called HCF (Wilson et al. 1993).

The POU domain is a bipartite DNA-binding domain. It consists of a POU-specific domain (POUs, approximately 75 amino acids long) linked to a 60-amino acid POU homeodomain (POUhd). For stimulation of Ad DNA replication as well as for site-specific DNA binding, the intact POU domain is required (Verrijzer et al. 1990b). The POUhd by itself recognizes the right half of the octamer, with a consensus recognition sequence TAATNA, as determined by a binding site selection procedure (Verrijzer et al. 1992a). This is very similar to the recognition sequence of classic homeodomain proteins. The POUs binds much weaker to DNA, but nevertheless is able to specifically recognize the sequence ATGCA, which is the other half of the octamer sequence. Thus both subdomains recognize half of the binding site. This unique mechanism of DNA recognition was recently further elucidated by determination of the solution structure of POUs employing multidimensional nuclear magnetic resonance (NMR) techniques (Dekker et al. 1993; Assa-Munt et al. 1993). POUs consists of a tetrahelical structure employing a helix–turn–helix motif to bind DNA very similar to the λ–repressor. Helix III recognizes the ATGCA sequence in the major groove. Since the POUhd also recognizes DNA by a helix–turn–helix motif, the POU domain represents the first example of two HTH motifs in one protein, together recognizing one consecutive series of bases in the major groove (Fig. 3).

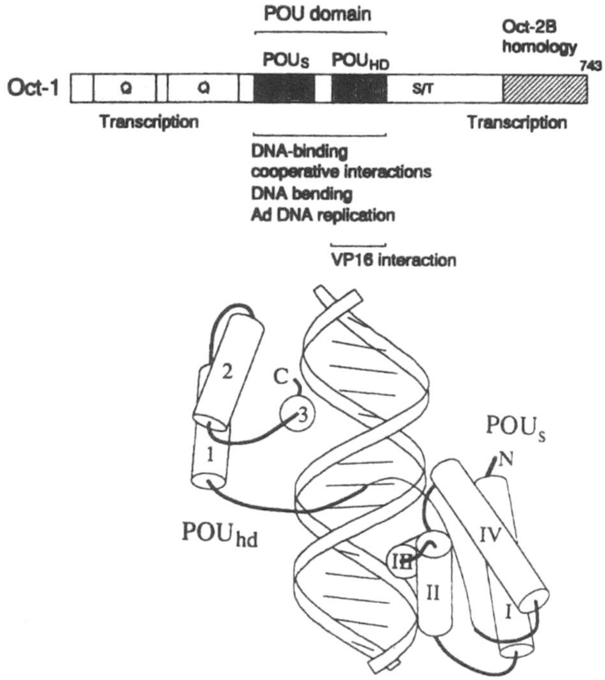

Fig. 3. Three-dimensional structure of the Oct-1 POU domain. *Top*, the various functional regions of Oct-1 are given. The bipartite POU domain suffices for stimulation of adenovirus (*Ad*) DNA replication and has several other properties, as indicated. *Bottom*, model of the POU–DNA interaction based upon the nuclear magnetic resonance (NMR) structure of the POU-specific domain (DEKKER et al. 1993; ASSA-MUNT et al. 1993) and the POU homeodomain (POU_{hd}; COX et al. 1995). The orientation of helix III, the DNA recognition helix of the POU-specific domain (POU_s), is arbitrary

3.2.1 How Does Nuclear Factor III Stimulate Adenovirus DNA Replication?

In trying to determine how NFIII/Oct-1 stimulates adenovirus DNA replication, one should keep in mind the fact that the position of the NFIII/Oct-1 binding site in the various Ad serotypes is even more conserved than its actual sequence. Insertion or deletion of more than one base between the NFI and Oct-1 sites leads to loss of stimulation, despite perfect binding of the protein to such a mutated origin (COENJAERTS et al. 1991). This suggests that the POU domain is involved in protein–protein contacts at the origin. No such contacts were detected between NFI and the POU domain and, despite their close proximity, both proteins bind DNA independently. Moreover, stimulation of initiation by NFI and NFIII/Oct-1 is additive (MUL et al. 1990), suggesting that they do not act through exactly the same target.

A possible mechanism was suggested by the observation that the POU domain is capable of DNA bending as shown by three independent techniques (VERRIJZER et al. 1991). DNA bending might facilitate the interaction between the various replication proteins, thereby enhancing the formation of an initiation complex or stabilizing such a complex. However, in contrast to NFI, stabilization of a preinitiation complex by the POU domain could not be observed by template challenge experiments or by direct band shift experiments (MUL and VAN DER VLIET 1993). An alternative explanation for the observed stimulation could be that the distortion of the DNA structure induced by bending leads to a facilitated opening of the origin, possibly aided by the energy stored in the bent protein–DNA complex. If this were true, it should be possible to exchange the Oct-1-binding site for binding sites for other DNA-binding transcription factors. We have inserted an AP1-binding site at several positions next to the NFI site, to replace the octamer sequences, and studied the effect of adding various combinations of c-*fos* and c-*jun*, but in no case was stimulation observed. The same was true for insertion of prebent DNA at that position. Thus, the effect of Oct-1, is presumably specific, in agreement with the observation that in none of the adenovirus origins have binding sites other than for NFI and Oct-1 been found. An attractive explanation is that the POU domain recognizes one of the viral proteins, possibly the pTP–pol complex. Such an interaction could also explain that the level of stimulation by the POU domain is dependent on the pTP–pol concentration, although less pronounced than with NFI (MUL et al. 1990). At low concentrations of pTP–pol, the POU domain stimulates initiation about sevenfold, whereas at high concentrations most of the stimulation disappears and only a twofold effect remains.

Direct evidence for an interaction between POU and pTP–pol was recently obtained by COENJAERTS et al. (1994). He employed a GST-POU fusion protein immobilized on glutathione-agarose beads and observed that the pTP–pol complex was specifically retained, using a replication assay to measure the amounts of pTP–pol. Binding by the POUhd was still measurable, but much lower than with the intact POU domain, while POUs was negative. Interestingly, NFI did not compete for this interaction, indicating that different domains of pTP–pol could be recognized by the two transcription factors. This agrees with the previous observations that stimulation by NFI and POU domain is additive and thus acts through slightly different targets.

In conclusion, a picture emerges in which both transcription factors touch the pTP–pol complex (see also Fig. 5). Whether this is just aiding correct positioning of pTP–pol to the origin or whether it may have other consequences for the function of the polymerase is presently under investigation. Since neither NFI nor the POU domain changes the K_m of the initiation reaction, the transcription factors do not seem to have a direct effect on the active site of the polymerase.

3.3 Nuclear Factor II

Although the three viral proteins, together with NFI and NFIII/Oct-1, can sustain initiation very efficiently, these five proteins are not sufficient for replication of the intact genome. In vitro, replication does not proceed beyond approximately 30%

(10 kb) unless a third cellular protein is present (NAGATA et al. 1983b). This protein was purified as a 30-kDa polypeptide containing topoisomerase I activity and was called NFII. Presumably NFII is a proteolytic product of HeLa topoisomerase I. Topoisomerase I is also required in vivo as shown by specific inhibitors (SCHAACK et al. 1990a). Thus topological stress accumulates during movement of the replication fork, suggesting that unwinding is limited despite the presence of free molecular ends. In vivo this may be caused by attachment of DNA through TP to the nuclear matrix (FREDMAN and ENGLER 1993; SCHAACK et al. 1990b), but the need for topoisomerase in vitro cannot be explained in this way. Possibly TP DNA circularizes by interactions between the two terminal proteins, thereby restricting the rotational freedom of the templates. This would explain the circles that were observed originally by EM and that led to the discovery of the TP (ROBINSON et al. 1973).

To investigate whether the presence of TP causes the limited elongation in vitro, we have analyzed the template activity of TP DNA obtained from Ad5*dl* 309 and digested with *Xba*I. This leads to two fragments of 1339 and 34.596 bp, respectively, each containing TP at one end. With these fragments as templates, replication of the short fragment was unimpeded, whereas the long fragment replicated only to about 25%, similar to intact TP DNA (D.C. ZIJDERVELD and P.C. VAN DER VLIET, unpublished work). This block could be relieved by adding topoisomerase I (Fig. 4). Since these templates initiate at the TP-containing origins and thus are TP free at the other molecular end, this means that there is no apparent reason for these molecules not to rotate freely. Therefore, we can conclude that the blockade is not caused by TP–TP interactions, but by the length of DNA. Possibly the proteins in the replication fork dissociate when positive supercoils accumulate and are not relieved in time. The reason for accumulation could be that the release of topological stress is too slow compared to the rate of fork movement, despite the presence of a free molecular end. In this respect it is interesting to note that several bacteriophages containing linear DNA such as T7 also require topoisomerase action during replication (ITOH and TOMIZAWA 1977).

4 Function of the Parental Terminal Protein

Proteins covalently bound to the molecular ends of DNA are not uncommon. They occur, in addition to adenovirus, in a number of bacteriophages (notably ϕ29 and PRD-1), as well as in several plasmids (SALAS 1991). Their function is clearly related to the protein-priming mechanism of initiation.

The adenovirus TP are highly conserved in the various serotypes. pTP becomes attached to the DNA as a consequence of protein priming and is proteolytically cleaved by the viral protease late in infection. Proteolytical cleavage of the free pTP leads to inactivation of the priming capacity, since N-terminal deletions are lethal. Thus proteolysis could be a signal to stop replication late in

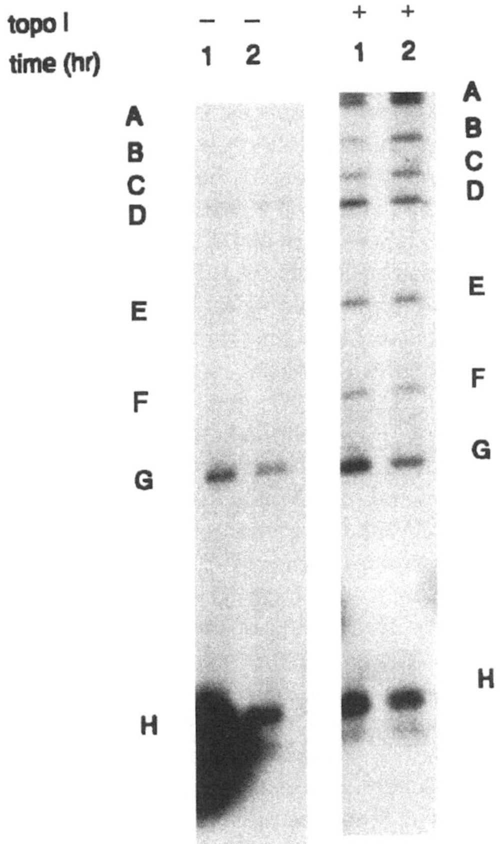

Fig. 4. Topoisomerase I (nuclear factor II), is required for replication of long stretches of adenovirus DNA irrespective of the presence of terminal protein (TP) at both ends. Ad5*dl* 309 (N.C. JONES and SHENK 1979) was digested with *XbaI*, leading to short TP-containing fragment H (1339 bp) and a 34; 5-kbp fragment. This mixture was used as a template for in vitro replication during 1 or 2 h and digested afterwards with *KpnI*. In the absence of topoisomerase (*lanes 1, 2*), replication only proceeded for about 3 kbp, leading to the labeling of fragments G, H, and part of D. When topoisomerase I (Wheatgerm) was added, all fragments were labeled, indicating a suppression of the block

infection. DNA-bound TP and pTP, however, appear not to differ significantly in template activity. This indicates that, in addition to priming and interaction with pol, the pTP harbors several other properties that are located in the TP moiety rather than in the precursor part of the molecule.

What are the functions of the parental DNA-bound TP? First of all, it protects the DNA against nucleolytic breakdown by 5'→3' exonucleases, such as pL (KENNY et al. 1988). It also prevents other end-binding proteins such as the NFIV/Ku protein to enter the DNA and block replication (DE VRIES et al. 1989). Also, TP-DNA has been shown to attach tightly to the nuclear matrix throughout the course of infection (BODNAR et al. 1989; SCHAACK et al. 1990b). Matrix association may localize the viral genome to particular nuclear compartments in which replication and transcription factors could be concentrated, as shown by the appearance of replication foci in infected cells. Thus, matrix association could be instrumented both in replication and transcription.

An important function of TP is the enhancement of infectivity and template efficiency. TP-DNA is a much more efficient template than protein-free DNA (TAMANOI and STILLMAN 1982; VAN BERGEN et al. 1983). Under certain conditions the difference can be as much as 100-fold (PRONK and VAN DER VLIET 1993). Only a part of the TP seems to be required for enhancement (PRONK et al. 1992). Since this enhanced template activity is observed in a reconstituted system, in which the nucleases or other end-binding proteins are absent and matrix association cannot occur, it must be an intrinsic property of TP itself.

How does TP enhance template activity in vitro? First of all, it enhances the binding of pTP–pol to the origin two- to threefold. This could be shown employing band shift experiments with a short, TP-containing origin fragment that was end labeled. The isolation of such a fragment, in which the TP remains functionally intact, was possible by anion exchange chromatography of restriction enzyme digests of virion DNA. During chromatography, the presence of TP changes the elution profile enough to enable separation of TP-DNA from the main TP-free fragments. Whereas binding of pTP–pol is slightly enhanced, the binding of NFI or NFIII/Oct-1 was not influenced (PRONK and VAN DER VLIET 1993). Increased binding of pTP–pol can only partly explain the enhanced template efficiency. Therefore, it was interesting to note that TP also has a direct effect on the structure of the origin. By DNAse I footprinting, a clear difference in sensitivity was observed between TP-DNA and protein-free DNA. Strong DNAse I hypersensitivity was found at positions 5 and 6, close to the molecular end, as well as some less pronounced hypersensitivity further away, at portions 41, 42, and 48 in the top strand (PRONK and VAN DER VLIET 1993). These subtle changes in the origin structure could well be instrumental in influencing the interaction of other replication proteins with the origin. Alternatively, these changes in the origin could indicate some form of twisting of the DNA as a result of TP, which may help in unwinding of the origin occurring after formation of a preinitiation complex.

5 An Integrated Model for Adenovirus DNA Replication

5.1 Initiation

Upon accumulation of the viral replication proteins and correct location inside infected nuclei, replication can start in vivo. The mechanism by which this occurs has been studied almost exclusively in vitro and will be described below. At sufficiently high concentrations, DBP forms a multimeric protein–DNA complex with viral DNA. This leads to subtle changes in the DNA configuration, as shown by hydroxyl radical footprinting (STUIVER et al. 1992), and leaves the binding sites for NFI and Oct-1 accessible. The pTP–pol complex, either bound to NFI or free, recognizes the core origin weakly by itself, but it is mainly guided to the correct position by NFI which binds the pTP–pol complex fairly tightly (CHEN et al. 1990; BOSHER et al. 1990; MUL and VAN DER VLIET 1992). Binding of NFI, or the pTP–pol–NFI complex, to its recognition sequence in the auxiliary origin is facilitated by DBP. When positioned correctly, pTP–pol induces structural changes in the core origin that are enhanced by NFI and possibly by the other proteins as well. Binding and positioning of pTP–pol is further enhanced by the POU domain of Oct-1 and by the parental TP. The latter effects could occur via a direct interaction with particular domains on pTP–pol other than the domain bound by NFI or could be caused indirectly by structural changes that are induced by the POU domain (bending) or by TP (DNase I hypersensitivity at several positions).

After assembly of this preinitiation complex (Fig. 5), the origin must be unwound or otherwise distorted to permit the template strand to enter the active site of the polymerase. How this occurs is not yet clear. Unwinding may be facilitated by DBP, which has the intrinsic ability to unwind DNA (ZIJDERVELD and VAN DER VLIET 1994; MONAGHAN et al. 1994), by TP, and by the POU domain through structural DNA changes. For this unwinding, a helicase activity is not required in contrast to other replication systems. Moreover, initiation is independent of ATP (PRONK et al. 1994) and occurs equally well in the presence of nonhydrolyzable ATP analogues. Since unwinding occurs at the molecular ends, it may be easier than in other systems with internal origins.

In a next step, the actual initiation reaction occurs, i.e., a serine residue of pTP is coupled to dCMP. This reaction requires Mg^{2+} (PRONK et al. 1994), but two other bivalent cations (Mn^{2+}, Ca^{2+}) can substitute for Mg^{2+}. The K_m for dCTP for pTP–dCMP formation is $3.7 \pm 0.9\,\mu M$ in the absence of DBP, whereas DBP lowers the K_m six- to eightfold (MUL and VAN DER VLIET 1993). This might occur through a better presentation of the terminal dG residue in the template strand to pol, resulting in a more efficient selection and binding of the dCTP substrate. Otherwise, DBP might induce a better fit for dCMP through structural changes in the polymerase active site. The two transcription factors NFI and NFIII/Oct-1 have no effect on the K_m for dCTP, but rather they increase the V_{max}, presumably by increasing the effective enzyme concentration and by stabilizing the initiation complex. Their recognition sequences in the auxiliary origin are critically spaced. This is in accordance with formation of a well-balanced multiprotein initiation complex stabilized by multiple protein–protein and protein–DNA interactions requiring precise ordering of all components involved.

5.2 From Initiation to Elongation

Several changes must occur in order to proceed from an initiation to an elongation mode. It is likely from mutagenesis studies that the same dNTP-binding site is used for initiation and elongation. Therefore, the region of pTP

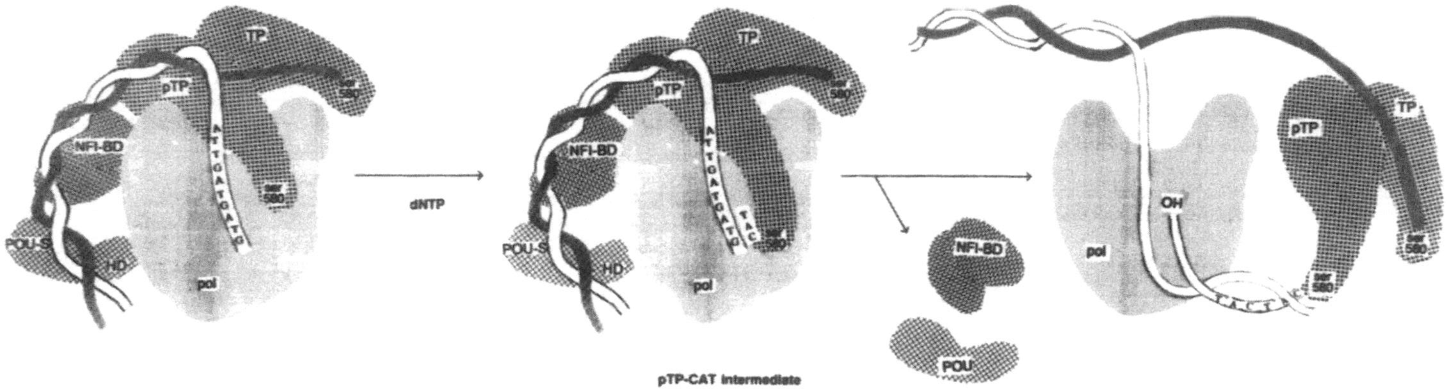

Fig. 5. Model for events during initiation of adenovirus type 5 (Ad5) DNA replication. The DNA polymerase (*pol*) is drawn according to the model for *Escherichia coli* pol I (Klenow fragment; BEESE et al. 1993), and the POU domain according to the solution structure (DEKKER et al. 1993) The other structures are hypothetical. A preinitiation complex is formed in which the template strand is present in cleft one and is separated from the other strand. The core sequence of the origin binds the precursor part of precursor terminal protein (*pTP*) (R.T. HAY, personal communication). Both nuclear factor I (*NFI*) and the POU domain touch the polymerase. The extent of unwinding is unknown. After formation of pTP–CAT, a kinetic barrier exists (MUL and VAN DER VLIET1993), suggesting that this product is an intermediate in initiation. This is caused by start at position 4 followed by jumping back to position 1 (KING and VAN DER VLIET 1994). Finally, elongation can start after dissociation of DNA from pTP to enable the DNA to pass freely along the polymerase. The transcription factors dissociate and possibly also pTP and pol. Whether pTP and TP stick together as drawn is not known. Note that, for the sake of clarity, DNA-binding protein (DBP) is not depicted. *NFI-BD*, NFI-binding domain; *dNTP*, deoxynucleoside triphosphate; *POU-S*, POU-specific domain; HO, homeodomain

containing Ser-580 would be positioned close to the active site of the polymerase (Fig. 5) and must either shift its position or dissociate completely from the polymerase in order to give access to the moving template strand. In this respect, it is noteworthy that free pTP inhibits the polymerizing activity of pol, at least on synthetic templates. On the other hand, antibodies against pTP or TP inhibit elongation (RIJNDERS et al. 1983), which could indicate that pTP or TP are still present at or nearby a replication fork. Moreover, pTP mutants exist in which elongation activity is uncoupled from initiation activity, suggesting a role for pTP after the initiation step (PETTIT et al. 1989; ROOVERS et al. 1993). Clearly this issue has not yet been settled. Despite this, it is likely that the active site of the polymerase changes its configuration when going from initiation to elongation, since K_m the for dCTP is different. Moreover, initiation can occur with several bivalent cations, whereas the step from initiation to elongation depends exclusively on Mg^{2+} (PRONK et al. 1994).

When elongation has started, the initiation complex presumably dissociates (Fig. 5). Dissociation of NFI occurs very early (COENJAERTS and VAN DER VLIET 1994). Upon going from initiation to elongation, several other events may occur even before dissociation of the initiation proteins. Recently we observed that, during elongation at low dCTP concentrations, part of the pTP does not participate in elongation but remains trapped in an early intermediate form. Since dATP and dTTP were added, we presume that this represents pTP containing just the first three nucleotides, pTP–CAT (MUL and VAN DER VLIET 1993). The accumulation of this product, a pTP–CAT intermediate (Fig. 5), indicates a kinetic barrier that inhibits proceeding beyond the fourth nucleotide, a C residue, despite, the presence of dCTP. This might be explained by a blockade at this position, due to the low dCTP concentrations, which delays further elongation and could lead to dissociation of the complex. Remarkably however, no such accumulation was observed at other C residues at positions 7, 17, or 18, and therefore this explanation is less likely. A more interesting explanation is that, in order to proceed from initiation to elongation, the polymerase employs a slideback mechanism similar to that described for other protein-priming systems such as ϕ29 and PRD-1 (MENDEZ et al. 1992; CALDENTEY et al. 1993). It could be that initiation actually occurs at G4 in the template strand, synthesizing pTP–CAT followed by a sliding or jumping back of the complex such that CAT can basepair with the first three nucleotides of the template strand. This process could be the rate-limiting step at low dCTP concentrations. This would agree with the presence of trinucleotide repeats in all adenovirus serotypes and is also in agreement with mutagenesis studies, showing that mutation of the first nucleotide in the single-stranded template has no effect, while mutating both G1 and G4 is lethal (DOBBS et al. 1990). Moreover, it explains that adenovirus can regenerate two nucleotides absent from both molecular ends (GRAHAM et al. 1989), presumably by making use of the information present in the repeat. Evidence for such a jumping-back mechanism was obtained by us recently after analyzing the length of elongated products made on mutated templates (KING and VAN DER VILET 1994).

5.3 From Elongation to Daughter Molecules

Once started, elongation can occur at a high rate and unabated, provided that DBP is present and that NFII removes topological stress. The polymerase can synthesize at least 30 kbp on primed poly-dT in one single association step in the presence of DBP (Field et al. 1984). Thus, DBP makes the polymerase highly processive, presumably by influencing the DNA structure, although a direct interaction with the polymerase is not excluded (Lindenbaum et al. 1986). DBP also facilitates unwinding as shown in direct assays (Zijderveld and Van der Vliet 1994; Monaghan et al. 1994), enables displacement synthesis, and protects the ssDNA. Presumably it forms a protein polymer growing from C terminus to N terminus around which the DNA is wound, starting at the 5' displaced end of the DNA (Tucker et al. 1994). Like initiation, elongation is ATP independent (De Jong et al. 1983). The replication fork then moves towards the end of the molecule, entering the inverted terminal repeat from the inside. It is unlikely that the inverted terminal repeat contains a specific termination signal, since virus mutants exist that contain multiple repeats of the inverted terminal repeat and nevertheless maintain their genome (Chen and Horwitz 1990). When arriving at the end of the molecule the polymerase most likely dissociates, but we cannot exclude the possibility that it remains bound to the end of the molecule and participates in initiation at the other end, assisted by NFI, Oct-1, and TP present at this end. If the polymerase dissociates, the displaced ssDNA is released as a rather stable ssDNA–DBP complex. In principle, this could function again as a template by formation of a panhandle, restoring a double-stranded origin which can be used for the same protein-primed initiation process (Leegwater et al. 1988). However, the intramolecular renaturation process required for panhandle formation is inhibited considerably by DBP, presumably because the DBP chain is rather rigid and prevents panhandle formation of the displaced ssDNA (Zijderveld et al. 1993). As an alternative, displaced strands with opposite polarity, arising from initiation at two different molecular ends, might renature to form a double-stranded daughter molecule (see Fig. 2). This interstrand renaturation process is considerably enhanced by DBP and might be the main pathway in vivo compared to intramolecular panhandle formation, although the latter process can occur in vivo (Stow 1981). In part of the molecules the need for complementary strand synthesis will be bypassed when both origins are used simultaneously. In that case both forks, when they meet, continue, while the two DNA molecules separate (Fig. 2). However, this process will occur only when the concentration of initiating proteins is high compared to the template concentration. This is less likely late in infection, when progeny DNA accumulates. Finally, the pTP-containing templates that originate from replication and that are as replication competent as TP-DNA will be processed during assembly by the viral protease. This enzyme is dependent for its activity on a peptide present in the viral protein PVI, which can function as a signal for the start of viral assembly (Webster et al. 1993; Mangel et al. 1993). Removal of the precursor part of free pTP by the protease will destroy its capacity to function as a primer and thus will stop replication, followed by assembly of progeny virions.

6 Perspectives

Our understanding of the molecular processes governing initiation of adenovirus DNA replication has come a long way since the development of the first, crude in vitro system employing nuclear extracts. All essential proteins have been cloned, overexpressed, and purified and the three-dimensional structure of two (DBP, POU domain) has largely been solved. Moreover, a number of protein–protein and protein–DNA interactions have been described. Despite these considerable efforts, several essential questions remain unanswered. We still do not understand if, how, and when the origin is opened during initiation. Moreover, we do not understand the conformational changes that must be the result of the various interactions and the dynamics of the process. Future studies will concentrate on these aspects now that large amounts of proteins can be produced. Also, intermediates in initiation will be defined, and site-directed mutagenesis will undoubtedly lead to a fine mapping of the interacting domains. In this respect, the adenovirus DNA replication system will remain one of the best-studied eukaryotic replication systems. On a longer timescale it may be possible to bring this reconstituted system back into permeabilized nuclei in order to study the effects of the interaction with nuclear matrix components. This may enable the study of replication in a more native environment mimicking control mechanisms that may exist in intact cells and might facilitate interactions due to high local concentrations in replication foci. An initial step in this direction has recently been made by the development of an immobilized replication system (Coenjaerts and Van der Vliet 1994).

Acknowledgements. I wish to thank Paul Tucker (EMBL, Heidelberg), Frank Coenjaerts, Rob Pronk, Diederik Zijderveld, Audrey King, Hans van Leeuwen, Job Dekker, Wouter de Laat, Michel Cox, and Rob Kaptein for unpublished results and Diederik Zijderveld for help with figures, in particular Fig. 4. This work was supported in part by the Netherlands Foundation for Chemical Research (SON) and the Medical Research Council (G-MW) with financial support from the Netherlands Organization for Scientific Research (NWO).

References

Adhya S, Shneidman PS, Hurwitz J (1986) Reconstruction of adenovirus replication origins with a human nuclear factor I binding site. J Biol Chem 261: 3339–3346

Anderson KP, Klessig DF (1984) Altered mRNA splicing in monkey cells abortively infected with human adenovirus may be responsible for inefficient synthesis of the virion fiber polypeptide. Proc Natl Acad Sci USA 81: 4023–4027

Assa-Munt N, Mortishire-Smith RJ, Aurora R, Herr W, Wright PE (1993) The solution structure of the Oct-1 POU-specific domain reveals a striking similarity to the bacteriophage 434 repressor DNA-binding domain. Cell 73: 193–205

Beese LS, Derbyshire V, Steitz TA (1993) Structure of DNA polymerase I Klenow fragment bound to duplex DNA. Science 260: 352–355

Bodnar JW, Hanson PI, Polvino-Bodnar IGM, Zempsky W, Ward DC (1989) The terminal regions of adenovirus and minute virus of mouse DNAs are preferentially associated with the nuclear matrix in infected cells. J Virol 63: 4344–4353

Bosher J, Clare Robinson E, Hay RT (1990) Interactions between the adenovirus type 2 DNA polymerase and the DNA binding of nuclear factor I. New Biol 2: 1083–1090
Bosher J, Leith IR, Temperley SM, Wells M, Hay RT (1991) The DNA-binding domain of nuclear factor I is sufficient to cooperate with the adenovirus type 2 DNA-binding protein in viral DNA replication. J Gen Virol 72: 2975–2980
Bosher J, Dawson A, Hay RT (1992) Nuclear factor I is specifically targeted to discrete subnuclear sites in adenovirus type 2-infected cells. J Virol 66: 3140–3150
Caldentey J, Blanco L, Bamford DH, Salas M (1993) In vitro replication of bacteriophage PRD1 DNA. Characterization of the protein-primed initiation site. Nucleic Acids Res 21: 3725–3730
Campbell JL (1986) Eukaryotic DNA replication. Annu Rev Biochem 55: 733–771
Challberg MD, Kelly TJ (1979) Adenovirus DNA replication in vitro. Proc Natl Acad Sci USA 76: 655–659
Challberg MD, Kelly TJ (1989) Animal virus DNA replication. Annu Rev Biochem 58: 671–717
Challberg MD, Rawlins DR (1984) Template requirements for the initiation of adenovirus DNA replication. Proc Natl Acad Sci USA 81: 100–104
Challberg MD, Desiderio SV, Kelly TJ (1980) Adenovirus DNA replication in vitro: characterization of a protein covalently linked to nascent DNA strands. Proc Natl Acad Sci USA 77: 5105–5109
Chase JW, Williams KR (1986) Single-stranded DNA-binding proteins required for DNA replication. Annu Rev Biochem 55: 103–136
Chen M, Horwitz MS (1989) Dissection of functional domains of adenovirus DNA polymerase by linker insertion mutagenesis. Proc Natl Acad Sci USA 86: 6116–6120
Chen M, Horwitz MS (1990) Replication of an adenovirus type 34 mutant DNA containing tandem reiterations of the inverted terminal repeat. Virology 179: 567–575
Chen M, Mermod N, Horwitz MS (1990) Protein-protein interactions between adenovirus DNA polymerase and nuclear factor-1 mediate formation of the DNA replication preinitiation complex. J Biol Chem 265: 18634–18642
Cleat PH, Hay RT (1989) Co-operative interactions between NFI and the adenovirus DNA binding protein at the adenovirus origin of replication. EMBO J 8: 1841–1848
Cleghon V, Klessig DF (1992) Characterization of the nucleic acid binding region of adenovirus DNA-binding protein by partial proteolysis and photochemical cross-linking. J Biol Chem 267: 17872–17881
Coenjaerts FEJ, De Vries E, Pruijn GJM, Van Driel W, Bloemers SM, Van der Lugt NMT, Van der Vliet PC (1991) Enhancement of DNA replication by transcription factor NFI and NFIII/Oct-1 depends critically on the positions of their binding sites in the adenovirus origin of replication. Biochim Biophys Acta 1090: 61–69
Coenjaerts FEJ, Van der Vliet PC (1994) Early dissociation of nuclear factor I from the origin during initiation of adenovirus DNA replication studied by origin immobilization. Nucl Acids Res 22: 5235–5240
Coenjaerts FEJ, Van Oosterhout JAWM, Van der Vliet PC (1994) The Oct-1 POU domain stimulates adenovirus DNA replication by a direct interaction between the viral precursor terminal protein-DNA polymerase complex and the POU homeodomain. EMBO J 13: 5401–5409
Cox M, Van Tilborg PJA, de Laat W, Van Leeuwen HC, Van der Vliet PC, Kaptein R (1995) Solution structure of the Oct-1 POU-homeodomain determined by NMR and restrained Molecular dynamics. J Biomol NMR (in press)
De Jong PJ, Kwant MM, Van Driel W, Jansz HS, Van der Vliet PC (1983) The ATP requirements of adenovirus type 5 DNA replication and cellular DNA replication. Virology 124: 45–58
Dekker N, Cox M, Boelens R, Verrijzer CP, Van der Vliet PC, Kaptein R (1993) Solution structure of the POU-specific DNA-binding domain of Oct-1. Nature 362: 852–855
DePamphilis ML (1993a) Eukaryotic DNA replication: anatomy of an origin. Annu Rev Biochem 62: 29–62
DePamphilis ML (1993b) How transcription factors regulate origins of DNA replication in eukaryotic cells. Trends Cell Biol 3: 161–167
De Vries E, Van Driel W, Van den Heuvel SJ, Van der Vliet PC (1987) Contact point analysis of the nuclear factor I recognition site reveals symmetric binding at one side of the DNA helix. EMBO J 6: 161–168
De Vries EW, Van Driel W, Bergsma WG, Arnberg AC, Van der Vliet PC (1989) HeLa nuclear protein recognizing DNA termini and translocating on DNA forming a regular DNA-multimeric protein complex. J Mol Biol 208: 65–78
Dobbs L, Zhao LJ, Scipad G, Padmanabhan R (1990) Mutational analysis of single-stranded DNA templates active in the in vitro initiation assay for adenovirus DNA replication. Virology 178: 43–51
Eagle PA, Klessig DF (1992) A zinc-binding motif located between amino acids 273 and 286 in the adenovirus DNA-binding protein is necessary for ssDNA binding. Virology 187: 777–787

Ensinger MG, Ginsberg HS (1972) Selection and preliminary characterisation of temperature sensitive mutants of type 5 adenovirus. J Virol 10: 328–339
Field J, Gronostajski RM, Hurwitz J (1984) Properties of the adenovirus DNA polymerase J Biol Chem 259: 9487–9495
Fredman JN, Engler JA (1993) Adenovirus precursor to terminal protein interacts with the nuclear matrix in vivo and in vitro. J Virol 67: 3384–3395
Fredman JN, Pettit SC, Horwitz MS, Engler JA (1991) Linker insertion mutations in the adenovirus preterminal protein that affect DNA replication activity in vivo and in vitro J Virol 65: 4591–4597
Freimuth PI, Ginsberg HS (1986) Codon insertion mutants of the adenovirus terminal protein. Proc Natl Acad Sci USA 83: 7816–7820
Friefeld BR, Korn R, De Jong PJ, Sninsky JJ, Horwitz MS (1985) The 140-kDa adenovirus DNA polymerase is recognized by antibodies to Escherichia coil-synthesized determinants predicted from an open reading frame on the adenovirus genome. Proc Natl Acad Sci USA 82: 2652–2656
Georgaki A, Strack B, Podust V, Hubscher U (1992) DNA unwinding activity of replication protein A. FEBS Lett 308: 240–244
Gounari F, De Francesco R, Schmidt J, Van der Vliet PC, Cortese R, Stunnenberg HG (1990) Amino terminal domain of NFI binds to DNA as a dimer and activates adenovirus DNA replication. EMBO J 9: 559–556
Graham FL, Rudy J, Brinkley P (1989) Infectious circular DNA of human adenovirus type 5: regeneration of viral DNA termini from molecules lacking terminal sequences. EMBO J 8: 2077–2085
Guggenheimer RA, Nagata K, Lindenbaum J, Hurwitz J (1984) Protein-primed replication of plasmids containing the terminus of the adenovirus genome. J Biol Chem 259: 7807–7814
Hatfield L, Hearing P (1993) The NFIII/Oct-1 binding site stimulates adenovirus DNA replication in vivo and is functionally redundant with adjacent sequences. J Virol 67: 3931–3939
Hay RT (1985) Origin of adenovirus DNA replication. Role of nuclear factor I binding site in vivo. J Mol Biol 186: 129–136
Hay RT, McDougall IM (1986) Viable viruses with deletions in the left inverted terminal repeat define the adenovirus origin of DNA replication. J Gen Virol 67: 321–332
Hay RT, Russell WC (1989) Recognition mechanisms in the synthesis of animal virus DNA. Biochem J 258: 3–16
Hay RT, Stow ND, McDougall IM (1984) Replication of adenovirus mini-chromosomes. J Mol Biol 175: 493–510
Herr W, Sturm RA, Clerc RG, Corcoran LM, Baltimore D, Sharp PA, Ingraham HA, Rosenfeld MG, Finney M, Ruvkun G, Horvitz HR (1988) The POU domain: a large conserved region in the mammalian pit-1, oct-1, oct-2 and Caenorhabditis elegans unc-86 gene products. Genes Dev 2: 1513–1516
Ingraham HA, Flynn SE, Voss JW, Albert VR, Kapiloff MS, Wilson L, Rosenfeld MG (1990) The POU-specific domain of pit-1 is essential for sequence-specific, high affinity DNA binding and DNA-dependent pit-1–pit-1 interactions. Cell 61: 1021–1033
Itoh T, Tomizawa J–I (1977) Involvement of DNA gyrase in bacteriophage T7 DNA replication. Nature 270: 79–80
Jones KA, Kadonaga JT, Rosenfeld PJ, Kelly TJ, Tjian R (1987) A cellular DNA-binding protein that activates eukaryotic transcription and DNA replication. Cell 48: 79–89
Jones NC, Shenk T (1979) Isolation of adenovirus type 5 host range deletion mutants defective for transformation of rat embryo cells. Cell 17: 683–689
Joung I, Engler JA (1992) Mutations in two cysteine-histidine-rich clusters in adenovirus type 2 DNA polymerase affect DNA binding. J Virol 66: 5788–5796
Joung I, Horwitz MS, Engler JA (1991) Mutagenesis of conserved region I in the DNA polymerase from human adenovirus serotype 2. Virology 184: 235–241
Kelly TJ (1984) Adenovirus DNA replication. In: Ginsberg HS (ed) The adenoviruses Plenum ,New York, pp 271–308
Kenny MK, Hurwitz J (1988) Initiation of adenovirus DNA replication II. Structural requirements using synthetic oligonucleotide adenovirus templates. J Biol Chem 263: 9809–9817
Kenny MK, Balogh LA, Hurwitz J (1988) Initiation of adenovirus replication I. Mechanism of action of a host protein required for replication of adenovirus DNA templates devoid of the terminal protein. J Biol Chem 263: 9801–9808
King AJ, Van der Vliet PC (1994) A precursor terminal protein-trinucleotide intermediate during initiation of adenovirus DNA replication: regeneration of molecular ends in vitro by a jumping back mechanism EMBO J 13: 5786–5792
Kitchingman GR (1985) Sequence of the DNA binding protein of a human subgroup E adenovirus (type 4): comparisons with subgroup A (type 12), subgroup B (type 7) and subgroup C (type 5). Virology 146: 90–101

Kruijer W, Van Schaik FMA, Sussenbach JS (1981) Structure and organization of the gene coding for the DNA binding protein of adenovirus type 5. Nucleic Acids Res 9: 4439–4457

Kuil ME, Van Amerongen H, Van der Vliet PC, Van Grondelle R (1989) Complex formation between the adenovirus DNA-binding protein and single-stranded poly (rA). Biochemistry 28: 9795–9800

Lally C, Dörper T, Gröger W, Antoine G, Winnacker E-L (1984) A size analysis of the adenovirus replicon. EMBO J 3: 333–337

Leegwater AJ, Rombouts RFA, Van der Vliet PC (1988) Adenovirus DNA replication in vitro: duplication of single-stranded DNA containing a panhandle structure. Biochim Biophys Acta 951: 403–410

Lindenbaum JD, Field J, Hurwitz J (1986) The adenovirus DNA-binding protein and adenovirus DNA polymerase interact to catalyze elongation of primed DNA templates. J Biol Chem 261: 10218–10227

Mangel WF, McGrath WJ, Toledo DL, Anderson CW (1993) Viral DNA and a viral peptide can act as cofactors of adenovirus virion proteinase activity. Nature 361: 274–275

Meisterernst M, Rogge L, Foechler R, Karaghiosoff M, Winnacker EL (1989) Structural and functional organization of a porcine gene coding for nuclear factor I. Biochemistry 28: 8191–8200

Mendez J, Blanco L, Esteban JA, Bernad A, Salas M (1992) Initiation of Ø29 DNA replication occurs at the second 3' nucleotide of the linear template: a sliding-back mechanism for protein-primed DNA replication. Proc Natl Acad Sci USA 89: 9579–9583

Mermod N, O'Neill EA, Kelly TJ, Tjian R (1989) The proline-rich transcriptional activator of CTF/NFI is distinct from the replication and DNA binding domain. Cell 58: 741–753

Monaghan A, Webster A, Hay RT (1994) Adenovirus DNA binding protein: helix destabilising properties. Nucl Acids Res 22: 742–748

Morin N, Delsert C, Klessig DF (1989) Nuclear localization of the adenovirus DNA-binding protein: requirement of two signals and complementation during viral infection. Mol Cell Biol 9: 4372–4380

Mul YM, Van der Vliet PC (1992) Nuclear factor I enhances adenovirus DNA replication by increasing the stability of a preinitiation complex. EMBO J 11: 751–760

Mul YM, Van der Vliet PC (1993) The adenovirus DNA binding protein effects the kinetics of DNA replication by a mechanism distinct from NFI or Oct-1. Nucleic Acids Res 21: 641–647

Mul YM, Van Miltenburg RT, De Clercq E, Van der Vliet PC (1989) The mechanism of inhibition of adenovirus DNA replication by the acyclic nucleoside triphosphate analogue (S)-HPMPApp: influence of the adenovirus DNA binding protein. Nucleic Acids Res 17: 8917–8929

Mul YM, Verrijzer CP, Van der Vliet PC (1990) Transcription factors NFI and NFIII/oct-1 function independently, employing different mechanisms to enhance adenovirus DNA replication. J Virol 64: 5510–5518

Nagata K, Guggenheimer RA, Enomoto T, Lichy JH, Hurwitz J (1982) Adenovirus DNA replication in vitro: identification of a host factor that stimulates synthesis of the preterminal protein-dCMP complex. Proc Natl Acad Sci USA 79: 6438–6442

Nagata K, Guggenheimer RA, Hurwitz J (1983) Adenovirus DNA replication in vitro: synthesis of full-length DNA with purified proteins. Proc Natl Acad Sci USA 80: 4266–4270

Nakano R, Zhao LJ, Padmanabhan R (1991) Overproduction of adenovirus DNA polymerase and preterminal protein in HeLa cells. Gene 105: 173–178

Novak A, Goyal N, Gronostajski RM (1992) Four conserved cysteine residues are required for the DNA binding activity of nuclear factor I. J Biol Chem 267: 12986–12990

O'Hare P (1993) The virion transactivator of herpes simplex virus. Semin Virol 4: 145–155

O'Neill EA, Fletcher C, Burrow CR, Heintz N, Roeder RG, Kelly TJ (1988) Transcriptional factor OTF-1 is functionally identical to the DNA replication factor NFIII. Science 241: 1210–1213

Paonessa G, Gounari F, Frank R, Cortese R (1988) Purification of a NFI-like DNA-binding protein from at liver and cloning of the corresponding cDNA. EMBO J 7: 3115–3123

Pettit SC, Horwitz MS, Engler JA (1988) Adenovirus preterminal protein synthesized in COS cells from cloned DNA is active in DNA replication in vitro. J Virol 62: 496–500

Pettit SC, Horwitz MS, Engler JA (1989) Mutations of the precursor to the terminal protein of adenovirus serotypes 2 and 5. J Virol 63: 5244–5250

Pronk R, Van der Vliet PC (1993) The adenovirus terminal protein influences binding of replication proteins and changes the origin structure. Nucleic Acids Res 21: 2293–2300

Pronk R, Stuiver MH, Van der Vliet PC (1992) Adenovirus DNA replication: the function of the covalently bound terminal protein. Chromosoma 102: 39–45

Pronk R, Van Driel W, Van der Vliet PC (1994) Adenovirus DNA replication is ATP-independent. FEBS Lett 337: 33–38

Pruijn GJM, Van Driel W, Van der Vliet PC (1986) Nuclear factor III, a novel sequence-specific DNA-binding protein from HeLa cells stimulating adenovirus DNA replication. Nature 322: 656–659

Pruijn GJM, Van Miltenburg RT, Claessens AJ, Van der Vliet PC (1988) Interaction between the octamer-binding protein nuclear factor III and the adenovirus origin of DNA replication. J Virol 62: 3092–3102
Pruijn GJM, Van der Vliet PC, Dathan NA, Mattaj IW (1989) Anti-OTF-1 antibodies inhibit NFIII stimulation of in vitro adenovirus DNA replication. Nucleic Acids Res 17: 1845–1863
Ramachandra M, Nakano R, Mohan PM, Rawitch AB, Padmanabhan R (1993) Adenovirus DNA polymerase is a phosphoprotein. J Biol Chem 268: 442–448
Rekosh DMK, Russell WC, Bellett AJD, Robinson AJ (1977) Identification of a protein linked to the ends of adenovirus. Cell 11: 283–295
Rekosh D, Lindenbaum J, Brewster J, Mertz LM, Hurwitz J, Prestine L (1985) Expression in Escherichia coli of a fusion protein product containing a region of the adenovirus DNA polymerase. Proc Natl Acad Sci USA 82: 2354–2358
Rijnders AWM, Van Bergen BGM, Van der Vliet PC, Sussenbach JS (1983) Specific binding of the adenovirus terminal protein precursor-DNA polymerase complex to the origin of DNA replication. Nucleic Acids Res 11: 8777–8789
Robinson AJ, Younghusband HB, Bellet AJD (1973) A circular DNA-protein complex for adenoviruses. Virology 56: 54–59
Roovers DJ, Overman PF, Chen XQ, Sussenbach JS (1991) Linker mutation scanning of the genes encoding the adenovirus type 5 terminal protein precursor and DNA polymerase. Virology 180: 273–284
Roovers DJ, Van der Lee FM, Van der Wees J, Sussenbach JS (1993) Analysis of the adenovirus type 5 terminal protein precursor and DNA polymerase by linker insertion mutagenesis. J Virol 67: 265–276
Rosenfeld PJ, O'Neill EA, Wides RJ, Kelly TJ (1987) Sequence-specific interactions between cellular DNA-binding proteins and the adenovirus origin of DNA replication. Mol Cell Biol 7: 875–886
Salas M (1991) Protein-priming of DNA replication. Annu Rev Biochem 60: 39–71
Santoro C, Mermod N, Andrews PC, Tjian R (1988) A family of human CAAT-box-binding proteins active in transcription and DNa replication: cloning and expression of multiple cDNAs. Nature 334: 218–224
Schaack J, Schedl P, Shenk T (1990a) Topoisomerase I and II cleavage of adenovirus DNA in vivo: both topoisomerase activities appear to be required for adenovirus DNA replication. J Virol 56: 78–85
Schaack J, Yew-Wai Ho J, Freimuth P, Shenk T (1990b) Adenovirus terminal protein mediates both nuclear matrix association and efficient transcription of adenovirus DNA. Gene 4: 1197–1208
Shu L, Horwitz MS, Engler JA (1987) Expression of enzymatically active adenovirus DNA polymerase from cloned DNA requires sequences upstream of the main open reading frame. Virology 161: 520–526
Steenbergh PH, Sussenbach JS, Roberts RJ, Jansz HS (1975) the 3'-terminal nucleotide sequences of adenovirus types 2 and 5 DNa. J Virol 15: 268–272
Stillman BW (1989) Initiation of eukaryotic DNA replication in vitro. Annu Rev Cell Biol 5: 197–245
Stillman BW, Topp WC, Engler JA (1982a) Conserved sequences at the origin of adenovirus DNA replication. J Virol 44: 530–537
Stillman BW, Tamanoi F, Mathews MB (1982b) Purification of an adenovirus-coded DNA polymerase that is required for initiation of DNA replication. Cell 31: 613–623
Stow ND (1981) The infectivity of adenovirus genomes lacking DNA sequences from their left-hand termini. Nucleic Acids Res 10: 5105–5119
Stuiver MH, Van der Vliet PC (1990) The adenovirus DNA binding protein forms a multimeric protein complex with double-stranded DNA and enhances binding of nuclear factor I. J Virol 64: 379–386
Stuiver MH, Bergsma WG, Arnberg AC, Van Amerongen H, Van Grondelle R, Van der Vliet PC (1992) Structural alterations of double-stranded DNA in complex with the adenovirus DNA-binding protein. Implications for its function in DNA replication. J Mol Biol 225: 999–1011
Stunnenberg HG, Lange H, Philipson L, Van Miltenburg RT, Van der Vliet PC (1988) High expression of functional adenovirus DNA polymerase and precursor terminal protein using recombinant vaccinia virus. Nucleic Acids Res 16: 2431–2444
Sussenbach JS, Van der Vliet PC (1983) The mechanism of adenovirus DNA replication and the characterization of replication proteins. In: Doerfler W (ed) The molecular biology of adenoviruses. Springer, Berlin Heidelberg New York, pp 53–73 (Current topics in microbiology and immunology, vol 109)
Sussenbach JS, Van der Vliet PC, Ellens DJ, Jansz HS (1972) Linear intermediates in the replication of adenovirus DNA. Nature New Biol 239: 47–49
Tamanoi F, Stillman BM (1982) Function of the adenovirus terminal protein in DNA replication. Proc Natl Acad Sci USA 79: 2221–2225

Temperley SM, Hay RT (1992) Recognition of the adenovirus type 2 origin of DNA replication by the virally encoded DNA polymerase and preterminal proteins. EMBO J 11: 761–768
Temperley SM, Burrow CR, Kelly TJ, Hay RT (1991) Identification of two distinct regions within the adenovirus minimal origin of replication that are required for adenovirus type 4 DNA replication in vitro. J Virol 65: 5037–5044
Tolun A, Aleström P, Pettersson U (1979) Sequence of inverted terminal repetitions from different adenoviruses: demonstration of conserved sequences and homology between SA-7 termini and SV40-DNA. Cell 17: 705–713
Tsernoglou D, Tucker AD, Van der Vliet PC (1984) Crystallization of a fragment of the adenovirus DNA binding protein. J Mol Biol 172: 237–239
Tsernoglou D, Tsugita A, Tucker AD, Van der Vliet PC (1985) Characterization of the chymotryptic core of the adenovirus DNA-binding protein. FEBS Lett 188: 248–252
Tucker PA, Tsernoglou D, Tucker AD, Coenjaerts FEJ, Leenders H, Van der Vliet PC (1994) Crystal structure of the adenovirus DNA binding protein reveals a hook-on model for cooperative DNA binding. EMBO J 13: 2994–3002
Van Amerongen H, Van Grondelle R, Van der Vliet PC (1987) The interaction between the adenovirus DNA binding protein and single-stranded polynucleotides studied by circular dichroism and ultraviolet absorption. Biochemistry 26: 4646–4652
Van Bergen BGM, Van der Ley PA, Van Driel W, Van Mansfeld ADM, Van der Vliet PC (1983) Replication of origin containing fragments that do not carry the terminal protein. Nucleic Acids Res 11: 1975–1989
Van der Vliet PC (1990) Adenovirus DNA replication in vitro. In: Straus P, Wilson SH (eds) The eukaryotic nucleus, vol 1, Telford, Caldwell, pp 1–32
Van der Vliet PC (1991) The role of cellular transcription factors in the enhancement of adenovirus DNA replication. Semin Virol 2: 271–280
Van der Vliet PC, Levine AJ (1973) DNA binding proteins specific for cells infected by adenovirus. Nature New Biol 246: 170–174
Van der Vliet PC, Levine AJ, Ensinger MJ,Ginsberg HS (1975) Thermolabile DNA binding proteins from cells infected with a temperature-sensitive mutant of adenovirus defective in viral DNA synthesis. J Virol 15: 348–354
Van der Vliet PC, Keegstra W, Jansz HS (1978) Complex formation between the adenovirus type 5 DNA-binding protein and single-stranded DNA. Eur J Biochem 86: 389–398
Van der Vliet PC, Claessens J, De Vries E, Leegwater PAJ, Pruijn GJM, Van Miltenburg RT (1988) Interaction of cellular proteins with the adenovirus origin of DNA replication. Cancer Cells 6: 61–70
Verrijzer CP, Kal AJ, Van der Vliet PC (1990a) The oct-1 homeo domain contacts only part of the octamer sequence and full oct-1 DNA binding activity requires the POU-specific domain. Genes Dev 4: 1964–1974
Verrijzer CP, Kal AJ, Van der Vliet PC (1990b) The DNA binding domain (POU domain) of transcription factor oct-1 sufficers for stimulation of DNA replication. EMBO J 9: 1883–1888
Verrijzer CP, Van Oosterhout JAWM, Van Weperen W, Van der Vliet PC (1991) POU proteins bend DNA via the POU-specific domain. EMBO J 10: 3007–3014
Verrijzer CP, Alkema MJ, Van Weperen WW, Van Leeuwen HC, Strating MJ, Van der Vliet PC (1992a) The DNA binding specificity of the bipartite POU domain and its subdomains. EMBO J 11: 4993–5003
Verrijzer CP, Strating M, Mul YM, Van der Vliet PC (1992b) POU domain transcription factors from different subclasses stimulate adenovirus DNA replication. Nucleic Acids Res 20: 6369–6375
Verrijzer CP, Van Oosterhout JAWM, Van der Vliet PC (1992c) The Oct-1 POU domain mediates interactions between Oct-1 and other POU proteins. Mol Cell Biol 12: 542–551
Voelkerding K, Klessig DF (1986) Identification of two nuclear subclasses of the adenovirus type 5-encoded DNA-binding protein. J Virol 60: 353–362
Vos HL, Van der Lee FM, Sussenbach JS (1988) The binding of in vitro synthesized adenovirus DNA-binding protein to single-stranded DNA is stimulated by zinc ions. FEBS Lett 239: 251–254
Wang K, Pearson GD (1985) Adenovirus sequences required for replication in vivo. Nucleic Acids Res 13: 5173–5178
Wang TS-F, Wong SW, Korn D (1989) Human DNA polymerase alpha: predicted functional domains and relationships with viral DNA polymerases. FASEB J 3: 14–21
Watson CJ, Hay RT (1990) Expression of adenovirus type 2 DNA polymerase in insect cells infected with a recombinant baculovirus. Nucleic Acids Res 18: 1167–1173
Webster A, Hay RT, Kemp G (1993) The adenovirus protease is activated by a virus-coded disulphide-linked peptide. Cell 72: 97–104

Wides RJ, Challberg MD, Rawlins DR, Kelly TJ (1987) Adenovirus origins of DNA replication: sequence requirements for replication in vitro. Mol Cell Biol 7: 864–874
Wilson AC, LaMarco K, Peterson MG, Herr W (1993) The VP16 accessory protein HCF is a family of polypeptides processed from a large precursor protein. Cell 74: 115–125
Zhao L-J, Padmanabhan R (1988) Nuclear transport of adenovirus DNA polymerase is facilitated by interaction with preterminal protein. Cell 55: 1005–1015
Zhao L-J, Irie K, Trirawatanapong T, Nakano R, Nakashima A, Morimatsu M, Padmanabhan R (1991) Synthesis of biologically active adenovirus preterminal protein in insect cells using a baculovirus vector. Gene 100: 147–154
Zijderveld DC, Van der Vliet PC (1994) Helix-destabilizing properties of the adenovirus DNA-binding protein. J Virol 68: 1158–1164
Zijderveld DC, Stuiver MH, Van der Vliet PC (1993) The adenovirus DNA binding protein enhances intermolecular renaturation but inhibits intramolecular DNA renaturation. Nucleic Acids Res 21: 2591–2598
Zorbas H, Rogge L, Meisterernst M,Winnacker EL (1989) Hydroxyl radical footprints reveal novel structural features around the NFI binding site in adenovirus DNA. Nucleic Acids Res 17: 7735–7748
Zorbas H, Rein T, Krause A, Hoffmann K, Winnacker E-L (1992) Nuclear factor I (NFI) binds to an NF I-type site but not to the CCAAT site in the human α-globin gene promoter. J Biol Chem 267: 8478–8484

Molecular Interactions During Adenovirus DNA Replication

R.T. Hay, A. Freeman, I. Leith, A. Monaghan, and A. Webster

1 Introduction

Over the past 20 years studies on the replication of adenovirus DNA have contributed not only to an understanding of the mechanics of adenovirus DNA replication, but have also shed light on basic processes such as the assembly of nucleoprotein complexes and virus–host interactions. This subject has been reviewed extensively (Hay and Russell 1989; Stillman 1989; Van der Vliet 1990; Salas 1991), but a number of recent findings have suggested that the time may be ripe for further evaluation of new developments in the field.

Within the virion the linear adenovirus genome is found in a highly condensed form associated with small basic proteins. Upon infection of a susceptible cell, the virion is uncoated and the viral DNA released into the nucleus of the cell. Early transcription leads to the expression of three viral gene products, pre-terminal protein (pTP), DNA polymerase (pol) and DNA-binding protein (DBP), that genetic evidence indicates are required for viral DNA replication. Initial work utilising the in vitro system developed by Challberg and Kelly (1979) quickly demonstrated that,

Irvine Building, School of Biological and Medical Sciences, University of St. Andrews, North Street, St. Andrews KY16 9AL, Scotland, UK

as well as the viral proteins, additional cellular factors were also required. While one of these factors was required for complete elongation of the genome and could be substituted for by calf thymus topoisomerase (NAGATA et al. 1982), the other factors were both sequence-specific DNA binding proteins that recognised DNA within the inverted terminial repeats (ITR) of the adenovirus genome (NAGATA et al. 1983; PRUIJN et al. 1986). The adenovirus genome is a linear, double-stranded DNA molecule of 35-36 kb with ITR of about 100bp, the exact size depending upon serotype. DNA replication initiates at either terminus of the linear 36-kbp viral genome by the formation of a covalent linkage between the α-phosphoryl group of the terminal residue, deoxycytidine monophosphate (dCMP), and the β-OH group of a serine residue in pTP. The 3'-OH group of the pTP–dCMP complex then serves as a primer for synthesis of the nascent strand by the viral pol, which proceeds by displacing the non-template strand. Displaced single strands can form partial duplexes by base pairing of the ITR, on which a second round of DNA synthesis may be initiated (HAY et al. 1984; STOW 1982). An alternative to the formation of panhandle structures is that displaced complementary strands simply reanneal to form double-stranded products. This suggestion is based on the in vitro properties of DBP, which promotes intermolecular reannealing but inhibits intramolecular reannealing (ZIJDERVELD et al. 1993).

As the adenovirus genome can be replicated in vitro by the action of three viral proteins and two cellular transcription factors, all of which have been cloned, overexpressed and purified, it represents one of the few systems in which all the components for DNA replication have been precisely defined. The objective of this review will be to describe the components of the adenovirus replicon and discuss how these components assemble into a nucleoprotein complex that is a requirement for the initiation of viral DNA replication.

2 Adenovirus Origins of DNA Replication

Located within the ITR are the *cis*-acting DNA sequences which define ori, the origin of DNA replication. Covalently attached to each 5' end of the DNA is a terminal protein (TP), which is likely to be an additional *cis*-acting component of ori. Comparison of DNA sequences which constitute origins of DNA replication from many different adenoviruses and extensive mutational analysis on such templates has defined four regions within the terminal 51 bp of the adenovirus type 2 (Ad2) genome that contribute to ori activity in vitro and in vivo. Like many other viral origins of DNA replication, that of adenovirus appears to consist of an essential core region and auxilliary regions that enhance the efficiency of DNA replication. The terminal 18 bp of the viral genome are regarded as the minimal replication origin and contain a 10-bp region that is perfectly conserved in all of the human adenoviruses sequenced (STILLMAN et al. 1982). While the integrity of this region is required for viral replication, in isolation it can support only a very limited basal level of initiation of DNA replication (CHALLBERG and RAWLINS 1984; GUGGENHEIMER et al. 1984; HAY 1985a; LALLY et al. 1984; TAMANOI and STILLMAN 1983; VAN BERGEN et al. 1983; WIDES et al. 1987). Separated from the "core" by a precisely defined spacer region (ADYHA et al. 1986; BOSHER et al. 1990; WIDES et al. 1987) is the recognition site for the cellular protein nuclear factor I (NFI) or CAAT transcription factor (JONES et al. 1987; MEISTERERNST et al. 1989; PAONESSA et al. 1988; SANTORO et al. 1988). Binding of NFI to this region of the genome increases the frequency of initiation of viral DNA replication both in vivo and in vitro (NAGATA et al. 1983; RAWLINS et al. 1984; GUGGENHEIMER et al. 1984; DE VRIES et al. 1985; HAY 1985a,b; SCHNEIDER et al. 1986). Immediately adjacent to the NFI-binding site is the recognition site for another cellular DNA binding protein, nuclear factor III (NFIII) or octamer-binding protein (Oct-1); O'NEILL and KELLY 1988; PRUIJN et al. 1986; ROSENFELD et al. 1987; STRUM et al. 1988; WIDES et al. 1987). Although dele-tion of this DNA sequence is without consequence in vivo (HAY and MCDOUGALL 1986), addition of NFIII to an Ad2 in vitro system results in the stimulation of DNA replication (MUL et al. 1990). In contrast to Ad2, the subgroup E virus Ad4 only requires the terminal 18 bp of the viral genome for efficient DNA replication in vivo (HAY 1985b) and in vitro (HARRIS and HAY 1988; TEMPERLEY and HAY 1991; TEMPERLEY et al. 1991). Thus Ad4 appears to have circumvented the need for the host factors NFI and NFIII. The inverted terminal repeat does not contain an NFI recognition site, and whilst it does have a binding site for NFIII neither factor is required for DNA replication in vivo or is capable of stimulating DNA replication in vitro (HAY et al. 1988)

3 Role of Pre-terminal Protein, Terminal Protein and Protease in the Replication of Viral DNA

It has been known for some time that a 55-kDa Tp is linked to the 5' termini of mature adenovirus DNA. However, early in vitro DNA replication studies indicated the presence of 80-kDa pTP covalently attached to the 5' ends of nascent DNA (CHALLBERG et al. 1980). Analysis of an Ad2 virus protease temperature-sensitive mutant, H2*ts*1, which can replicate efficiently at its restrictive temperature, but whose progeny virions are not infectious, revealed that the 80-kDa pTP was processed to the mature TP late in infection (STILLMAN et al. 1981). Subsequent studies demonstrated that during the initiation of adenovirus DNA replication, dCMP was transferred onto pTP to act as the primer for viral DNA synthesis. This protein priming reaction was catalysed by the intimately associated adenovirus DNA polymerase. In adenovirus-infected cells, pTP and pol form a stable heterodimer that can only be dissociated under strong denaturing conditions such as 1.7 *M* urea (LICHY et al. 1982; STILLMAN et al. 1982). Studies on pTP have been facilitated by the cloning of cDNA copies of the gene (PETTIT et al. 1988) into vaccinia virus (STUNNENBERG et al. 1988) or baculovirus (BOSHER et al. 1990) vectors for high-level expression in human or insect cells.

Adenovirus pTP is processed to TP via an intermediate form by the action of the virus-coded protease, which is produced at high levels late in the infectious cycle and is essential for viral infectivity. Sequence alignments and site-directed mutagenesis of the protease indicate that the enzyme may represent a new subclass of cysteine protease (RANCOURT et al. 1994). Recently, it has been shown that adenovirus protease is activated by a disulphide-linked peptide (MANGEL et al. 1993; WEBSTER et al. 1993) derived from the C terminus of the virus structural protein pVI, with the suggested activation mechanism being one of thiol–disulphide interchange (WEBSTER et al. 1993). Given the fact that the protein primer for DNA replication, pTP, is processed by adenovirus protease, it was suggested that DNA was a co-factor in the protease reaction (MANGER et al. 1993). However, other studies have demonstrated that, while DNA can stimulate protease activity under certain defined conditions, it is clearly not required for protease activity (WEBSTER et al. 1993, 1994). It has been demonstrated that Ad2 pTP is cleaved at three sites; MRGF-G and MGGR-G, separated by eight amino acids, to generate the intermediate TP (iTP), and MTGG-V to give TP. These sites were predicted previously based on synthetic peptide studies and are in accord with the substrate specificity of the adenovirus protease (WEBSTER et al. 1989). Furthermore, sites giving rise to both iTP and TP are conserved in all serotypes in which pTP has been sequenced to date, suggesting that the cleavage of pTP, via an intermediate, to TP, plays a role in the infectious cycle.

The major approach to mapping the functional domains of pTP has been linker/ insertion mutagenesis, and although this approach has provided some useful information, the major conclusion appears to be that the virus is very sensitive to mutations in pTP, with changes throughout the 671 amino acids proving lethal (FREDMAN et al. 1991; FREIMUTH and GINSBERG 1986; ROOVERS et al. 1993). In terms of activity in in vitro replication assays, pTP is fractionally more tolerant to mutagenesis, but it has been shown that a number of regions spanning the N-terminal 250 amino acids are absolutely required for initiation of DNA replication (PETTIT et al. 1989; ROOVERS et al. 1993). In a recent study (WEBSTER et al. 1994) the natural processing of pTP, by adenovirus protease, was used to define the regions of pTP that bind to adenovirus pol (Adpol) and the DNA at the origin of replication. Immunoprecipitation experiments demonstrated that pTP and iTP, but not TP, bound to Adpol, suggesting that the amino acids critical for interaction with Adpol reside between the iTP and TP sites. Further evidence that Adpol binds to pTP in proximity to iTP sites comes from the observation that the rate of digestion to iTP is much slower for the Adpol–pTP complex than for free pTP. Studies on the DNA-binding properties of pTP and its digestion products show that only the pTP binds to DNA, indicating that the pTP probably contacts DNA in the region of the iTP cleavage sites. The importance of the amino acids in the region of the iTP cleavage sites in DNA replication is emphasised by studies of a mutant *in*179, in which an insertion has been introduced 48 amino acids to the N terminus of iTP site. It has been shown that the mutant virus has a delayed onset of DNA replication and that it has a *trans*-dominant negative effect in that *in*179 pTP inhibits wild-type DNA replication in a dose-dependent manner (FREIMUTH and GINSBERG 1986).

In considering the role of the pTP processing by protease in DNA replication, free pTP and pTP/TP which is covalently attached to the 5' end of DNA should be regarded as distinct functional entities. Clearly, the role of free pTP is to act as the protein primer for DNA replication, whilst the significance of the covalently attached pTP/TP remains to be fully elucidated; however, it has been known for some time that in transfection experiments the infectivity of TP-DNA is orders of magnitude greater than naked DNA.

Roles for the covalently attached pTP/TP that have been suggested include protection of viral DNA from exonucleases, attachment of virus DNA to nuclear matrix (SCHAAK et al. 1990) and unwinding of the DNA duplex at the origin of replication. Although long suspected, it has only recently been demonstrated that the genome-bound TP serves to stabilise binding of the incoming pTP–pol heterodimer at the origin of DNA replication (PRONK and VAN DER VLIET 1993). In the same study the authors also presented evidence indicating that TP-bound origin DNA adopted a different structure to free origin DNA. However, it is yet to be established whether the altered DNA structure of origin DNA is responsible for the increased binding of pTP–pol or whether it is an entirely unrelated phenomenon. To date, however, few attempts have been made to distinguish between TP-DNA and pTP-DNA, and in most cases the terms have been used interchangeably. This is perhaps surprising, given that in vivo the template for early transcription and the first round of DNA replication is TP-DNA, whilst the template for subsequent rounds of replication will be pTP-DNA, irrespective of whether infections are carried out with wild-type Ad2 or Ad2*ts*1 at the permissive or non-permissive temperatures. One role for the digestion of pTP-DNA to TP-DNA by the protease that has been suggested is that the cleavage is required to release the TP-DNA from the nuclear matrix, prior to assembly within the virus particles (FREDMAN and ENGLER 1993). The fact the pTP-DNA is packaged within virus particles during Ad2*ts*1 infections at non-permissive temperatures would tend to argue against this (WEBER 1990). It cannot be ruled out, however, that the nuclear matrix binding properties of TP-DNA and pTP-DNA are distinct and that this has some bearing on the transcription/replication sites of viral DNA. Another possibility is that the purpose of the processing of pTP-DNA to TP-DNA is to create a different template for either early transcription or the first round of DNA replication. As pTP binds to DNA and Adpol while the mature TP does not (WEBSTER et al. 1994), it would not be surprising to find differences between the properties of TP-DNA and pTP-DNA as templates for DNA replication.

4 Adenovirus DNA Polymerase

Purification of an activity required for viral replication from an Ad2-infected cytosolic extract yielded two polypeptide species: the previously described 80-kDa pTP and a 140-kDa protein with a unique DNA polymerase activity distinguishable from cellular DNA polymerases (ENOMOTO et al. 1981). The viral origin of the DNA polymerase was first suggested from complementation experiments with a group of temperature-sensitive mutants of Ad5 (group N mutants) defective in both the initiation and elongation of DNA replication in vivo (STILLMAN et al. 1982; VAN BERGEN et al. 1983). Nuclear extracts prepared from cells infected with these mutant viruses were unable to support initiation of DNA replication, but activity was restored by addition of the 140-kDa pol (STILLMAN et al. 1982).

The presence of DBP was found to have a profound effect on both the pol and exonuclease activities of the Adpol (FIELD et al. 1984). DNA synthesis on poly-dT:oligo-rA was stimulated ten- to 100-fold by the presence of DBP due to an increase in processitivity of the Adpol. This effect was template specific, as DNA synthesis on activated DNA and poly-dC:oligo-dG were not effected by DBP. Other single-stranded binding proteins (SSB) such as *Escherichia coli* SSB were unable to substitute for adenovirus DBP. Alternatively, it was found that adenovirus DBP did not substitute for human SSB in stimulating the activity of human cellular polymerase-α. These results are consistent with a specific cooperative interaction between DBP and the Adpol (LINDENBAUM et al. 1986) in which DBP is thought to act by stabilising the interaction between Adpol and the DNA template. Indirect evidence for the formation of a physical complex was provided by LINDENBAUM et al. (1986), who observed an increase in the thermostability of Adpol in the presence of an excess of DBP. Although physical evidence of functional complexes between SSB and pol have been found in a variety of procaryotes such as T4 and T7 (HUBERMAN and KORNBERG 1971; REUBEN and GEFTER 1973) and eucaryotes such as human SSB (KENNY et al. 1989, 1990), it has not yet been possible to isolate an Adpol–DBP complex.

In addition to its polymerase function Adpol also possesses an intrinsic 3'–5' exonuclease activity, common to many procaryotic and eucaryotic DNA polymerases (BERNAD et al. 1989). This exonuclease, which is thought to have a role in the proofreading of nascent DNA during elongation, was found to be inhibited up to sevenfold by the presence of DBP. The nature of inhibition by DBP was shown to be due to a direct effect upon the pol rather than binding to and protecting the DNA (LINDENBAUM et al. 1986)

Expression of pol cDNA (SHU et al. 1987) in HeLa cells and insect cells using vaccinia virus (STUNNENBERG et al. 1988) and baculovirus vectors (WATSON and HAY 1990) has allowed structure function analysis studies of pol to be initiated. Protein affinity chromatography, co-immunoprecipitation and cross-linking experiments have all demonstrated that pol can interact directly with the cellular protein NFI, which has been shown to stimulate the initiation of adenovirus DNA replication between five- and 30-fold (BOSHER et al. 1990; CHEN et al. 1990; MUL and VAN DER VLIET 1990). This protein–protein interaction is thought to play an important role in assembly of the pre-initiation nucleoprotein complex at the adenovirus DNA origin of replication (discussed below).

The Adpol shares regions of amino acid sequence homology with a large number of DNA polymerases including procaryotic phage polymerases (bacteriophages T4 and ϕ29), eucaryotic polymerases (human DNA polymerase α) and viral polymerases (adenovirus, herpes simplex and vaccinia virus). Four conserved amino acid regions designated I–IV (I being the most conserved between species) have been found in the Adpol. The most conserved regions (I–II) are present in the C-terminal domain of the protein and have been implicated in metal binding (region I) and substrate binding (regions II and III). The importance of these conserved regions (BERNAD et al. 1987) is suggested by comparison with regions in mutant herpes simplex virus pol that confer altered sensitivity to nucleotide analogues such as aphidicolin and acyclovir (LARDER et al. 1987; MARCY et al. 1990). Mutations in conserved region I of pol have defects both in their ability to participate in the initiation of adenovirus DNA replication in vitro and in pol catalytic activity. Further mutational analysis has identified two potential metal-binding domains consisting of cysteine–histidine clusters in the Ad2 pol that affect its DNA-binding and catalytic properties. N-terminal domain mutations in this enzyme had a moderate effect on both DNA synthesis and elongation, but failed to make the pTP–dCMP complex or bind DNA. C-terminal mutants had the greatest effect on both DNA synthesis, DNA binding and pTP–dCMP complex formation (JOUNG and ENGLER 1992). Recent experiments using the active site label pyridoxal phosphate have suggested the involvement of a specific lysine residue in the catalytic mechanism of the pol enzyme (MONAGHAN and HAY, submitted for publication). While a candidate lysine residue is present in a domain involved in template-primer binding deoxynucleoside triphosphate (dNTP) selection, positive identification of the modified lysine must await microchemical analysis.

Intriguingly, a role for phosphorylation in the activity of pol has been suggested (RAMACHANDRA et al. 1993) by the finding that pol is phosphorylated on S67 by a stably associated histone H1 kinase which exhibits properties similar to the *cdc*2 family of kinases (RAMACHANDRA and PADMANABHAN 1993). Dephosphorylation of the protein altered its ability to participate in the initiation reaction.

5 Adenovirus DNA-Binding Protein

The abundance of DBP during the adenovirus infectious cycle (up to 5×10^6 molecules per cell) helped make it become the first of the viral non-structural proteins to be identified and subsequently purified (VAN DER VLIET and LEVINE 1973). Although DBP has an apparent molecular weight of 72 kDa, this was subsequently shown to be an aberrant electrophoretic mobility, with its true molecular weight as predicated from amino acid composition, being around 59 kDa (KRUIJER et al. 1981).

The essential role of DBP in viral infectivity was demonstrated genetically by isolation of the temperature-sensitive mutations Ad5 H5*ts*125 and H5*ts*107 in the DBP gene. Nuclear extracts from both of these mutants were found to be defective for full-length adenovirus DNA replication on both exogeneous and endogeneous templates (FRIEFELD et al. 1983). In both cases activity was restored by the addition of purified wild-type DBP.

Although the role of DBP in elongation has been well documented over the years by both in vivo and in vitro studies, its role in the initiation of adenovirus DNA replication has been more contentious. However, more recent studies using purified components have observed a stimulatory effect on the level of initiation when DBP is present (KENNY and HURWITZ 1988; CLEAT and HAY 1989; MUL and VAN DER VLIET 1993). One effect of DBP is a consequence of a functional interaction with NFI, a cellular DNA binding protein which has also been found to stimulate the initiation of DNA replication. This was initially suggested by the observation that the abilityof NFI to stimulate initiation in vitro was influenced by the concentration of DNA binding protein (DE VRIES et al. 1985). Subsequent experiments demonstrated that DBP co-operatively increased the affinity of NFI for its recognition site in the adenovirus origin of DNA replication (CLEAT and HAY 1989; STUIVER and VAN DER VLIET 1990). However, no evidence for a direct interaction between the two proteins was detected, and it was suggested that the observed effect was a consequence of DBP′s ability to alter the structure of bound DNA. Using a combination of electron microscopy, hydroxyl radical footprinting and circular dichromism, this contention was confirmed when it was demonstrated that DBP can remove the tertiary structure of double-stranded DNA fragments upon binding (STUIVER et al. 1992). Thus the structure of the DBP–DNA complex may alter the relative positions of hydrogen bond donor and acceptor groups in the major grooves, causing a fine tunning of the contacts between NFI and DNA that leads to a higher-affinity interaction. It has also been demonstrated that DBP can stimulate initiation of DNA replication by decreasing the K_m of the polymerase for the initiatior nucleotide dCTP (MUL and VAN DER VLIET 1993).

Partial chymotryptic digestion of purified DBP yields a C-terminal fragment of around 40 kDa and a highly phosphorylated 27-kDa N-terminal fragment (KLEIN et al. 1979). The C-terminal portion of the molecule can substitute for the full-length molecule during adenovirus DNA replication in vitro and contains a number of conserved domains that are present in all adenovirus serotypes (KITCHINGMAN 1985; VOS et al. 1988). Mutations in each of the three highly conserved domains alter the affinity of DBP for single-stranded DNA (NEALE and KITCHINGMAN 1990). Further investigation revealed the presence of a zinc-binding motif within this domain (between amino acids 273 and 286) which, when mutated, destroyed all of DBP's functions (EAGLE and KLESSIG 1992). More recent studies using limited proteolysis and photo-cross-linking techniques have revealed that the two residues, Met-299 and Phe-418, also play an important in DBP's ability to bind single-stranded DNA (CLEGHORN and KLESSIG 1992).

Adenovirus DBP has been shown to possess the properties of a helix-destabilising protein (MONAGHAN et al. 1994; ZIJDERVELD and VAN DER VLIET 1994). When templates contain a large amount of single-stranded DNA, the double-stranded portion is efficiently unwound in a highly co-operative reaction. Completely double-stranded templates are also unwound, but this reaction is restricted by the length and G+C content of the DNA fragment (MONAGHAN et al. 1994). Like other helix-destabilising proteins, DBP-promoted unwinding requires neither adenosine triphosphate (ATP) nor $MgCl_2$. In fact, the latter is inhibitory to the process, as are other agents, such as NaCl, which increase the stability of duplex DNA molecules. Very similar properties have recently been ascribed to both the calf thymus and herpes simplex virus-coded single-strand specific DBP which appear to be involved in cellular and viral DNA replication (GOERGAKI et al. 1992; GOERGAKI and HUBSCHER 1993; BOEHMER and LEHMAN 1993). Two possibilities exist to explain the ability of DBP to unwind completely double-stranded DNA. In one model DBP could utilise its ability to first bind to double-stranded DNA (CLEAT and HAY 1989; STUIVER and VAN DER VLIET 1990), invade the DNA duplex and bind in a stable fashion to the exposed single strands. An alternative model is that DBP binds to transiently single-stranded regions of DNA that are exposed during "breathing" of short, double-stranded DNA molecules. What is clear, however, is that when the double-stranded DNA fragments are tighltly bound by either NFI or NFIII at their cognate recognition sites DBP is unable to unwind the DNA. This could be due to the bound proteins stabilising the DNA duplex or alternatively the bound proteins may interfere with the ability of DBP to form a continuous protein chain on the DNA. Unlike DNA helicases which translocate unidirectionally on DNA, there does not appear to be a strict directionality to the unwinding reaction catalysed by DBP. Determination of the three-dimensional structure of DBP by X-ray crystallography (TUCKER et al. 1994, and discussed in detail by P.C. VAN DER VLIET, this volume) has suggested a mechanism for the formation of a protein chain of DBP molecules on DNA. The ability of DBP to impose a rigid structure on DNA has recently been shown to be responsible for the ability of DBP to also promote renaturation of complementary single strands (ZIJDERVELD et al. 1993). However, whereas the denaturation reaction described here is inhibited by high concentrations of monovalent and divalent cations, the renaturation reaction described above is highly resistant to their presence (ZIJDERVELD et al. 1993).

The ability of DBP to destabilise double-stranded DNA duplexes could be utilised at a number of points in the viral replicative cycle. While DBP has been shown to stimulate the initiation of Ad2 DNA replication in vitro by decreasing the K_m for transfer of dCMP onto pTP and by increasing the binding of NFI to the replication origin, neither of these reactions are likely to involve the melting of duplex DNA. However, DBP may stimulate the initiation of DNA replication by additional means, and it has previously been demonstrated that Ad4 DBP dramatically stimulates initiation of DNA replication. In this case the extent of stimulation is independent of the concentrations of NFI and dCTP (TEMPERLEY and HAY 1991). One possibility is that DBP may participate with other replication proteins in the unwinding of the DNA double helix that is expected to take place at the termini of the genome prior to initiation (see Fig. 1). Numerous experiments have indicated that TP (PRONK et al. 1992; PRONK and VAN DER VLIET 1993), NFI (DE VRIES et al. 1987; MUL and VAN DER VLIET 1992) and NFIII (VERRIJZER et al. 1991) can all distort adenovirus genomic DNA to some extent. Along with DBP's ability to distort and unwind short, fully duplex DNA, this could provide a mechanism by which the adenovirus replication proteins can destabilise and eventually open the origin DNA. The requirement for DBP during progress of the replication fork has been well established and it is likely to be a consequence of DBP stabilising

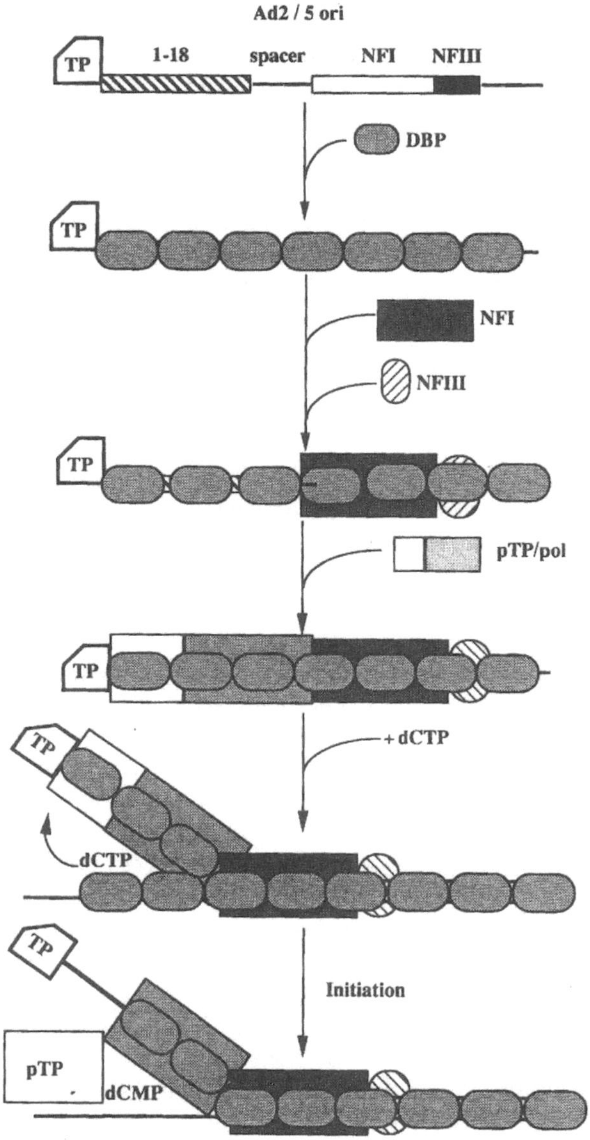

Fig. 1. Initiation of adenovirus type 2 (Ad2) DNA synthesis showing formation of the pre-initiation complex and helix opening at the origin of DNA replication. *TP*, terminal protein; *NF*, nuclear factor; *pTP*, precursor TP; *dCTP*, deoxycytidine triphosphate; *dCMP*, deoxycytidine monophosphate; *pol*, DNA polymerase; *DBP*, DNA-binding protein; *ori*, origin of DNA replication

displaced single strands and altering the properties of the viral pol. In the latter case, DBP has been shown to convert pol into a form that is capable of strand displacement and highly processive DNA synthesis (Field et al. 1984; Lindenbaum et al. 1986). Both of these activities may well be a consequence of the ability of DBP to transiently destabilise double-stranded DNA at the advancing replication fork.

6 Cellular Proteins Involved in Adenovirus DNA Replication

Limited initiation of DNA replication in vitro can occur in the presence of only the virally encoded pTP, pol and DBP proteins. However, the addition of a crude nuclear extract of uninfected HeLa cells was observed to restore efficient levels of initiation. This stimulatory effect was due to the presence of two cellular transcription factors, NFI and NFIII. As both of these factors will be discussed elsewhere (P.C. VAN DER VLIET et al., this volume) only information that is relevant to the formation of a pre-initiation complex will be discussed here.

6.1 Nuclear Factor I

Early in vitro DNA replication studies by both IKEDA et al. (1982) and LICHY et al. (1982) demonstrated the need for a component in uninfected HeLa cell nuclear extract for optimal initiation and elongation in the presence of infected cytosol or purified pTP–pol heterodimer and DBP. Such an activity, purified from the nuclear extract, contained a single major protein species with a molecular weight of 47 kDa which was termed NFI (NAGATA et al. 1982). Early characterisation of NFI concentrated on its ability to specifically bind the sequence 5'-TGGC (N_6) GCCAA-3' present in a number of adenovirus origins of DNA replication (LEEGWATER et al. 1985). On the basis of immunological cross-reactivity data, amino acid composition and proteolytic cleavage patterns, it appears that the NFI proteins are indistinguishable from CCAAT-binding transcription factors (CTF), a family of proteins involved in cellular gene transcription. Analysis of the cDNA from various human NFI/CTF mRNA have shown that they all originate from a single gene, giving rise to multiple mRNA transcripts by differential splicing of a precursor molecule (SANTORO et al. 1988; MEISTERERNST et al. 1989). The NFI proteins had virtually indistinguishable DNA-binding activities and stimulated Ad2/Ad5 initiation of DNA replication in vitro. Mutagenesis studies on the cDNA of the largest NFI protein (CTF-1) identified two functional domains: a highly conserved N-terminal domain, which contains the functions for DNA binding, dimerisation and DNA replication (SANTORO et al. 1988; MERMOD et al. 1989; GOUNARI et al. 1990; BOSHER et al. 1991) and a less highly conserved, proline-rich C-terminal domain which has a transcriptional activation function (MERMOD et al. 1989).

The mechanism of stimulation of adenovirus DNA replication by NFI in vitro is complex (MUL et al. 1990). The degree of stimulation appears to be strongly dependent on the concentration of pTP–pol. At low pTP–pol concentrations, NFI or the DNA-binding domain of NFI (NFI-BD) stimulated replication up to 50-fold, while at high concentrations of pTP-pol stimulation was less than twofold. This demonstrates that in vitro the need for NFI can be overcome by high pTP–pol concentrations. This result provided an explanation for the different levels of stimulation by NFI reported previously (ADHYA et al. 1986; DE VRIES et al. 1985) as well as a strong indication for a direct interaction between NFI and pTP–pol. The orientation and spacing between the NFI site and the 1–18 core origin sequence is critical in adenovirus DNA replication. Insertion of additional sequence between these two sequence regions abolished NFI-mediated stimulation of DNA replication in vitro (ADHYA et al. 1986; WIDES et al. 1987) and in vivo (BOSHER et al. 1990). This suggested that a strict constraint on the spatial arrangement between the 1–18 core sequence and the NFI site existed which was neccessary to allow specific protein–protein interactions between NFI and other replication proteins. A direct interaction between NFI and pol was subsequently demonstrated (BOSHER et al. 1990; CHEN et al. 1990; MUL et al. 1990), the function of which was to stabilise the relatively weak interaction between the pTP–pol heterodimer and its recognition site in the adenovirus origin of DNA replication.

6.2 Nuclear Factor III

The host encoded cellular factor NFIII was initially identified in HeLa nuclear extracts through its ability to stimulate the initiation of adenovirus DNA replication in vitro in the presence of NFI (PRUIJN et al. 1986). Purified NFIII has a molecular weight of 92 kDa and, when bound to its recognition site in the adenovirus origin, stimulates the level of initiation in vitro three- to sevenfold (O'NEILL and KELLY 1988). A combination of DNase I footprinting and methylation protection studies on Ad2 identified the core binding site as 5'-TATGATAAT-3', which is situated between nucleotides 39 and 48 in the Ad2 ITR. NFIII was recognised as a member of the octamer transcription factor family by O'NEILL et al. (1988) and PRUIJIN et al. (1989), who demonstrated that NFIII was indistinguishable from octamer transcriptiion factor (OTF-1). Many more octamer-binding proteins have been identified which contain a conserved DNA-binding region known as the POU domain. NFIII contains this domain, which like the N-terminal domain of NFI is sufficient for the in vitro stimulation of Ad2 DNA replication (VERRIJZER et al. 1990). While NFIII clearly stimulates Ad2 DNA replication in vitro, the mechanism by which it does so has yet to be determined. In vivo the situation is less clear, as mutant Ad2 viruses containing genomes with a deleted NFIII site are as infectious as wild-type virus with an intact NFIII site (HAY and McDOUGALL 1986). However, transfection assays on plasmids containing only the 1–18 core origin of replication linked to an NFIII site (and no NFI site) found that these templates could replicate more efficiently than plasmids containing only the 1–18 core origin of replication (HAY et al. 1988). This confirmed in vitro replication studies by MUL et al. (1990), who found that both the NFIII and NFI-binding sites were needed for optimal stimulation of replication. Since the binding sites of the two proteins overlapped, this suggested that they could interact in a co-operative manner to stimulate DNA replication. However, this does not appear to be the case, it seems that both proteins can, by binding to their respective recognition sites, stimulate initiation independently (MUL et al. 1990). It has been suggested that NFIII-induced bending of the DNA at the origin of DNA replication may stimulate DNA replication by promoting interactions between the various components in the pre-initiation complex (VERRIJZER et al. 1991).

7 A Model for the Initiation of Adenovirus DNA Replication

Based on the observations described above, a model outlined in Fig. 1 has been formulated to describe events that lead to the initiation of adenovirus DNA replication. Ad2 DBP, originally classified as a single-stranded DBP, but which also binds to double-stranded DNA, is produced in such large amounts that it is likely that in vivo all template molecules are coated with the protein. This has the effect of increasing the affinity of NFI for its recognition site in the Ad2 ori and thus ensuring that the NFI-binding site in ori is fully occupied (CLEAT and HAY 1989, STUIVER and VAN DER VLIET 1990). A direct protein–protein interaction between NFI and pol then recruits the pTP–pol heterodimer into the preinitiation complex (BOSHER et al. 1990; CHEN et al. 1990; MUL et al. 1990). Correct positioning of the pTP–pol heterodimer at the origin is accomplished by an interaction between pTP–pol and the DNA sequence from base pairs 9–18 that is perfectly conserved in all human adenoviruses sequenced to date and is regarded as the core of the Ad2 ori (MUL and VAN DER VLIET 1992; TEMPERLEY and HAY 1992). Further stabilisation of this complex may be accomplished by interactions between the incoming pTP–pol and genome-bound TP (PRONK and VAN DER VLIET 1993). At this stage the origin must unwind to expose the single-stranded DNA that is the template for DNA synthesis. The helix-destabilising properties of DBP may be particularly important in catalysing this reaction. Chemical modification experiments have recently demonstrated that this unwinding takes place in the highly conserved region within the core of the viral origin of DNA replication (I. LEITH and R.T. HAY, unpublished work).

References

Adhya S, Schneidman Hurwitz J (1986) Reconstruction of adenovirus replication origins with a human nuclear factor 1 binding site. J Biol Chem 261: 3339–3346

Bernad A, Zaballos A, Salas M, Blanco L (1987) Structural and functional relationships between procaryotic and eukaryotic DNA polymerases. EMBO J 6: 4219–4225

Bernad A, Blanco L, Lazaro JM, Martin G, Salas M (1989) A conserved 3'–5' exonuclease active site in prokaryotic and eukaryotic DNA polymerases. Cell 59: 219–228

Boehmer PE, Lehman IR (1993) Herpes simplex virus type I ICP8: helix-destabilising properties. J Virol 67: 711–715
Bosher J, Robinson EC, Hay RT (1990) Interactions between the adenovirus type 2 polymerase and the DNA binding domain of nuclear factor I. New Biol 2: 1083–1090
Bosher J, Leith IR, Temperley SM, Wells M, Hay RT (1991) The DNA-binding domain of nuclear factor I is sufficient to cooperate with the adenovirus type 2 DNA-binding protein in viral replication. J Gen Virol 72: 2975–2980
Challberg MD, Kelly TJ (1979) Adenovirus DNA replication in vitro. Proc Natl Acad Sci USA 76: 655–659
Challberg MD, Rawlins DR (1984) Template requirements for the initiation of adenovirus DNA replication. Proc Natl Acad Sci USA 81: 100–104
Challberg MD, Desiderio SV, Kelly TJ (1980) Adenovirus DNA replication in vitro: characterisation of a protein covalently linked to nascent DNA strands. Proc Natl Acad Sci USA 77: 5105–5109
Chen M, Mermod N, Horwitz MS (1990) Protein-protein interactions between adenovirus DNA polymerase and nuclear factor I mediate formation of the DNA replication pre-initiation complex. J Biol Chem 265: 18634–18642
Cleat PH, Hay RT (1989) Co-operative interactions between NF-I and the adenovirus DNA binding protein at the origin of DNA replication. EMBO J 8: 1841–1848
Cleghorn V, Klessig DF (1992) Characterisation of the nucleic acid binding region of adenovirus DNA binding protein by partial proteolysis and photochemical cross-linking. J Biol Chem267: 17872–17881
De Vries E, van Driel W, Tromp M, van Boom J, Van der Vliet PC (1985) Adenovirus DNA replication in vitro: site directed mutagenesis of the nuclear factor I binding site of the Ad. origin. Nucleic Acids Res 13: 4943–4952
De Vries E, van Driel W, van den Heuvel SJL, Van der Vliet PC (1987) Contact point analysis of the HeLa nuclear factor I recognition site reveals symmetrical binding at one side of the helix. EMBO J 6: 161–168
Eagle PA, Klessig DF (1992) A zinc-binding motif located between amino acids 273 and 286 in the adenovirus DNA-binding protein is necessary for ssDNA binding. Virology 187: 777–787
Enomoto T, Lichy JH, Ikeda JE, Hurwitz J (1981) Adenovirus replication in vitro: Purification of the terminal protein in a functional form. Proc Natl Acad Sci USA 78: 6779–6783
Field J, Gronostajski RM, Hurwitz J (1984) Properties of the adenovirus DNA polymerase. J Biol Chem 259: 9487–9495
Fredman JN, Engler JA (1993) Adenovirus precursor to terminal protein interacts with the nuclear matrix in vivo and in vitro. J Virol 67: 3384–3395
Fredman JN, Pettit SC, Horwitz MS, Engler JA (1991) Linker insertion mutations in the adenovirus preterminal protein that affect DNA replication in vivo and in vitro. J Virol 65: 4591–4597
Freimuth PI, Ginsberg HS (1986) Codon insertion mutants of the adenovirus terminal protein. Proc Natl Acad Sci USA 83: 7816–7820
Friefeld BR, Krevolin MD, Horwitz MS (1983) Effect of the adenovirus H5ts125 and Hts107 DNA binding proteins on DNA replication in vitro. Virology 124: 380–389
Georgaki A, Hubscher U (1993) DNA unwinding by replication factor A is a property of the 70KDa subunit and is facilitated by phosphorylation of the 32KDa subunit. Nucleic Acids Res 21: 3659–3665
Georgaki A, Strack B, Podust V, Hubscher U (1992) DNA unwinding activity of replication protein A. FEBS Lett 308: 240–244
Gounari F, De Francesco R, Schmitt J, Van der Vliet PC, Cortese R, Stunnenberg H (1990) Amino terminal domain of NF-I binds to DNA as a dimer and activates adenovirus DNA replication. EMBO J 9: 559–566
Guggenheimer RA, Nagata K, Kenny M, Hurwitz J (1984) Protein primed replication of plasmids containing the terminus of the adenovirus genome 2: purification and characterisation of a host protein required for the replication of DNA templates devoid of the terminal protein. J Biol Chem 259: 7815–7825
Harris MPG, Hay RT (1988) DNA sequences required for the initiation of adenovirus type 4 DNA replication J Mol Biol 201: 57–67
Hay RT (1985a) The origin of adenovirus DNA replication: minimal DNA sequence in vivo. EMBO J 4: 421–426
Hay RT (1985b) The origin of adenovirus DNA replication: role of the nuclear factor I site in vivo. J Mol Biol 186: 129–136
Hay RT, McDougall IM (1986) Viable viruses with deletions in the left inverted terminal repeat define the origin of Ad2 DNA replication. J Gen Virol 6: 321–332

Hay RT, Russell WC (1989) Recognition mechanisms in the synthesis of animal virus DNA's. Biochem J 258: 3–16

Hay RT, Stow ND, McDougall IM (1984) Replication of adenovirus mini chromosomes. J Mol Biol 175: 493–510

Hay RT, Clark L, Cleat PH, Harris MPG, Robertson EC, Watson CJ (1988) Requirements for the initiation of adenovirus types 2 and 4 DNA replication. Cancer Cells 6: 71–75

Huberman JA, Kornberg A (1971) Stimulation of T4 bacteriophage DNA polymerase by the protein product of gene 32. J Mol Biol 62: 39–52

Ikeda J-E, Enomoto T, Hurwitz J (1982) Adenovirus protein primed initiation of DNA chains in vitro. Proc Natl Acad Sci USA 79: 2442–2446

Jones KA, Kadonaga JT, Rosenfeld PJ, Kelly TJ, Tjian R (1987) A cellular DNA binding protein that activates eukaryotic transcription and DNA replication. Cell 48: 79–89

Joung I, Engler JA (1992) Mutations in two cysteine-histidine-rich clusters in adenovirus type 2 DNA polymerase affect DNA binding. J Virol 66: 5788–5796

Kenny MK, Hurwitz J (1988) Initiation of adenovirus DNA replication. II. Structural requirements using synthetic oligonucleotide adenovirus templates. J Biol Chem 263: 9801–9808

Kenny MK, Balogh LA, Hurwitz J (1988) Initiation of adenovirus DNA replication. I. Mechanism of action of a host protein required for replication of adenovirus DNA templates devoid of the terminal protein. J Biol Chem 263: 9809–9817

Kenny MK, Lee S-H, Hurwitz J (1989) Multiple functions of human single-stranded DNA binding protein in simian virus 40 DNA replication: single-strand stabilisation and stimulation of DNA polymerases α and δ. Proc Natl Acad Sci USA 86: 9757–9761

Kenny MK, Schlegel U, Furneaux H, Hurwitz J (1990) The role of human single-stranded DNA binding protein and its individual subunits in simian virus 40 DNA replication. J Biol Chem 265: 7693–7700

Kitchingman GR (1985) Sequence of the DNA binding protein of a human subgroup E adenovirous (type 4): comparisons with subgroup A (type 12), subgroup B (type 7) and subgroup C (type 5). Virology 146: 90–101

Klein H, Maltzman W, Levine AJ (1979) Structure function relationships of the adenovirus DNA binding protein. J Biol Chem 254: 11051–11060

Kruijer W, van Schaik FAM, Sussenbach JS (1981) Structure and organisation of the gene coding for the DNA binding protein of adenovirus type 5. Nucleic Acids Res 9: 4439–4450

Lally C, Dorper T, Groger W, Antoine G, Winnacker E-L (1984) A size analysis of the adenovirus replicon. EMBO J 3: 333–337

Larder BA, Kemp SD, Darby G (1987) Related functional domains in virus DNA polymerases. EMBO J 6: 169–175

Leegwater PAJ, van Driel W, Van der Vliet PC (1985) Recognition site of nuclear factor I, a sequence specific DNA binding protein from HeLa cells that stimulates adenovirus DNA replication. EMBO J 4: 1515–1521

Lichy JH, Field J, Horwitz MS, Hurtwitz J (1981) Formation of a covalent complex between the 80kDa adenovirus terminal protein and 5'-dCMP in vitro. Proc Natl Acad Sci USA 78: 2678–2682

Lichy JH, Field J, Horwitz MS, Hurwitz J (1982) Separation of the adenovirus terminal protein precursor from its associated DNA polymerase: role of both proteins in the initiation of adenovirus DNA replication. Proc Natl Acad Sci USA 79: 5225–5229

Lindenbaum JO, Field J, Hurwitz J (1986) The adenovirus DNA binding protein and adenovirus DNA polymerase interact to catalyse elongation of primed DNA templates. J Biol Chem 261: 10218–10227

Mangel WF, McGrath WJ, Toledo D, Anderson CW (1993) Viral DNA and a viral peptide are cofactors of adenovirus virion proteinase activity. Nature 361: 274–275

Marcy AI, Yager DR, Coen DM (1990) Engineered herpes simplex virus DNA polymerase point mutants: the most highly conserved region shared among α-like DNA polymerases is involved in substrate recognition. J Virol 64: 2208–2216

Meisterernst M, Rogge L, Foeckler R, Karaghiosoff M, Winnacker E-L (1989) Structural and functional organisation of a porcine gene coding for nuclear factor I. Biochemistry 28: 8191–8200

Mermod N, O'Neill EA, Kelly TJ, Tjian R (1989) The proline rich transcriptional activator of CTF/NF-I is distinct from the replication and DNA binding domain. Cell 58: 741–753

Monaghan A, Hay RT (1994) Pyridoxal phosphate inhibition of the adenovirus DNA polymerase. (submitted for publication)

Monaghan A, Webster AG, Hay RT (1994) Adenovirus DNA binding protein: helix destabilising property. Nucleic Acids Res 5: 742–748

Mul YM, Van der Vliet PC (1992) Nuclear factor I enhances Adenovirus DNA replication increasing the stability of a preinitiation complex. EMBO J 11: 751–760
Mul YM, Van der Vliet PC (1993) The adenovirus DNA binding protein effects the kinetics of DNA replication by a mechanism distinct from NF-I or Oct-1. Nucleic Acids Res 21: 641–647
Mul YM, Verrijzer CP, Van der Vliet PC (1990) Transcription factors NF-I and NF-III/Oct-1 function independently, employing different mechanisms to enhance adenovirus DNA replication. J Virol 64: 5510–5518
Nagata K, Guggenheimer RA, Enomoto T, Lichy JH, Hurwitz J (1982) Adenovirus DNA replication in vitro: identification of a host factor that stimulates synthesis of the preterminal protein-dCMP complex. Proc Natl Acad Sci USA 79: 6438–6442
Nagata K, Guggenheimer RA, Hurwitz J (1983) Specific binding of acellular DNA replication protein to the origin of replication of adenovirus DNA. Proc Natl Acad Sci USA 80: 6177-6181
Neale GAM, Kitchingman GR (1990) Conserved region 3 of the adenovirus type 5 DNA binding protein is important for interaction with single-stranded DNA. J Virol 64: 630–638
O'Neill EA, Kelly TJ (1988) Purification and characterisation of nuclear factor III (origin recognition protein C), a sequence specific DNA binding protein required for efficient initiation of adenovirus DNA replication. J Biol Chem 263: 931–937
O'Neill EA, Fletcher C, Burrow CR, Heintz N, Roeder RG, Kelly TJ (1988) The transcription factor OTF-1 is functionally identical to the adenovirus DNA replication factor NF-III. Science 241: 1210–1213
Paonessa G, Gounari F, Frank R, Cortese R (1988) Purification of an NFI-like binding protein from rat liver and cloning of the corresponding cDNA. EMBO J 7: 3115–3123
Pettit SC, Horwitz MS, Engler JA (1988) Adenovirus preterminal protein synthesised in COS cells from cloned DNA is active in DNA replication in vitro. J Virol 62: 496–500
Pettit SC, Horwitz MS, Engler JA (1989) Mutations in the precursor to the terminal protein of adenovirus serotypes 2 and 5. J Virol 63: 5344–5350
Pronk R, Van der Vliet PC (1993) The adenovirus terminal protein influences binding of replication proteins and changes the origin structure. Nucleic Acids Res 21: 2293–2300
Pronk R, Stuiver MH, Van der Vliet PC (1992) Adenovirus DNA replication: the function of the covalently bound terminal protein. Chromosoma 102, 39–45
Pruijin GJM, Van Driel W, Van der Vliet PC (1986) A novel sequence specific DNA binding protein from HeLa cells stimulating adenovirus DNA replication. Nature 322: 656–659
Pruijin GJM, Van der Vliet PC, Dathan NA, Mattaj IW (1989) Anti-OTF-1 antibodies inhibit NF-III stimulation in vitro adenovirus DNA replication. Nucleic Acids Res 17: 1845–1863
Ramachandra M, Padmanabhan R (1993) Adenovirus DNA polymerase is phosphorylated by a stably associated histone H1 kinase. J Biol Chem 268: 17448–17456
Ramachandra M, Nakano R, Mohan PM, Rawitch AB, Padmanabhan P (1993) Adenovirus DNA polymerase is a phosphoprotein. J Biol Chem 268: 442–448
Rancourt C, Tihanyi K, Bourbonniere M, Weber JM (1994) Identification of active-site residues of the adenovirus endopeptidase. Proc Natl Acad Sci USA 91: 844–847
Rawlins DR, Rosenfeld PJ, Wides RJ, Challberg MD, Kelly TJ (1984) Structure and function of the adenovirus origin of DNA replication. Cell 37: 309–319
Reuben RC, Gefter MC (1973) A DNA-binding protein induced by bacteriophage T7. Proc Natl Acad Sci USA 70: 1864–1870
Roovers DJ, van der Lee FM, van der Wees J, Sussenbach JS (1993) Analysis of the adenovirus type 5 terminal protein precursor and DNA polymerase by linker insertion mutagenesis. J Virol 67: 265–276
Rosenfeld PJ, O'Neil, EA, Wides RJ, Kelly TJ (1987) Sequence specific interactions between cellular DNA-binding proteins and the adenovirus origin of replication. Mol Cell Biol 7:875–886
Salas M (1991) Protein-priming of DNA replication. Annu Rev Biochem 60: 39–71
Santoro C, Mermod N, Andrews PC, Tjian R (1988) A family of human CCAAT box binding proteins active in transcription and DNA replication: cloning and expression of multiple cDNA's. Nature 334: 218–224
Schaak J, Yew-Hai Ho W, Freimuth P, Shenk T (1990) Adenovirus terminal protein mediates both nuclear matrix association and efficient transcription of adenovirus DNA. Genes Dev 4: 1197–1208
Schneider R, Gander I, Muller U, Mertz R, Winnacker EL (1986) A sensitive and rapid gel retention assay for nuclear factor I and other DNA binding proteins in crude nuclear extracts. Nucl Acid Res 14: 1303–1317
Shu L, Horwitz MS, Engler JA (1987) Expression of enzymatically active adenovirus DNA polymerase from cloned cDNA requires sequences upstream of the main open reading frame. Virology 161: 520–526

Stillman B (1989) Initiation of eukaryotic DNA replication in vitro. Annu Rev Cell Biol 5: 197–245
Stillman B, Lewis JB, Chow LT, Mathews MB, Smart JE (1981) Identification of the gene and mRNA for the adenovirus terminal protein precursor. Cell 23: 497–508
Stillman B, Tamanoi F, Matthews MB (1982) Purification of an adenovirus-coded DNA polymerase that is required for initiation of DNA replication. Cell 31: 613–623
Stow ND (1982) The infectivity of adenovirus genomes lacking DNA sequence from their left hand termini. Nucleic Acids Res 10: 5105–5119
Strum RA, Herr W (1988) The POU domain is a bipartite DNA-binding structure. Nature 336: 601–604
Strum RA, Das G, Herr W (1988) The ubiquitous octamer binding protein oct-1 contains a POU domain and a homeobox subdomain. Genes Dev 2: 1582–1599
Stuiver MH, Van der Vliet PC (1990) Adenovirus DNA binding protein forms a multimeric protein complex with double-stranded DNA and enhances binding of nuclear factor I. J Virol 65: 379–389
Stuiver MH, Bergsma WG, Arnberg AC, Van Amerongen H, Van Grondelle R, Van der Vliet PC (1992) Structural alterations of double-stranded DNA in a complex with the adenovirus DNA-binding protein: implications for its function in DNA replication. J Mol Biol 225: 99–1011
Stunnberg HG, Lange H, Philipson L. Van Miltenburg RT, Van der Vliet PC (1988) High expression of functional adenovirus DNA polymerase and precursor terminal protein using recombinant vaccinia virus. Nucleic Acids Res 16: 2431–2444
Tarmanoi F, Stillman B (1982) Function of adenovirus terminal protein in the initiation of DNA replication. Proc Natl Acad Sci USA 79: 2221–2225
Tamanoi F, Stillman B (1983) Initiation of adenovirus DNA replication in vitro requires a specific DNA sequence. Proc Natl Acad Sci USA 80: 6446–6450
Temperley SM, Hay RT (1991) Replication of adenovirus type 4 DNA by a purified fraction from infected cells. Nucleic Acids 19: 3243–3249
Tempereley SM, Hay RT (1992) Recognition of the adenovirus type 2 origin of replication by the virally encoded DNA polymerase and preterminal proteins. EMBO J 11: 761–768
Temperely SM, Burrow CR, Kelly TJ, Hay RT (1991) Identification of two distinct regions within the adenovirus minimal origin of replication that are required for adenovirus type 4 DNA replication in vitro. J Virol 65: 5037–5044
Tucker PA, Tsernoglou D, Tucker AD, Coenjaerts FEJ, Leenders H, Van der Vliet PC (1994) X-Ray crystal structure of the 39KDa C-terminal adenovirus DNA binding protein. EMBO J 13: 2994–3002
Van Bergen BGM, Van der Ley PA, Van Driel W, Van Mansfeld ADM, Van der Vliet PC (1983) Replication of origin containing adenovirus DNA fragments that do not carry the terminal protein. Nucleic Acids Res 11: 1975–1989
Van der Vliet PC (1990) Adenovirus DNA replication in vitro. In: Straus P, Wilson SH (eds) The eukaryotic nucleus. Telford, Caldwell
Van der Vliet PC, Levine AJ (1973) DNA binding proteins specific for cells infected by adenovirus. Nature New Biol 246: 170–174
Van der Vliet PC, Sussenbach JS (1975) An adenovirus type 5 gene function required for initiation of viral DNA replication. Virology 67: 415–426
Van der Vliet PC, Zandberg J, Jansz HS (1977) Evidence for the function of the adenovirus DNA binding protein in initiation of DNA synthesis as well as in elongation on nascent DNA chains. Virology 80: 98–110
Verrijzer CP, Kal AJ, Van der Vliet PC (1990) The DNA binding domain (POU domain) of transcription factor Oct-1 suffices for stimulation of DNA replication. EMBO J 9: 1883–1888
Verrijzer CP, Van Oosterhout JAW, Van Weperen WW, Van der Vliet PC (1991) POU Proteins bend DNA via the POU-specific domain. EMBO J 10: 3007–3014
Verrijzer CP, Van Oosterhout JAW, Van der Vliet PC (1992) The Oct-1 POU domain mediates interactions between Oct-1 and other POU proteins. Mol Cell Biol 12: 542–551
Vos HL, van der Lee FM, Reemst AMCB, van Loon AE, Sussenbach JS (1988) The genes encoding the DNA binding protein and the 23 K protease of adenovirus types 40 and 41. Virology 163: 1–10
Watson CJ, Hay RT (1990) Expression of adenovirus type 2 DNA polymerase in insect cells infected with a recombinant baculovirus. Nucleic Acids Res 18: 1167–1173
Weber J (1990) The adenovirus proteinases. Semin Virol 1: 379–384
Webster A, Russell S, Talbot P, Russell WC, Kemp GD (1989) Characterisation of the adenovirus protease: substrate specificity J Gen Virol 70: 3225–3234
Webster A, Hay RT, Kemp G (1993) The adenovirus protease is activated by a virus-coded disulphide-linked peptide. Cell 72: 97–104
Webster A, Leith I, Hay RT (1994) Activation of the adenovirus coded protease and processing of the preterminal protein. J Virol (in press)

Wides RJ, Challberg MD, Rawlins DR, Kelly TJ (1987) Adenovirus origin of replication: sequence requirements for replication in vitro Mol Cell Biol 7: 864–874

Zijderveld DC, Van der Vliet PC (1994) Helix-destablizing properties of the adenovirus DNA-binding protein. J Virol 68: 1158–1164

Zijderveld DC, Stuiver MH, Van der Vliet PC (1993) The adenovirus DNA binding protein enhances intermolecular renaturation but inhibits intramolecular DNA renaturation. Nucleic Acids Res 21: 641–647

Expression, Nuclear Transport, and Phosphorylation of Adenovirus DNA Replication Proteins

M. RAMACHANDRA and R. PADMANABHAN

Department of Biochemistry and Molecular Biology, University of Kansas Medical Center, Kansas City, KS 66160-7421, USA

1 Introduction

DNA tumor viruses have contributed immense wealth of knowledge in the past few years regarding the eukaryotic cellular processes involving replication, transcription, and translation. Adenoviruses (Ad) in particular have played a pioneering and significant role in the understanding of the mechanisms of many of these biological processes mainly due to the interaction of viral proteins with the host proteins during the virus life cycle. The development of the first cell-free system to study Ad DNA replication (CHALLBERG and KELLY 1979; for reviews, see CHALLBERG and KELLY 1989; STILLMAN 1989; HAY and RUSSELL 1989) was pivotal to our current understanding of eukaryotic DNA replication.

The Ad genome contains a linear double-stranded DNA of 36 kb in length, and at each 5' end a 55-kDa protein (terminal protein, TP) is covalently linked through the β-hydroxyl group of a serine via a phosphodiester bond (REKOSH et al. 1977; CHALLBERG et al. 1980; RONINSON and PADMANABHAN 1980; STILLMAN 1981; DESIDERIO and KELLY 1981; CHALLBERG and KELLY Jr 1981; SMART and STILLMAN 1982). The viral DNA replication takes place in the nucleus of the host cell by a strand displacement mechanism through the interaction of three viral proteins, preterminal protein (pTP), DNA polymerase (AdPol), and the DNA-binding protein (DBP), and at least three cellular proteins, nuclear factors (NF) I, II, and III, as well as the origin of DNA replication (ori DNA). The ori DNA is located at both ends of the viral genome within the first 51 bp of the inverted terminal repeat sequences, and depending on the Ad serotype the length of the inverted terminal repeat varies from 102 to 162 bp (ARRAND and ROBERTS 1979, SHINAGAWA and PADMANABHAN 1979, 1980; TOLUN et al. 1979; TOKUNAGA et al. 1982; STILLMAN et al. 1982). The sequences required for Ad replication initiation include the minimal core origin between nucleotides 1 and 18 (domain A in WIDES et al. 1987) and an auxiliary region located between nucleotides 19 and 51 (domain B between 19 and 40, domain C between 41 and 51) containing binding sites for the host transcription factors, NFI (or CCAAT-box transcription factor, CTF; NAGATA et al. 1982; RAWLINS et al. 1984; LEEGWATER et al. 1985; JONES et al. 1987; WIDES et al. 1987; DE VRIES et al. 1989; GOUNARI et al. 1990; MUL and VAN DER VLIET 1992), and NFIII (Oct-1; PRUIJN et al. 1986, 1987; O'NEILL and KELLY 1988; O'NEILL et al. 1988).

The viral as well as the host proteins that participate in initiation and elongation reactions and the template requirements for initiation have been identified using the in vitro system (for reviews, see CHALLBERG and KELLY 1989; STILLMAN 1989; HAY and RUSSELL 1989). The DNA replication starts by a novel protein-priming mechanism in which AdPol catalyzes the covalent linkage of the 5'-terminal nucleotide deoxycytidine monophosphate (dCMP) to the β-OH of a serine residue of the pTP which then serves as a primer for elongation reaction. Studies have shown that NFI binding recruits the AdPol-pTP complex to the minimal ori through specific protein–protein and protein–DNA interactions (KENNY and HURWITZ 1988; CHEN et al. 1990; MUL and VAN DER VLIET 1992) to form the preinitiation complex. It has been suggested that a single-stranded region might be generated resulting from the interaction of NFI with its binding site followed by unwinding of the duplex DNA at the core origin sequence (NAGATA et al. 1983b; CHALLBERG and RAWLINS 1984). In support of this model, a single-stranded oligodeoxynucleotide complementary to the template core origin sequence was found to inhibit the initiation of DNA replication, suggesting that formation of single-stranded region involving the core origin sequence is a prerequisite for the initiation reaction (DOBBS et al. 1990).

Ad2- (or Ad5-) encoded DBP is thought to increase the affinity of NFI for its binding site (DE VRIES et al. 1985; KENNY and HURWITZ 1988; CLEAT and HAY 1989; STUIVER and VAN DER VLIET 1990; BOSHER et al. 1991) and stimulates the NFI-dependent formation of pTP-dCMP initiation complex (NAGATA et al. 1982; DE VRIES et al. 1985). DBP binds preferentially in a nonsequence-specific manner to single-stranded DNA (VAN DER VLIET and LEVINE 1973; LEVINE et al. 1976; LINNE et al. 1977; SUGAWARA et al. 1977; SCHECHTER et al. 1980) and plays an essential function in the AdPol-catalyzed elongation step of viral DNA replication (VAN DER VLIET et al. 1977; HORWITZ 1978; KEDINGER et al. 1978; KAPLAN et al. 1979; FRIEFELD et al. 1983; OSTROVE et al. 1983; VAN BERGEN and VAN DER VLIET 1983; for reviews, see HORWITZ 1990; VAN DER VLIET 1990; K.R. WILLIAMS and CHASE 1990). The rate of DNA synthesis as well as the processivity of AdPol increases in the presence of DBP during DNA chain elongation (FIELD et al. 1984; LINDENBAUM et al. 1986). In addition to its DNA replication function, DBP is involved in early (CARTER and BLANTON 1978; BABICH and NEVINS 1981; HANDA et al.1983) as well as late viral gene expression (KLESSIG and GRODZICKER 1979) and in transformation (GINSBERG et al. 1974).

Synthesis of full-length Ad DNA by AdPol in the in vitro DNA replication system requires, in addition to DBP, the NFII, a type I DNA topoisomerase (NAGATA et al. 1983a, b). More recently, the requirement for both topoisomerases I and II in Ad DNA replication was demonstrated by using camptothecin and VM26 as the respective inhibitors of these enzymes (SCHAACK et al. 1990a). Inhibition of topoisomerase I activity by camptothecin resulted in an immediate block of DNA replication, whereas the inhibition of topoisomerase II activity by VM26 occurred only after completion of approximately one additional round of DNA replication (SCHAACK et al. 1990a).

2 Expression of Adenovirus DNA Replication Proteins

A detailed biochemical characterization of AdPol and pTP has been difficult because of their low levels of synthesis in Ad-infected cells. In Ad2-infected cells, AdPol and pTP exist as a 1:1 stoichiometric complex, and attempts to purify this complex from Ad-infected cells resulted in low yields. For example, ENOMOTO et al. (1981) described a procedure for the purification of this protein complex about 1000-fold starting from Ad2-infected HeLa (2.5×10^{10}) cells through several steps with an overall yield of about 7% of total activity and less than 110 μg protein. In order to carry out a detailed biochemical characterization and identify the functional domains of the Ad replication proteins, it was necessary to express these proteins using eukaryotic expression systems.

2.1 Transcription Unit of E2 Region

The mRNA for AdPol, pTP, and DBP are transcribed early from a single promoter at genome coordinate 75. The 59-kDa DBP is encoded by E2A region which shares common RNA leader sequences near genome coordinates 75 and 68 with mRNA for the pTP and AdPol. The pTP and the AdPol proteins are encoded by the E2B region of the viral genome and the mRNA share a common exon at genome coordinate 39 (STILLMAN et al. 1981; GINGERAS et al. 1982). These short exons are spliced to the main body of the open reading frames (m-ORF) for pTP and AdPol at genome coordinates 28.9 and 24.1, respectively.

2.2 Expression of Adenovirus DNA Polymerase and Preterminal Protein in Biologically Active Forms Using Transient Expression Systems

Several lines of experimental evidence suggested that the region upstream of the first translation initiation codon ATG of pTP and AdPol m-ORF might be important for biological activity. First, it had been suggested that the mRNA leader at coordinate 39 may encode the N-terminal amino acids of pTP, based on the observation that the sequence upstream of the first ATG of pTP and AdPol m-ORF is highly conserved among several serotypes (SMART and STILLMAN 1982; SHU et al. 1986). Second, a linker scanning mutation that was mapped to the region upstream of the first ATG codon of pTP m-ORF was conditionally defective for DNA replication (FREIMUTH and GINSBERG 1986). Third, when the m-ORF of AdPol between genome coordinates 22.9 and 14.2, containing the first ATG at residue 8367, was placed under the control of *lac* promoter, the resulting expression construct produced a protein of 120 kDa in size in *Escherichia coli* upon induction with isopropyl-beta-D-thiogalacto-pyranoside (IPTG). However, the rabbit polyclonal antibodies raised against this *E. coli*-produced protein recognized a 140-kDa protein in the nucleus of the Ad2-infected cells at late phase of Ad2 infection (SASAGURI et al. 1987). Since the size of the *E. coli*-expressed AdPol on sodium dodecyl sulfate polyacrylamide gel electrophoresis (SDS-PAGE) was smaller than the native protein, it was suggested that the 140-kDa native AdPol contained additional amino acids coded by the short upstream exon (SASAGURI et al. 1987). Evidence that this upstream exon is important for the biological activity of AdPol and pTP came from the experiments in which the eukaryotic expression plasmids containing the *HindIIIJ* fragment encoding the exon at genome coordinate 39 cloned upstream of the AdPol m-ORF (SHU et al. 1987) or the pTP m-ORF (PETTIT et al. 1988) produced biologically active AdPol and pTP in transient expression systems. Independently, it was observed that synthetic linkers providing the translation initiation codon inserted upstream of the pTP and AdPol m-ORF also produced pTP and AdPol proteins in the transfected cells which were biologically active in the in vitro initiation of DNA replication, whereas the proteins expressed from the first ATG of the m-ORF of AdPol and pTP were not (ZHAO and PADMANABHAN 1988). The precise sequence of this exon from the genome coordinate 39 revealed that this exon, encoding three amino acids residues M-A-L, was spliced to the m-ORF of pTP and AdPol (SHU et al. 1988).

2.3 Expression of Adenovirus DNA Polymerase and Preterminal Protein Using Recombinant Vaccinia Virus Expression Systems

The DBP is expressed abundantly both early and late during the Ad life cycle (VAN DER VLIET and LEVINE 1973; LEWIS et al. 1976; SABORIO and OBERG 1976; FLINT and SHARP 1976; LEVINSON and LEVINE 1977), and therefore the amounts produced in the Ad-infected cells are sufficient for detailed biochemical analyses. For this reason, considerably more information is available regarding the function of DBP than the other two replication proteins. In contrast, the amounts of AdPol and pTP produced in Ad-infected cells are very low, and in the transient expression systems they are still not sufficient in order to launch a detailed biochemical characterization of these proteins. STUNNENBERG et al. (1988) reported successful expression of pTP and AdPol using recombinant vaccinia virus expression system. The vaccinia expression vector pATA-18 containing a mutated 11k late promoter (–100 to +8) was used. Both AdPol and pTP were produced at levels at least 30-fold higher compared to extracts from Ad-infected cells and were active in the in vitro replication assay.

AdPol and pTP proteins have been overproduced using the vaccinia expression vector pTM1 (described in MOSS et al. 1990), in which the wild-type coding sequences of AdPol and pTP were cloned under the control of the bacteriophage T7 promoter and the 5'-untranslated leader sequences of the encephalomyocarditis virus (EMCV 5'-UTR) placed between the promoter and the gene (NAKANO et al. 1991). It was shown in earlier studies that in the absence of EMCV 5'-UTR sequences, although the RNA transcribed from the T7 promoter amounted to about 30% of the total steady state RNA in the cytoplasm of the infected cells, only a small fraction (about 5%–10%) of these transcripts were capped and methylated, which restricted the amount of translatable RNA in the cytoplasm (FUERST and MOSS 1989). By the placement of the EMCV 5'-UTR immediately downstream from the T7 promoter, the translation of the transcripts was made cap independent, which increased the level of expression five- to ten-fold to about 10% of the total cell protein (ELROY-STEIN et al. 1989). The levels of expression of AdPol and pTP were higher in this T7 hybrid system (NAKANO et al. 1991) than the previously used vaccinia expression vectors, and routinely at least 1–2 mg protein could be obtained from one liter HeLa cells infected with the recombinant vaccinia virus. However, much of the protein produced in the recombinant vaccinia virus-infected cells was found to be associated with the nuclear pellet fraction, possibly due to aggregation resulting from over production. The protein in the pellet fraction could be solubilized using buffers of high ionic strengths (1 *M* NaCl).

This recombinant vaccinia virus/T7 promoter hybrid expression system offers some advantages over the other vaccinia expression systems used earlier in that site-directed mutants on pTP and AdPol can be constructed in the pTM1-based expression plasmids and their biological activities can be assayed rapidly using the "infection-transfection" protocol (Moss et al. 1990) without the necessity of constructing a recombinant vaccinia virus for each mutant. The amounts of heterologous protein produced in this transient expression system are sufficiently high to carry out a number of biochemical analyses.

2.4 Expression of Adenovirus DNA Polymerase and Preterminal Protein in Insect Cells Using a Baculovirus Vector

The expression in insect cells using the baculovirus *Autographa californica* nuclear polyhedrosis virus (AcNPV) is currently one of the most widely used heterologous gene expression systems for structure and function studies of eukaryotic proteins, because the proteins produced using this system have been shown to undergo proper folding and post-translational processing and thus maintain biological activity (Miller 1988; Lucknow and Summers 1988). Using the baculovirus transfer vector pACRP23 (Matsuura et al. 1987), the AdPol was expressed in insect cells to the level of 2–3 mg/l infected cells (2×10^9 cells). The recombinant AdPol was biologically active in the in vitro DNA replication assay (Watson and Hay 1990). A different baculovirus transfer vector was constructed in which the polyhedrin promoter was assembled from synthetic oligodexynucleotides, and a transcription unit for the expression of the *lac*Z gene under the control of RSV-LTR promoter was inserted in the opposite orientation to that of the polyhedrin promoter. The coexpression of *lac*Z allowed easy identification of the recombinant plaques generated from in vivo recombination. Using this modified transfer vector both pTP and AdPol were expressed in insect cells at high levels which could be detected by Coomassie blue staining of the SDS-PAGE on which crude extracts were fractionated (Zhao et al. 1991; Nakano et al. 1991). Both pTP and AdPol were biologically active in the in vitro DNA replication assay (Zhao et al. 1991; Nakano et al. 1991). Thus both AdPol and pTP have been overproduced using the heterologous eukaryotic expression systems, which should facilitate their structure and function studies in Ad2 DNA replication.

3 Nuclear Transport of Adenovirus Replication Proteins

3.1 Adenovirus DNA Polymerase, Preterminal Protein, and DNA-Binding Protein Are Associated with Distinct Intranuclear Foci or Active Centers During Viral DNA Replication

Ad DNA replication takes place in the nucleus of the infected cells producing 10^4–10^5 progeny virus particles. Using in situ hybridization with biotinylated Ad DNA probes, replicating viral genomes were localized as discrete foci within the nuclei of infected cells (BRIGATI et al. 1983). When the localization of Ad5 DBP was examined in the infected HeLa cells, VOELKERDING and KLESSIG (1986) obtained evidence for the presence of two nuclear subclasses of DBP. The first subclass of DBP exhibited diffuse intranuclear staining and was extractable by a buffer containing 1% Nonidet P-40 and 150 m*M* NaCl. The second subclass of DBP, which required 2 *M* NaCl for extraction, appeared to be associated with distinct centers of active viral DNA replication and to be very similar to those distinct foci of replicating viral genomes observed by BRIGATI et al. (1983). In fact, the replicating Ad5 DNA and DBP were colocalized in those globular structures (VOELKERDING and KLESSIG 1986). It has been suggested that the DBP globuler structures may represent sites of Ad DNA replication (SUGAWARA et al. 1977; GINSBERG et al. 1977; MCPHERSON et al. 1982; MURTI et al. 1990).

In addition to the DBP, the replicating Ad2 DNA and AdPol were colocalized in these globular sites in the nucleus by in situ hybridization using biotinylated Ad2 DNA probe and by indirect immunofluorescence microscopy using rabbit polyclonal antibodies raised against a 120-kDa AdPol produced in *E. coli* (see Sect. 2.2; SASAGURI et al. 1987). Association of DBP and AdPol with these globular sites, which are presumably the centers for DNA replication, occurred only during viral DNA replication. When DNA replication was inhibited by treatment with hydroxyurea, or when the cells were infected with the temperature-sensitive AdPol mutant Ad5*ts*149 and the infected cells were incubated at nonpermissive temperature, only diffuse distribution of DBP and AdPol in the nucleus was observed (VOELKERDING and KLESSIG 1986; SASAGURI et al. 1987). Moreover, pTP was also localized in the nucleus of the Ad2-infected cells by subcellular fractionation (GREEN et al. 1981). Using immunofluorescence microscopy, pTP (ZHAO 1990; MURTI et al. 1990) as well as NFI (BOSHER et al. 1992) have been localized in distinct globular sites similar to those of AdPol within the nuclei of Ad2-infected cells. Interestingly, infection with Ad4, which does not require NFI for its replication, did not result in the localization of NFI in these globular sites, in contrast to the localization of Ad4 DBP (BOSHER et al. 1992).

The targeting of these replication proteins and the viral DNA to these foci or centers for DNA replication is not unique to Ad. The association of the herpesvirus DBP with discrete foci has also been observed during viral DNA replication (QUINLAN et al. 1984; WILCOCK and LANE 1991). The host proteins p53 and retinoblastoma susceptibility protein (Rb), which are presumably involved in the control of eukaryotic cell cycle and in the chromosomal DNA replication, have also been colocalized with the herpesvirus DBP, suggesting a possibility that these are the same sites utilized for chromosomal DNA replication (WILCOCK and LANE 1991). Moreover, by labeling with bromodeoxyuridine followed by indirect immunofluorescence microscopy, the replicating eukaryotic chromosome has been shown to be associated with distinct foci (NAKAMURA et al. 1986).

3.2 Identification of Nuclear Localization Signals for Preterminal Protein

Nuclear proteins larger than 40–60 kDa are transported into the nucleus by a receptor-mediated process. This process is energy dependent (NEWMEYER and FORBES 1988; RICHARDSON et al. 1988) and requires the presence of nuclear localization signal (NLS) sequence (DE ROBERTIS et al. 1978) on the protein to be transported through the nuclear pore complex. The molecular mechanism involved in the nuclear import of proteins and export of RNA from the nucleus through the pore complex is an active area of research, and several excellent reviews have been published recently (DINGWALL and LASKEY 1986; GOLDFARB 1989; ROBERTS 1989; STARR and HANOVER 1990; SILVER 1991; DINGWALL 1991; NIGG et al. 1991; DINGWALL and LASKEY 1991, 1992; GARCIA-BUSTOS et al. 1991; HANOVER 1992). There is evidence that small proteins could freely diffuse into the nucleus and the rate of diffusion is inversely proportional to their size (reviewed in PETERS 1986), although very few nuclear proteins enter the nucleus by free diffusion (GOLDFARB and MICHAUD 1991).

The best-studied example of the NLS sequence is that of SV40 large T antigen. The sequence PKK^{128}KRKV has been shown to be necessary and sufficient for the nuclear targeting of the SV40 large T antigen. Mutation of the lysine-128 to threonine or asparagine was shown to abolish the nuclear targeting function of this sequence (KALDERON et al. 1984a,b; LANFORD and BUTEL 1984). This sequence was also functional in targeting a non-nuclear protein, chicken pyruvate kinase, to the nucleus in the form of a fusion protein (KALDERON et al. 1984b). A number of NLS sequences have been identified based on the initial search for the presence of a motif similar to the SV40 large T antigen NLS with respect to the presence of one contiguous cluster of basic amino acid residues. In many cases the mutational analysis of the putative NLS motif and its ability to target a heterologous non-nuclear protein to the nucleus fulfilled the two criteria for the definition of an NLS that the sequence was necessary and sufficient for the nuclear targeting of the protein.

In pTP, the NLS motif RLPVRRRRRRVP (between residues 380–391; Fig. 1), based on its similarity to the SV40 large T antigen NLS, was identified for further analysis. Several in- frame deletions were introduced into the coding sequence of pTP cloned into an expression vector under the control of RSV-LTR

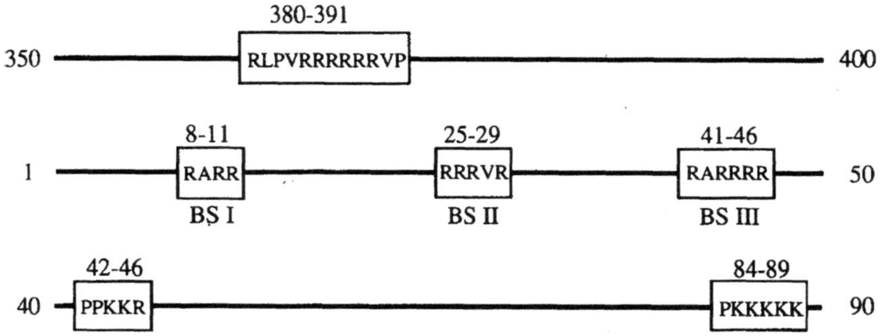

Fig. 1. Nuclear localization signals (NLS) of adenovirus DNA replication proteins: *top,* adenovirus type 2 (Ad2) preterminal protein (Ad2pTP); *middle,* Ad2 DNA polymerase (Ad2Pol); *bottom,* Ad5 DNA-binding protein (Ad5DBP). The residues in the NLS are shown within *boxes*

promoter and SV40 processing signals. These deletion constructs were tested by transient expression in monkey kidney (CV-1) cells and localization of the mutant proteins by immunofluorescence and subcellular fractionation. Mutant proteins which contained this motif were localized in the nucleus, whereas those which lacked this motif were localized predominantly in the cytoplasm, with one exception: the deletion construct encoding the C-terminal 213 amino acids devoid of this motif was localized in the nucleus (ZHAO and PADMANABHAN 1988). It was suggested that this mutant polypeptide was transported to the nucleus by passive diffusion, possibly because of its small size and lack of an NLS, and was subsequently retained in the nucleus because of its affinity to some nuclear component (ZHAO and PADMANABHAN 1988). It is especially interesting to test this possibility in view of recent reports that both TP and pTP have affinities for nuclear matrix (SCHAACK et al. 1990b; FREDMAN and ENGLER 1993), and the domain of TP or pTP involved in nuclear matrix attachment remains to be characterized. The motif RLPVRRRRRRVP of pTP was also functional in targeting *E. coli* β-galactosidase as a fusion protein into the nucleus and thus fulfilling both criteria for a functional NLS.

3.3 Nuclear Transport of Adenovirus DNA Polymerase Is Facilitated by Interaction with Preterminal Protein

In the Ad-infected cells pTP forms a stoichiometric complex with AdPol which is required for initiation of DNA replication as shown by in vitro assays. This complex can be dissociated by treatment with urea (1.7 *M*), suggesting that this complex is extremely stable (LICHY et al. 1982). For the synthesis of AdPol and pTP and thier complex formation, which is required for activity in the initiation reaction, the sequences upstream of the first ATG of the m-ORF of both pTP and AdPol are essential (SHU et al. 1987, 1988; PETTIT et al. 1988; ZHAO and PADMANABHAN 1988).

The proteins expressed from the first ATG of the m-ORF of AdPol and pTP genes (AdPol1 and pTP1) have a deletion of 142 N-terminal amino acids residues of AdPol and 18 residues of pTP, respectively. AdPol1 was localized exclusively in the cytoplasm, whereas the pTP1 was predominantly localized in the nucleus. These proteins were not biologically active in the in vitro DNA replication assay. When the expression plasmids encoding AdPol1 and pTP1 were cotransfected into CV-1 cells and the subcellular distribution of AdPol1 and pTP1 was examined by immunofluorescence and immunoprecipitation experiments, it was found that the distribution of AdPol1 dramatically changed from 95% cytoplasmic to about 40% nuclear location (ZHAO and PADMANABHAN 1988). These results indicated that AdPol1 was transported from the cytoplasm to the nucleus by a "piggyback" mechanism, and this observation, along with that of others (DINGWALL et al. 1982; MORELAND et al. 1987), provides a model for nuclear transport of a protein lacking an NLS as a complex with another protein with a functional NLS.

The biologically active proteins produced from plasmids constructed by inserting synthetic linkers containing the translational start sites upstream of the first ATG of the AdPol and pTP m-ORF were examined for subcellular localization. The AdPol, when expressed alone, was almost equally distributed between cytoplasm and nucleus, and this ratio changed to predominantly nuclear distribution for AdPol1 when its functional partner pTP was coexpressed (ZHAO and PADMANABHAN 1988). Two conclusions were reached from these experiments. First, a comparison of the subcellular localization of AdPol1 expressed from the first ATG and that of the biologically active AdPol, which included the upstream region of the m-ORF, revealed that the NLS of AdPol1 is located within this region. Second, the nuclear transport of AdPol is facilitated, irrespective of the presence of its NLS, by interaction with pTP, and this complex is transported more efficiently than AdPol alone. It is possible that a conformational change resulting from the interaction of AdPol and pTP might render the AdPol NLS present within its N-terminal region more competent for efficient nuclear transport. Another explanation for the efficient nuclear transport of the AdPol–pTP complex is due to the cumulative effects of the NLS sequences of individual proteins. This notion is consistent with the observation that some nuclear proteins have more than one NLS for efficient nuclear transport. For example, polyomavirus large T antigen has two NLS sequences, PPKKARED and VSKRPRP, for its nuclear localization and only when both NLS were mutated was its nuclear location abolished (RICHARDSON et al. 1986). Other examples of proteins containing two NLS include influenza virus NS1, yeast MATα_2, glucocorticoid receptor, and yeast ribosomal protein L29 (UNDERWOOD and FRIED 1990; see also review by GARCIA-BUSTOS et al. 1991). It is possible that when more than one NLS is present in a protein or a protein complex, the rate of nuclear transport may be additive (LANDFORD et al. 1986). It has recently been shown that AdPol is tightly associated with a *cdc2*-related histone H1 kinase, which consists of p33 or p34 catalytic subunit and a regulatory cyclin component, and this complex with AdPol does not exclude pTP (RAMACHANDRA and PADMANABHAN 1993; see Sect. 4.2.3). Thus, AdPol probably exists as a multimeric protein complex in Ad-infected HeLa cells, and the efficient nuclear transport of this complex may provide a way of regulating their coordinate arrival and function in the nucleus during Ad DNA replication. Many of the biological processes in a eukaryotic cell nucleus are carried out by protein complexes, and even if all the individual proteins of these complexes may not have NLS sequences, their efficient transport is accomplished by those that do have these sequences.

3.4 Nuclear Transport of Adenovirus DNA-Binding Protein and Adenovirus DNA Polymerase Is Mediated by Bipartite Nuclear Localization Signals

A DBP mutant Ad5*dl*802rl containing a deletion of amino acids from 23 to 105 in the coding region of DBP failed to localize in the nucleus, which suggested that this region contained the NLS of DBP (CLEGHON et al. 1989). Within this region two short clusters of basic amino acid residues, $P^{84}KKKKKR^{90}$ and $P^{42}PKKR^{46}$ (Fig. 1), were identified as potential NLS based on their similarity to SV40 NLS. Mutational analysis of each of these basic regions followed by subcellular localization of the mutant proteins indicated that both domains were required for the nuclear transport of DBP (MORIN et al. 1989a). Thus, these two NLS sequences are not redundant like those of polyomavirus large T antigen or influenza virus NS1 protein (GREENSPAN et al. 1988), but they are interdependent or the NLS is bipartite in nature. These two basic domains are separated by a spacer region of 37 amino acid residues. Interestingly, the mutant DBP which was localized in the cytoplasm when expressed transiently in human 293 cells was targeted to the nucleus when the cells were infected with the Ad virus encoding the mutant DBP, suggesting that the mutant DBP was transported possibly through an interaction with a virally encoded or induced factor in the infected cells (MORIN et al. 1989a). Other examples of nuclear proteins with bipartite NLS are the influenza virus polymerase basic protein (BP1) and the *Xenopus* nucleoplasmin. The NLS of BP1 consists of two clusters of four basic amino acids, RKRR and KRKQR, separated by a spacer of 16 amino acids (NATH and NAYAK 1990), and the NLS of nucleoplasmin has two interdependent basic domains, $K^{155}R$ and $K^{167}KKK$, which are separated by a spacer region of ten amino acids (ROBBINS et al. 1991).

The N-terminal region of AdPol contains three clusters between residues 4 and 52 which could function as NLS for the transport of AdPol (Fig.1). Deletion analysis and subcellular localization of mutant proteins showed that basic sequences (BS) I and II were both important for nuclear localization of AdPol. The deletion of either BSI or BSII resulted in cytoplasmic localization of AdPol. In contrast, the nuclear localization of mutant AdPol in which the BSIII was deleted was similar to the parent AdPol distributed equally between nucleus and cytoplasm, suggesting that BSIII was dispensible for the nuclear targeting of AdPol when expressed alone (ZHAO and PADMANABHAN 1991). When the region containing all three clusters of basic amino acids was fused to *E. coil* β-galactosidase, the fusion protein was predominantly localized in the nucleus and nucleolous, whereas the control (unfused) β-galactosidase was localized only in the cytoplasm. The analysis of the subcellular distribution of β-galactosidase fusion proteins containing an individual basic domain or pairs of basic domains revealed that the nuclear targeting of the β-galactosidase fusion protein required the interdependent interaction of either the BSI and BSII pair or BSII and BSIII, but not the BSI and BSIII pair. Point mutations of two or more arginine residues in either cluster severely interfered with the nuclear localization of the fusion protein. Although a single mutation of a basic amino acid residue in either cluster had no effect in the bipartite NLS function, the presence of two single mutations, one for each cluster, reduced the nuclear localization significantly (ZHAO and PADMANABHAN 1991). Similar results were obtained from the mutational analysis of the nucleoplasmin NLS (ROBBINS et al. 1991).

The function of BSII and BSIII as a very efficient bipartite NLS in the nuclear targeting of β-galactosidase is interesting, because BSIII is dispensible for the nuclear targeting of the AdPol protein. From these studies it appears that the three basic regions of AdPol constitute two overlapping bipartite NLS which interact differentially with the cellular transport system for the nuclear targeting of AdPol and the *E. coli* β-galactosidase. The presence of two bipartite NLS in the AdPol seems redundant for the nuclear targeting of AdPol when expressed alone. However, AdPol is normally complexed with pTP and NFI/CTF in the infected cell for its function in DNA replication. Moreover, as mentioned in Sect. 3.3, the pTP–AdPol complex also includes a histone H1 kinase in the Ad2-infected cells (RAMACHANDRA and PADMANABHAN 1993).

It is interesting that, although AdPol has mutliple basic clusters for nuclear localization, it is still distributed equally between the nuclear and cytoplasmic compartments. One possible explanation is that AdPol may have some cytoplasmic anchor in addition to an NLS. Its nuclear localization would only occur when the cytoplasmic anchor sites are saturated or no longer functional in retaining the AdPol due to conformational changes resulting from the interaction with pTP and other cellular proteins. In addition, when AdPol is complexed with pTP, NFI, and the histone H1 kinase, conformational changes, including those in the AdPol NLS, may in part account for the increased nuclear localization of AdPol that was observed earlier (ZHAO and PADMANABHAN 1988). Conformational changes can be induced by protein–protein, protein–nucleic acid or protein–ligand interaction or by protein phosphorylation (see Sect. 4.5). A conformational change resulting from the ligand binding to the C-terminal 256-amino acid domain has been invoked to explain the increased nuclear accumulation of the glucocorticoid receptor (PICARD and YAMAMOTO 1987). Thus, although BSIII is dispensible for the nuclear localization of AdPol alone, it may contribute to the nuclear localization of the multimeric AdPol complex.

The spacer region between BSI and BSII and between BSII and BSIII is 12 and 11 amino acids, respectively (ZHAO and PADMANABHAN 1991). Spacer mutations between BSII and BSIII were created which included a deletion of the first six amino acid residues or the entire spacer or an insertion of five amino acids into the spacer. All the mutant NLS *E. coli* β-galactosidase fusion proteins were predominantly localized in the nucleus, suggesting that the primary structure of the spacer is not critical for the bipartite NLS function (ZHAO and PADMANABHAN 1991). These results are consistent with the report that the nuclear localization of a spacer mutant of the influenza virus BP1 protein was not affected when seven amino acids of the spacer were deleted, as well as with the finding that altered spacer lengths of the nucleoplasmin NLS by insertion of repeats of SPGG sequence did not affect the nuclear localization, although insertion of a single copy of a hydrophobic sequence QPWL in the spacer reduced the nuclear accumulation of the fusion protein (ROBBINS et al. 1991).

The frequency of occurence of the nucleoplasmin-like bipartite NLS motif in nuclear proteins is significantly higher (about 50%) compared to the SV40-like NLS (17%), as determined by computer search of the SwissProt data base (DINGWALL and LASKEY 1991). If a common mechanism can be invoked in the bipartite signal-mediated nuclear transport of these proteins, then it is possible that the secondary and tertiary structure of a nuclear protein would play a role in briging these basic clusters into close proximity to one another, thus allowing their recognition by the cellular factor(s) as a single unit. First, this notion is supported by the observations that there are variations in spacer lengths in the bipartite motifs of nuclear proteins so far characterized. Second, insertion, deletion, and point mutations within the spacer do not seem to affect the nuclear targeting function (NATH and NAYAK 1990; ROBBINS et al. 1991; ZHAO and PADMANABHAN 1991). Third, the three-dimensional structure of the region containing the DNA-binding domain and the bipartite NLS of the glucocorticoid receptor was determined recently using two-dimensional nuclear magnetic resonance (NMR) and distance geometry, and the data indicates that this domain may be in a flexible region (HARD et al. 1990; reviewed by DINGWALL and and LASKEY 1991). Two functional domains in a protein could be brought into close proximity in different ways, such as by simple folding, via α-helical structure, by the presence of one or more β-turns, or by tertiary interactions in a protein molecule. Chou-Fasman predictive analysis (PREVELIGE and FASMAN 1989) of protein secondary structure in the vicinity of the bipartite NLS sequences of AdPol, Ad DBP, and nucleoplasmin support this model, in which protein conformation in the spacer region may play an important role in bringing these clusters of basic amino acids into close proximity (ZHAO and PADMANABHAN 1991). Further work is necessary to have a better understanding of how bipartite NLS function in nuclear transport.

It has recently been reported that the heat-shock protein or its cognate is required for nuclear transport of a protein (FINLAY et al. 1991; IMAMOTO et al. 1992; SHI and THOMAS 1992). DINGWALL and LASKEY (1992) proposed a model in which two sequential receptors are involved in protein import into the nucleus. First, the Hsp70 (or its cognate) binds to and stabilizes a locally unfolded NLS and then presents it to a second receptor. Experimental evidence for this model will shed light on the mechanism of protein import into the nucleus.

4 Phosphorylation of Adenovirus DNA Replication Proteins

Protein phosphorylation provides an excellent mechanism by which structure and function of a protein can be altered. Modulation of protein function by phosphorylation-dephosphorylation has been demonstrated for a number of proteins involved in DNA and RNA synthesis, and among them SV40 T antigen is one of the best-studied examples (for reviews, see PRIVES 1990; FANNING 1992). SV40 T antigen is phosphorylated on multiple sites, and depending upon the site of phosphorylation its DNA binding affinity and function are regulated positively or negatively.

Among the Ad-encoded DNA replication proteins, DBP has been studied extensively for its phosphorylation modification. Recently, the other two virus-encoded proteins, AdPol and pTP, have also been shown to be phosphorylated. This review summarizes our present state of knowledge concerning phosphorylation of Ad-encoded DNA replication proteins. Possible roles of phosphorylation in modulation of the activities of proteins involved in Ad DNA replication are discussed.

4.1 Phosphorylation of DNA-Binding Protein

DBP has been known to undergo phosphorylation modification (ROSENWIRTH et al. 1976; LEVINSON et al. 1977; JENG et al. 1977; RUSSEL and BLAIR 1977; LINNE et al. 1977; AXELROD 1978). DBP can be isolated from the infected cells as a phosphoprotein at early as well as late times after infection, and the degree of phosphorylation varies with the infection time (LINNE and PHILIPSON 1980). Isoelectric focusing of purified DBP resolves into as many as 15 subspecies, of which at least some are the result of differing degrees of phosphorylation (KLEIN et al. 1979; LINNE and PHILIPSON 1980). In Ad2 and Ad5, a moelcular weight of 59 kDa has been estimated for DBP based on sequence data, but SDS-PAGE analysis shows an apparent molecular weight of 72 kDa. Such a slow migration of the protein on an SDS polyacrylamide gel was initially thought to be the result of extensive phosphorylation. However, studies have later attributed the slower electrophoretic mobility of DBP to a relatively high proline content and to its highly asymmetric configuration (SUGAWARA et al. 1977; VAN DER VLIET et al. 1978). This conclusion was also supported by studies in which even after extensive dephosphorylation using calf intestinal alkaline phosphatase the apparent molecular weight of the protein was reduced by only 2 kDa (LINNE and PHILIPSON 1980).

4.1.1 DNA-Binding Protein Is Phosphorylated on Multiple Sites and Mostly on the N-Terminal Domain

Genetic and mutational analyses have indicated that DBP contains two domains that appear to be functionally distinct (RICE and KLESSIG 1984). These two domains can be separated by mild chymotrypsin treatment (KLEIN et al. 1979; LINNE and PHILIPSON 1980), and depending on the viral serotype chymotrypsin treatment yields products of 36–48 kDa arising from the C-terminal region and an N-terminal fragment of approximately 27 kDa. Occurence of such a cleavage within infected cells has also been reported (ASSELBERGS et al. 1983; BRANTON et al. 1985). Among these two domains, the N-terminal fragment is heavily phosphorylated on serine and threonine residues and contains most, if not all, of the estimated one to 13 phosphates bound per DBP molecule (KLEIN et al. 1979; LINNE and PHILIPSON 1980). Many of the phosphorylation sites have been preliminarily mapped to serine residues at positions 31, 33, 35, 70, 76, 92, 100, and 107 and threonine residues at positions 12 and 75 (C.W. ANDERSON et al. 1986; MANN 1987) of Ad2 DBP, which are also conserved in Ad5 DBP (MORIN et al.1989b).

Although serine and threonine are the major phosphoamino acids present in DBP, the occurrence of a phosphotyrosine, possibly at position 195, has also been reported (RUSSELL et al. 1989). Identification of a phosphotyrosine was possible only when cells were grown and infected in the presence of sodium vanadate, a known phosphatase inhibitor, suggesting that the phosphotyrosine may be readily susceptible for phosphatase action (RUSSELL et al. 1989). The tyrosine phosphorylation occurs earlier in the infection than the bulk of the phosphorylations on serine and threonine residues and therefore is probably not readily detected at the later times of infection (RUSSELL et al. 1989).

4.1.2 In Vitro Phosphorylation of DNA-Binding Protein

Studies have shown that DBP as well as two other Ad-infected cell-specific proteins can be phosphorylated in vitro by an enzme activity present in the nuclei of Ad-infected cells (LEVINSON et al. 1977; POSTEL et al. 1978). In addition, this protein kinase activity phosphorylates on identical peptides of DBP in vitro that are phosphorylated in vivo (POSTEL et al. 1978). In a later study, KLEIN et al. (1979) purified a chromatin-associated protein kinase from uninfected HeLa cell extracts, which in addition to exhibiting phosphorylation of histones H1, H2A, and H4 was able to phosphorylate DBP in a dose-dependent manner. Furthermore, this purified kinase phosphorylated all but one phosphopeptide obtained in vivo, suggesting that it may be the major enzyme responsible for the in vivo phosphorylation of DBP.

In vitro studies have also demonstrated the association or copurification of kinase (s) capable of phosphorylating DBP. CAJEAN-FEROLDI et al. (1981) found that at least two protein kinase activities of cellular origin, both capable of phosphorylating DBP, were coeluted with DBP on a DNA-cellulose affinity chromatography. Unlike the DBP kinase reported by KLEIN et al. (1979), the kinase that copurified with DBP did not phosphorylate histones. In vitro phosphorylation of DBP by an associated kinase which could also phosphorylate histone H3 was observed by BRANTON et al. (1985). DBP purified either by immunoprecipitation or single-stranded DNA-cellulose column chromatography contained a kinase activity that specifically phosphorylated the N terminus. In addition, using photoaffinity labeling with 8-azido-[α^{32}P] ATP, they were able to show that DBP contains an ATP-binding site at the N terminus (Branton et al. 1985). However, it was not clear whether the kinase activity is intrinsic to DBP or present in an associated enzyme. When the temperature-sensitive Ad DBP mutant H5*ts*125-infected cells were incubated at nonpermissive temperature, DBP was thermo-labile, whereas the associated kinase activity was still present, suggesting that the kinase activity is probably due to a DBP-associated enzyme (Branton et al. 1985).

Taken together, the results from the in vitro phosphorylation studies suggest that probably two or more kinases of cellular origin that are associated with DBP may be involved in the in vivo phosphorylation of DBP. The nature of the kinases and the effect of in vitro phosphorylation on DBP function remain to be identified and further characterized.

4.1.3 Functional Importance of DNA-Binding Protein Phosphorylation

It has long been proposed that the various phosphorylated forms of DBP may carry out different roles of this multifunctional protein. Several attempts have been made to address the functional importance of DBP phosphorylation, but the results from earlier studies have been inconclusive. Among the two distinct domains of DBP, N-terminal domain exhibits limited conservation among different serotypes and does not appear to play a direct role during DNA replication. Most of the estimated one to 13 phosphates bound per molecule of DBP are present in this N-terminal domain (Klein et al. 1979; Linne and Philipson 1980). The carboxyl part that is highly conserved among different adenovirus serotypes has the abilty to bind to DNA and RNA (Ariga et al. 1980; Cleghon and Klessig 1986, 1992; Neale and Kitchingman 1990) and to bind zinc (Eagle and Klessig 1992) and is required for viral DNA replication at the level of DNA chain elongation (Van der Vliet and Sussenbach 1975; Friefeld et al. 1983; Nagata et al. 1982). Characterization of adenovirus temperature-sensitive mutants and their revertants have also suggested a role for the C-terminal domain in the regulation of early viral gene expression (Carter and Blanton 1978; Babich and Nevins 1981; C.W. Anderson et al. 1983; Brough et al. 1985) as well as viral assembly (Nicolas et al. 1983). In contrast, the heavily phosphorylated N-terminal domain participates in late gene expression at both transcriptional and post-transcriptional levels (Johnston et al. 1985; Klessig and C.W. Anderson 1975; Klessig and Grodzicker 1979 K.P. Anderson and Klessig 1983), and in nuclear transport of the protein (Morin et al. 1989a), and it also may have a role in enhancement of its own expression (Morin et al. 1989b). Studies with host range mutants have revealed that a mutation within the N-terminal domain of DBP allows the efficient growth of Ad in nonpermissive monkey cells (Klessig and Grodzicker 1979). Even after cleaving the C-terminal fragment from the highly phosphorylated N-terminal domain using chymotrypsin, the C-terminal fragment still retains the ability to bind to single-stranded DNA (Klein et al. 1979; Linne and Philipson 1980) and function in an in vitro DNA replication assay (Ariga et al. 1980; Friefeld et al. 1983; Tsernoglou et al. 1985). In addition, Linne and Philipson (1980) found that removal of the majority of the phosphates by alkaline phosphatase did not alter the ability of DBP to bind to single-stranded DNA or the ends of double-stranded DNA. Based on these observations, it was believed that phosphorylation of DBP may not play any role in Ad DNA replication. Such conclusions were further supported by the in vitro studies, in which the 44-kDa C-terminal fragment of the wild-type DBP was able to complement the temperature-sensitive replication defect of the H5*ts*125 DBP (Ariga et al. 1980).

However, several other findings suggest a role for phosphorylation of DBP in DNA replication. In their studies, Klein et al. (1979) found that the less phosphorylated species of DBP bound single-stranded DNA more tightly than the phosphorylated form, suggesting that although DNA binding is mediated by the C-terminal fragment, the overall phosphorylation status is important for DNA binding activity. A role for phosphorylation of DBP in DNA replication was also suggested by the observation that the degree of DBP phosphorylation varies with infection time, and upon blocking viral DNA replication by cytosine arabinoside the extent of DBP phosphorylation doubles (Linne and Philipson 1980). It was also shown that when cells infected with H5*ts*125 mutant adenovirus at permissive temperature were shifted to nonpermissive temperature, phosphorylation of the DBP decreased and viral DNA replication ceased, suggesting that phosphorylation of DBP may play a role in viral DNA replication (Levinson et al. 1977).

In order to clarify the role of DBP phosphorylation, Morin et al. (1989b) systematically substituted cysteine or alanine for up to ten phosphorylation sites of the Ad5 DBP, which were preliminarily mapped on the very closely related Ad2 DBP (C.W. Anderson et al. 1986; Mann 1987). The mutant genes with one or more altered sites were introduced into the viral genome by in vivo recombination. Alteration of one or a few of these sites had little effect on the viability of virus containing the mutated DBP. However, when eight or more sites were altered, viral growth decreased significantly, suggesting that the overall phosphorylation state of the protein is more important. The reduction in growth correlated with both depressed DNA replication and expression of late genes. Moreover, although the stability of the mutated DBP was not affected, DBP synthesis and the level of its mRNA were depressed five- to tenfold for the under-phosphorylated protein, implying that DBP enhances its own expression and phosphorylation is important for this function.

The functional interactions between the N- and C-terminal domains were also indicated by other studies (Krevolin and Horwitz 1987; Brough et al. 1993). Brough et al. (1993) examined several viruses carrying lesions in the N-terminal region of DBP and found that accumulation of viral DNA and infectious virions was drastically reduced. Characterization of one of the mutants indicated that the N-terminal region (residues 2–38) affects viral DNA synthesis in vivo. The reduction in DNA synthesis was not due either to change in nuclear transport or disruption of DBP's role in early gene expression. In addition, the mutant also had single-stranded DNA-binding activities and participated efficiently in DNA elongation assays comparable to that of wild type, which further confirmed earlier studies that the C-terminal domain of DBP is responsible for these two functions.

Together these results suggest that, although DBP has two separate functional domains participating in DNA replication, for full activity of the protein both domains of the proteins are necessary.

In the case of intact DBP, phosphorylation on multiple sites within the N-terminal domain may possibly lead to a change in the conformation in such a way that it could affect the function mediated by the C-terminal domain. Similar to DBP, many other proteins including SV40 T antigen (SIMMONS et al. 1986), cellular transcription factors, c-Jun (BOYLE et al. 1991), and c-Myb (BIEDENKAPP et al. 1988) contain phosphorylation sites that are located outside the DNA-binding domain, and phosphorylation has been shown to negatively regulate their DNA-binding activities. In the case of SV40 T antigen, it has been shown that in vivo phosphorylation of the N-terminal region decreases the binding of the protein to the DNA origin, and the effect is reversed by in vitro dephosphorylation (SIMMONS et al. 1986; MOHR et al. 1987) or by proteolysis which removes the highly phosphorylated N-terminal arm of the polypeptide (SIMMONS et al. 1986). Enzymatic dephosphorylation of serine residues of SV40 T antigen also increases its activity in the initiation of viral DNA replication (MOHR et al. 1987), whereas in vitro phosphorylation of threonine-124 mediated by $p34^{cdc2}$ kinase enhances both DNA binding and replication initiation activities (McVEY et al. 1989). In contrast, casein kinase I-mediated in vitro phosphorylation of SV40 T antigen on many of the sites that are modified in vivo inhibits both origin unwinding and DNA replication (CEGIELSKA and VIRSHUP 1993). In the case of DBP, if an overall phosphorylation at the N terminus results in a physical or chemical constraint for DNA binding, then it would be logical to expect that the C-terminal fragment generated by enzymatic treatment or the dephosphorylated form of DBP would still bind to DNA, as shown previosly (KLIEN et al. 1979; LINNE and PHILIPSON 1980). Similar to SV40 T antigen, an overall phosphorylation of DBP may result in negative regulation of its DNA binding and, depending upon a specific site, the effect could be either positive or negative.

4.2 Phosphorylation of Adenovirus DNA Polymerase

Investigations of the phosphorylation of AdPol and pTP have been difficult because of their low levels of synthesis in Ad-infected cells. Recently, however, both pTP and AdPol have been overproduced using high-level recombinant vaccinia virus (STUNNENBERG et al. 1988; NAKANO et al. 1991) and baculovirus (WATSON and HAY 1990; ZHAO et al. 1991) systems. Several studies have suggested that the vaccinia virus expression system (for a review, see MOSS 1991) is particularly suitable to analyze phosphorylation of proteins that are normally synthesized in mammalian cells (HOSS et al. 1990; TEMPLETON 1992). The initial detection of the phosphorylation of AdPol was possible using the recombinant vaccinia virus expression system (RAMACHANDRA et al. 1993). A large-scale metabolic labeling of HeLa cells infected with Ad later confirmed that AdPol is also phosphorylated in the native system. In vivo labeling of insect cells infected with recombinant baculovirus also confirmed the phosphoprotein nature of AdPol-synthesized in insect cells using that system. Phosphoamino acid analysis of ^{32}P-AdPol indicated the presence of only phosphoserine independent of the source of AdPol. It is noteworthy that the extent of phosphorylation of AdPol was at least 60-fold lower in insect cells infected with the recombinant baculovirus compared with that in HeLa cells infected with the recombinant vaccinia virus. A similar observation was made regarding the phosphorylation state of Rb produced in insect cells compared with the Rb expressed in Hela cells (TEMPLETON 1992).

4.2.1 Adenovirus DNA Polymerase Is Phosphorylated on Multiple Sites

It was important to establish that in the recombinant vaccinia virus expression system, AdPol was modified at the same sites as in Ad2-infected cells. Comparison of tryptic peptide maps of native and recombinant AdPol revealed that most of the tryptic phosphopeptides, with the exception of two, were common in AdPol isolated from either Ad2-infected cells or recombinant vaccinia virus expression system, indicating that probably same kinase or kinases are involved in the in vivo phosphorylation of both native and recombinant vaccinia virus-expressed AdPol. There were nine spots in the tryptic phosphopeptide map of AdPol from Ad2-infected cells, indicating that phosphorylation occurs on multiple sites. Two-dimensional maps of AdPol expressed with or without pTP using recombinant vaccinia viruses showed that majority of the phosphopeptides were present in both preparations, suggesting that AdPol gets phosphorylated before its association with pTP or the interaction does not mask the major phosphorylation sites. AdPol isolated from nuclear and cytoplasmic fractions showed identical phosphopeptides on peptide maps, which suggests that AdPol gets phosphorylated on its potential sites in the cytoplasm and no other sites are modified in the nucleus (RAMACHANDRA et al. 1993).

4.2.2 Serine-67 Is the Major Adenovirus DNA Polymerase Phosphorylation Site

Since all the sites that were phosphorylated in native AdPol were also modified when the protein was expressed using the recombinant vaccinia virus expression sytem, the recombinant AdPol was used for mapping the phosphorylation sites using metabolic labeling, reverse-phase high-performance liquid chromatography (HPLC) fractionation of tryptic peptides, two-dimensional tryptic peptide mapping, and microsequence analysis techniques. One of the major phosphorylation sites was mapped at serine-67 (RAMACHANDRA et al. 1993).

4.2.3 In Vitro Phosphorylation of Adenovirus DNA Polymerase by a Stably Associated *cdc2*-Related Protein Kinase

Based on our observation (RAMACHANDRA et al. 1993) that serine-67 occurs at a site which is within the substrate recognition sequence of the families of serine/

threonine kinases such as Pro-X-Ser/Thr-Pro for mitogen-activated protein-2 kinase (ALVAREZ et al. 1991; CLARK-LEWIS et al. 1991; for reviews, THOMAS 1992) or Ser-Pro for cyclin-dependent protein kinases (MORENO and NURSE 1990; TSAI et al. 1991), the ability of $p34^{cdc2}$ kinase to phosphorylate AdPol in vitro was examined (RAMACHANDRA and PADMANABHAN 1993). In the in vitro kinase assays, AdPol was phosphorylated using purified $p34^{cdc2}$ kinase or by a protein kinase that coimmunoprecipitated with AdPol. Phosphorylation occured only on serine residues and on identical peptides in vivo as well as in vitro either by $p34^{cdc2}$ kinase or by the kinase that coimmunoprecipitated with AdPol. In addition to demonstrating the in vitro phosphorylation of AdPol by *cdc*2 kinase, these results indicated an association of a *cdc*2 or a closely related protein kinase that may be responsible for the in vivo phosphorylation of AdPol. Similar to *cdc*2 kinase, the AdPol-associated kinase was able to phosphorylate histone H1 and was recognized by the anti-$p34^{cdc2}$ antibodies in the immunoprecipitation and immunoblot experiments.

Similar to AdPol phosphorylated in vivo, serine-67-containing peptide was the major phosphopeptide in the case of AdPol modified in vitro by $p34^{cdc2}$ kinase or the AdPol-associated kinase. As mentioned above, this site is also within the substrate recognition sequence of mitogen-activated protein (MAP) kinase and cyclin-dependent kinases. These kinases are closely related proteins, and studies have implied similar modes of regulation and substrate specificity for these cell-cycle-regulated enzymes (PELECH and SANGHERA 1992). Because of the overlapping phosphorylation site requirements, some of the proteins that have been proposed as targets for cyclin-dependent kinase are also substrates for MAP kinases (PELECH and SANGHERA 1992). Phosphorylation of AdPol at serine-67 by MAP kinase may occur depending upon its steady state level at the time of synthesis of AdPol in the virus-infected cells. Studies with synthetic peptide analogues of the myelin basic protein have revealed that a proline residue present two residues upstream of the phosphoacceptor amino acid, as in substrate recognition sequence for MAP kinase, optimizes phosphorylation by $p34^{cdc2}$ (PELECH and SANGHERA 1992). Among the seven potential cyclin-dependent kinase phosphorylation sites within AdPol, only serine-67 is within the consensus recognition site for both cyclin-dependent kinase and MAP kinases. The other potential cyclin-dependent kinase phosphorylation sites in AdPol include serine residues at positions 51, 136, 387, 901, 926, and 1069.

Serine/threonine kinases such as 34-kDa *cdc*2, 33-kDa *cdk*2, *cdk*3, and *cdk*4 associate with different cyclins in order to be catalytically active. Among these protein kinases, $p34^{cdc2}$/cyclin B kinase is active in G2->M (PAGANO et al. 1992a; ROSENBLATT et al. 1992), whereas $p33^{cdc2}$/cyclin A kinase activity is maximum at G1->S transition (DRAETTA and BEACH 1988; PINES and HUNTER 1989). The levels of *cdc*2 kinase remain constant throughout the cell cycle (DRAETTA and BEACH 1988), and the interaction with cyclins and phosphorylation/dephosphorylation of critical residues regulate their activity at different phases of the cell cycle (for reviews, see DRAETTA 1990; MALLER 1990; MORENO and NURSE 1990). Recently, in addition to cyclins A and B, cyclins C–E and G (KOFF et al. 1991; LEOPALD and O'FARRELL 1991; LEW et al.1991; MATSUSHIME et al. 1991; XIONG et al. 1991; TAMURA et al. 1993) and a large family of *cdc*2-like kinases (PARIS et al. 1991; TSAI et al. 1991; MEYERSON et al. 1992), which may be active at different phases of cell cycle, have been discovered. All these cyclin-dependent kinases appear to recognize Ser/Thr-Pro motifs within their substrates, and except cyclin/*cdk*4 kinase other kinases utilize histone H1 as a substrate. Therefore, it is possible that in addition to p34^{cdc2}/cyclin B complexes, several members of the *cdc*2 kinase family could phosphorylate AdPol at a given point in the cell cycle. For the phosphorylation of a viral protein such as AdPol, cell cycle-dependent phosphorylation by a specific *cdc*2/cyclin complex may not be important. As has been shown in the case of cellular protein such as the Rb, the in vitro phosphorylation on the same sites could be mediated by more than one cyclin-dependent protein kinases, including p34^{cdc2} (LEES et al. 1992) and p33^{cdk2} kinases (AKIYAMA et al. 1992).

The AdPol-associated kinase was recognized by anti-p34^{cdc2} antibodies in the immunoprecipitation and immunoblot experiments, indicating the presence of p34^{cdc2} in the AdPol immunocomplex. However, we do not know whether p34^{cdc2} is the only cyclin-dependent kinase present in association with AdPol. Moreover, the nature of cyclin(s) in the AdPol immunocomplex remains to be determined. It is reported that in nonsynchronized cells p34^{cdc2}-cylin B is the major *cdc*2 kinase activity (DUTTA and STILLMAN 1992). The *cdc*2 kinase associates with cyclin B and cyclin A (MEYERSON et al. 1992), whereas, *cdk*2 and *cdk*4 interact with cyclins A, D, and E (TSAI et al. 1991; PAGANO et al 1992b; FAHA et al. 1993).

The histone H1 kinase activity capable of phosphorylating AdPol not only coimmunoprecipitates, but also copurifies with AdPol on a size exclusion column, further confirming a stable association of the kinase with AdPol. Copurification of other kinase activities with cellular DNA replication proteins, such as the 56-kDa protein kinase with DNA polymerase-α (PECK et al. 1993) and the casein kinase II with DNA topoisomerase II (CARDENAS et al. 1993), have recently been reported. However, stable associations of *cdc*2 kinases have previously been reported only with cell growth control proteins such as Rb (AKIYAMA et al. 1992; HU et al. 1992; KITAGAWA et al. 1992; LEES et al. 1992; R.T. WILLIAMS et al. 1992) and p53 (KRAISS et al. 1990; STURZBECHER et al. 1990; for review, WEINBERG 1991) and viral oncoproteins such as Ad E1A (GIORDANO et al. 1991; HERRMANN et al. 1991; KLEINBERGER and SHENK 1991; TSAI et al. 1991) and bovine papilloma virus E7 (TOMMASINO et al. 1993). It is believed that the negative or positive effects of these proteins on cellular proliferation are exerted through their interactions with other proteins. Interaction with a cyclin-dependent kinase is postulated to divert the normal function of these enzymes, either by altering their substrate specificity or by making them unavailable for the phosphorylation of certain cellular proteins. Interaction of AdPol with *cdc*2 might reflect a requirement for the virus to interfere with the cellular DNA synthesis apparatus in order to replicate its own genome.

4.2.4 Functional Importance of Adenovirus DNA Polymerase Phosphorylation

To evaluate the role of phosphorylation in the modulation of the enzyme activity in the in vitro replication initiation assay, unlabeled AdPol was treated with agarose-immobilized calf intestinal alkaline phosphatase. Dephosphorylation of AdPol with calf intestinal alkaline phosphatase resulted in a significant decrease in its activity in the in vitro DNA replication initiation assay and a moderate decrease in the DNA polymerase activity, suggesting that phosphorylation is important for its biological activity (RAMACHANDRA et al. 1993). Unlike DNA polymerase activity, the DNA replication initiation requires the specific interaction of AdPol with pTP and NFI as well as with the DNA sequences at the replication origin. Our recent preliminary experiments show that dephosphorylation of AdPol does not affect AdPol's interaction with pTP, but significantly diminishes the ability of AdPol to specifically recognize the DNA templates containing the replication origin sequences (M. RAMACHANDRA and R. PADMANABHAN, unpublished results). In AdPol, there are two Cys-His-rich sequences that are conserved between different Ad serotypes that could potentially be folded into zinc finger motifs capable of binding to DNA (CHEN and HORWITZ 1989). Though these Cys-His-rich sequences do not precisely fit the consensus for a zinc finger, mutations in the regions affect DNA synthesis, DNA binding, and in vitro replication, but show little or no effect on binding to other proteins including pTP (CHEN et al. 1990; JOUNG and ENGLER 1992). It is interesting to note that the potential *cdc*2 phosphorylation sites in AdPol are present in the vicinity of these two Cys-His-rich regions (Fig. 2A). Therefore, it is likely that phosphorylation may play a role in DNA binding of AdPol mediated through these Cys-His-rich sequences.

In order to address the functional importance of the major in vitro and in vivo phosphorylation site of AdPol, serine-67 was mutagenized to alanine and the mutant protein was transiently expressed in HeLa cells. Unlike the wild-type AdPol, the mutant was inactive in the DNA replication initiation assays. Our recent studies indicate that substitution of alanine for serine at position 67 affected overall stability of the protein, which could either be due to the absence of phosphate at this position or simply due to the presence of alanine at this position (M. RAMACHANDRA and R. PADMANABHAN, unpublished results). Attempts are underway to construct mutants with amino acids other than alanine in place of serine at position 67 to determine the importance of phosphorylation at this state. There are reported instances where mutation of a particular phosphoylation site affects the phosphorylation at one or more distant sites, as in the case of SV40 T antigen (SCHEIDTMAN et al. 1991). In the light of our results, it is interesting to note that in SV40 T antigen a mutant bearing a substitution of alanine for threonine at residue 124 was unable to replicate SV40 DNA in monkey cells and in the in vitro DNA replication assays (SCHNEIDER and FANNING 1988). It has been established that the phosphorylation of threonine-124 is mediated by p34^{cdc2} kinase (McVEY et al. 1989;

Fig. 2a–d. Location of the potential cyclin-dependent kinase sites and the known functional domains of proteins participating in adenovirus (Ad) DNA replication. **a** Ad DNA polymerase (AdPol). **b** Preterminal protein (pTP). **c** DNA-binding protein (DBP). **d** Nuclear factor I (NFI). *P*, phosphorylation site (confirmed in vivo phosphorylation sites are indicated by an *asterisk*); *NLS*, nuclear localization signal; *Zn*$^{++}$, zinc-binding region; *aa*, amino acid; *CR2* and *CR3*, regions of DBP conserved in different Ad serotypes. The zinc-binding region as well as CR2 and CR3 are implicated in DNA binding of DBP

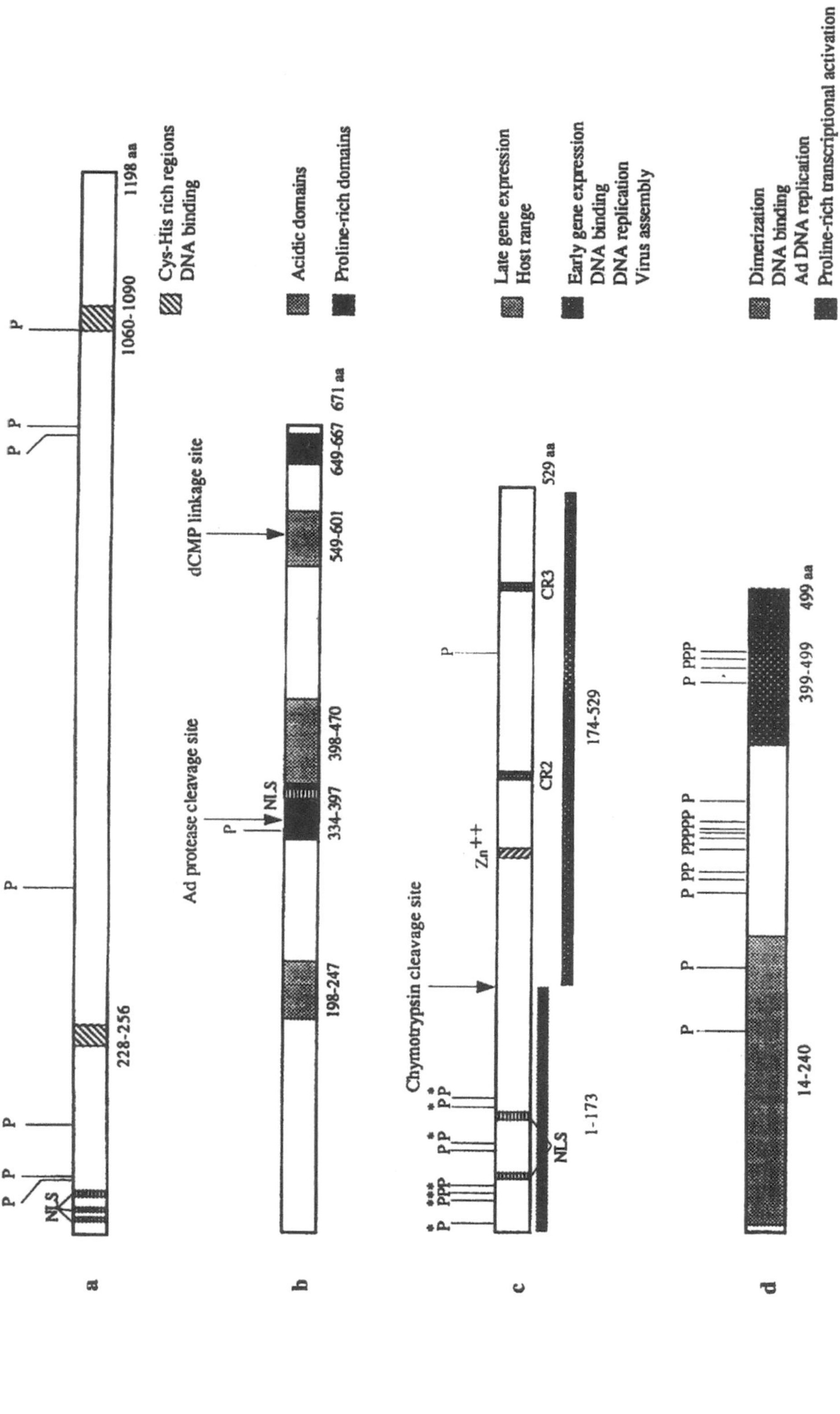
a
NLS
P
228-256
1060-1090
1198 aa
Cys-His rich regions
DNA binding
b
Ad protease cleavage site
dCMP linkage site
NLS
198-247
334-397
398-470
549-601
649-667
671 aa
Acidic domains
Proline-rich domains
c
Chymotrypsin cleavage site
Zn++
CR2
CR3
NLS
529 aa
1-173
174-529
Late gene expression
Host range
Early gene expression
DNA binding
DNA replication
Virus assembly
d
14-240
399-499
499 aa
Dimerization
DNA binding
Ad DNA replication
Proline-rich transcriptional activation

Hoss et al. 1990; Moarefi et al. 1993). It is also noteworthy that serine-67 of AdPol, which could potentially be phosphorylated by cyclin-dependent kinase, is also conserved in Ad5 (Chroboczek et al. 1992) and is not located within the conserved regions present in all the DNA-dependent DNA polymerases (Blanco et al. 1991), suggesting that phosphorylation of this residue is relevant only for adenovirus DNA replication.

Mapping all the phosphorylation sites of AdPol is a prerequisite in order to assign roles for individual phosphorylation sites and to identify the critical site(s) involved in AdPol function by site-directed mutagenesis. So far by using in vivo labeled AdPol we could identify only the major site of phosphorylation. The major drawback of the in vivo labeling approach is the difficulty in obtaining a relatively large quantity of purified, labeled AdPol for tryptic digestion followed by fractionation and microsequence analysis. Instead, in vitro phosphorylation of AdPol or synthetic peptides representing all the Ser-Pro motifs by $p34^{cdc2}$ kinase or by the associated kinase would aid in identification of the other phosphorylation sites. Similar in vitro approaches using synthetic peptides have proven to be successful in identifying the phosphorylation sites of other proteins (Sturzbecher et al. 1990; Lees et al. 1992).

4.3 Phosphorylation of Preterminal Protein

Similar to AdPol, phosphorylation of pTP was initially detected using the recombinant vaccinia virus expression system (Kusukawa et al. 1994). Subsequently, it was shown that pTP is also phosphorylated in Ad2-infected cells by a large-scale metabolic labeling experiment. In both these cases, pTP was phosphorylated only on serine residues, but on multiple sites as indicated by ten spots on two-dimensional tryptic peptide maps. Peptide maps of pTP expressed with or without AdPol using recombinant vaccinia viruses were similar, suggesting that an association with AdPol does not alter the phosphorylation state of pTP. It remains to be determined whether the same sites of pTP are phosphorylated in the vaccinia virus expression system as in Ad2-infected cells.

In order to determine the effect of phosphorylation on the biological activity of pTP, pTP preparations were dephosphorylated using calf intestinal alkaline phosphatase (Kusukawa et al. 1994). Characterization of the dephosphorylated pTP showed that the in vitro DNA replication initiation activities of pTP were significantly reduced, suggesting that phosphates that are critical for the activity were removed under these conditions. Addition of nuclear extract prepared from uninfected HeLa cells did not restore the activity, which again suggests that the decrease in activity was specifically due to dephosphorylation of pTP.

Studies have shown that phosphorylation can affect protein function by inducing conformational changes (Sprang et al. 1988) as well as by electrostatic repulsion effects (Hurley et al. 1990). Such mechanisms may result in altered protein–protein and protein–DNA interactions (Hunter and Karin 1992). Our experiments show that as in the case of AdPol, dephosphorylation of pTP did not alter the ability to form a tight pTP–AdPol complex in vitro. This result is not surprising, because many earlier mutational analyses of pTP and of AdPol suggest that the interaction of AdPol and pTP consists of multiple contact sites (CHEN and HORWITZ 1989; JOUNG and ENGLER 1992; ROOVERS et al. 1993).

A specific recognition of pTP in the absence of AdPol with the Ad DNA replication origin can be demonstrated in gel shift assays, using double-stranded DNA sequences within the terminal 1–18 bp of the Ad2 genome (TEMPERLEY and HAY 1992; KUSUKAWA et al. 1994). Gel electrophoresis-based DNA mobility shift assays of dephosphorylated pTP indicated a reduction in specific interactions with the replication origin upon dephosphorylation, suggesting that the decreased binding to the template DNA may the major reason for its loss of replication initiation activity (KASUKAWA et al. 1994). DNase I footprinting analysis has shown that the binding of pTP–AdPol heterodimer protects base pairs 8–17 (TEMPERLEY and HAY1992), which is within the conserved nucleotides 9–18 present in all human Ad. Domains within pTP that are required for binding to the origin of DNA replication have not been precisely mapped. Further identification and characterization of phosphorylation sites in pTP might be valuable in precise mapping of the DNA-binding domains.

The activities of many other proteins involved in DNA and RNA synthesis are modulated as a result of altered DNA binding. In the case of DNA polymerase-α, the phosphorylated form exhibits a lower affinity to DNA (NASHEUER et al. 1991). As noted earlier, in the case of SV40 T antigen the overall phosphorylation inhibits its DNA replication activities (SIMMONS et al. 1986; MOHR et al. 1987; CEGIELSKA and VIRSHUP 1993), whereas specific phosphorylation of threonine-124 (SCHNEIDER and FANNING 1988) mediated by $p34^{cdc2}$ kinase enhances both DNA binding and replication initiation activities (MCVEY et al. 1989, 1993; MOAREFI et al. 1993). Similarly, depending upon the site of phosphorylation, the biological activities of pTP and AdPol may be affected.

Similar to AdPol, pTP is phosphorylated on multiple serine residues. In cells infected with Ad or coinfected with recombinant vaccinia viruses encoding AdPol and pTP, the *cdc2*-related protein kinase exists in a multimeric complex that also contains pTP and AdPol (RAMACHANDRA and PADMANABHAN 1993). Therefore, it is likely that *cdc2*-related kinases also phosphorylate pTP. However, unlike AdPol, pTP contains only one site, serine-337, that could potentially be phosphorylated by cyclin-dependent kinases (Fig. 2B). Two-dimensional peptide maps of in vivo labeled pTP show multiple phosphopeptides, suggesting that even if *cdc2* or related kinases are involved in phosphorylation of pTP they are probably not the major kinases. pTP also contains several other serine residues that occur within the substrate recognition sequences of other well-characterized protein kinases such as casein kinase II and DNA-dependent protein kinases. However, the number and location of phosphoserines on pTP as well as the protein kinase(s) responsible for the phosphorylation of individual sites remain to be determined.

It is also well documented that phosphorylation plays an important role in the modulation of the activities of transcription factors (for review, see BOHMAN 1990; HUNTER and KARIN 1992). Although no trancriptional activation role has been directly ascribed to pTP, mutational analysis suggests a role for pTP in efficient transcription of adenovirus DNA (SCHAAK et al. 1990b). In addition, pTP contains sequence characteristics in common with known activation domains of cellular transcription factors. As previously noted by SCHAACK et al. (1990b), pTP contains domains (amino acids 198–247, 13 of 50 acidic residues; amino acids 398–470, 26 of 73 acidic residues; amino acids 549–601, 15 of 53 acidic residues) whose negative charge is comparable to the acidic activating domains in transcription factors such as GCN4 (HOPE and STRUHL 1986) and Gal4 (MA and PTASHNE 1987). These known acidic activating domains are capable of forming an amphipathic α-helical structure, and computer analysis indicated that the two of the three acidic domains in pTP exhibit this characteristic (SCHAAK et al. 1990b). pTP also contains two proline-rich domains (amino acids 334–397, 13 of 64 proline residues; amino acids 649–667, ten of 19 proline residues) similar to proline-rich activating domain of CTF/NFI (MERMOD et al. 1989; Fig. 2B). It remains to be seen whether phosphorylation of pTP has any effect in pTP/TP-mediated adenoviral transcription.

4.4 Possible Involvement of Cyclin-Dependent Protein Kinases in Regulation of Adenovirus DNA Replication

Several studies have implied that the *cdc*2 kinases target and regulate cellular DNA replication machinery. Among the four subunits of DNA polymerase-α, two polypeptides, the 180-kDa and the 70-kDa subunits, have been reported to be phosphorylated in a cell cycle-dependent manner and in vitro evidence indicates the involvement of $p34^{cdc2}$ kinase in the mitotic phosphorylation of DNA polymerase-α (NASHEUER et al. 1991). Replication protein A has been shown to be phosphorylated in S and G2 phases of the cell cycle, and the phosphorylation is mediated by the cyclin B–*cdc*2 complex as well as cyclin A-associated *cdc*2-like kinases (DUTTA and STILLMAN 1992; NASHEUER et al. 1992). Recently, it has been shown that phosphorylation of replication protein A by cyclin-dependent kinase facilitates DNA unwinding (GEORGAKI and HUBSCHER 1993). In addition, *cdc*2 kinase stimulates DNA replication in extracts purified from G1 phase cells (DUTTA and STILLMAN 1992). D-Type cyclins associate with cellular DNA replication and repair factor, proliferating cell nuclear antigen, PCNA (XIONG et al. 1992), and cyclin A lcoalizes at the sites of cellular DNA replication (SOBCZAK-THEPOT et al. 1993). The observations that phosphorylation of replication proteins such as AdPol and SV40 large T antigen by *cdc*2 is important for their function suggest that *cdc*2 kinases target not only cellular replication apparatus, but also the replication machinery of viruses multiplying in eukaryotic cells. Use of this kinase function may be a way by which the virus diverts a normal cellular mechanism for activation of its DNA replication.

One of the widely used approaches to determine the potential kinase(s) responsible for phosphorylation of a specific protein is to map the in vivo sites of phosphorylation and examine whether they occur within the consensus substrate recognition sequences of known protein kinases. As mentioned before, pTP contains only one site (serine-337) that could potentially be phosphorylated by cyclin-dependent protein kinases. It remains to be determined whether phosphorylation of this site is essential for the replication function of pTP. Among the ten confirmed DBP phosphorylation sites that are common to both Ad2 and Ad5 (MORIN et al. 1989a), seven are present within the substrate recognition sequences (S/T-P) for cyclin-dependent protein kinases (Fig. 2C). In addition to the confirmed phosphorylation sites, DBP contains two other sites (threonine residues at 68 and 416) that could potentially be phosphorylated by cyclin-dependent protein kinases. It is interesting to note that all these cyclin-dependent kinase sites are present in both Ad2 and Ad5, and these sites, except threonine-416, are present in the N-terminal domain that are known to be phosphorylated in vivo. Although it has not been demonstrated, these observations strongly suggest that cyclin-dependent protein kinases are involved in the phosphorylation of DBP. It is interesting to note that mutational analysis have defined a region (residues 2–38) in the N terminus of DBP that is required for efficient in vivo viral DNA synthesis (BROUGH et al. 1993), and the sequence S-P-S-P-S-P (residues 31–36) within this region is conserved in four serotypes, Ad2, Ad4, AD5, and Ad7. Serine residues within this sequence are phosphorylated in vivo and cyclin-dependent protein kinases may be involved in the phosphorylation.

Search of the primary sequence of NF I/CTF revealed that there are about 15 residues (serine or threonine) which are within the consensus substrate recognition sequences for cyclin-dependent protein kinases. Most of these sites are located outside of NFI's DNA-binding domain (MERMOD et al. 1989; Fig. 2D). Therefore, one obvious possibility is that cyclin-dependent protein kinases may also target and phosphorylate NFI in vivo. This possibility is further supported by a recent study by KAWAMURA et al. (1993), in which they demonstrated the in vitro phosphorylation of NFI by cyclin-dependent protein kinases. However, in vivo phosphorylation of NFI and its impact on NFI's function have not, to our knowledge, been reported. It appears that the phosphorylation of NFI is not critical for its DNA binding and replication activities, as suggested by the ability of bacterially expressed NFI to carry out these reactions (MERMOD et al. 1989). Furthermore, the N-terminal DNA-binding domain of NFI (residues 14–240) is sufficient for dimerization as well as for stimulation of adenovirus DNA replication (MERMOD et al. 1989; MUL et al. 1990). Based on the observation that serine and threonine residues that may be subject to phosphorylation in NFI/CTF-1 are absent in CTF-2 and CTF-3, which are nevertheless highly active in transcription, MERMOD et al. (1989) proposed that post-translational modification may not be essential for transcriptional activity, but did not rule out a possible regulation of NFI activity in transcription and DNA replication by such modification. Even though phosphorylation is not necessary for the DNA-binding and replication activities of NFI, phosphorylation on multiple sites at the C-terminal part mediated by cyclin-dependent protein kinases could result in a conformational change of the protein, which in turn could regulate NFI's DNA-binding and replication activities.

The cellular transcription factor Oct-1/NFIII is known to be a phosphoprotein (TANAKA and HERR 1990) and it is phosphorylated in a cell cycle-specific manner (SEGIL et al. 1991a, b; Roberts et al. 1991). The cell cycle-dependent phosphorylation of NFIII is thought to involve multiple kinases including p34^{cdc2}-related kinase as well as phosphatases (Roberts et al. 1991). A mitosis-specific phosphorylation in vitro by protein kinase A was shown to correlate with the inhibition of NFIII DNA binding activity (Segil et al. 1991b). Similar to NFI, the Ad DNA-binding region (POU domain) of NFIII is sufficient for enhancement of Ad DNA replication (Sturm and Herr 1988; Verrijzer et al. 1990). Since phosphorylation negatively regulates the DNA-binding activity (Segil et al. 1991b), it is likely that phosphorylation could also have an impact on NFIII's ability to enhance Ad DNA replication.

Ad DNA replication initiation is highly specific, and it is believed that the specificity is due to orderly coordination of protein–protein and protein–DNA interactions at the replication origin. From various studies, we now know that during initiation of DNA replication, DBP forms a multimeric protein complex with the double-stranded DNA and enhances the binding of NFI with the specific sequences at the replication origin (Cleat and Hay 1989; Stuiver and Van der Vliet 1990; Mul et al. 1993). Because of the direct interaction between NFI and AdPol, pTP–AdPol complex is positioned at the origin, which results in the binding of pTP–AdPol to the origin sequence and stabilization of the preinitiation complex (Bosher et al. 1990; Chen et al. 1990; Mul et al. 1990; Temperley and Hay 1992; Mul and Van der Vliet 1992). A functional interaction between DBP and AdPol has also been demonstrated in which DBP increases the processivity of AdPol (Field et al. 1984; Lindenbaum et al. 1986). The above-mentioned specific protein–protein and protein–DNA interactions, with the exception of functional interaction between AdPol and DBP, probably are disrupted during the elongation of the primed template. However, the elements controlling the specific interactions at the replication origin are not known. In light of our observation that *cdc2* or related kinase stably associates with AdPol, one possibility is that phosphorylation mediated by the AdPol-associated kinase may play a role in modulating these specific protein–protein and protein–DNA interactions. If indeed pTP, NFI, and DBP are the physiological substrates for cyclin-dependent protein kinases, the AdPol-associated kinase may be able to phosphorylate pTP, NFI, and DBP, which are also present along with AdPol in the preinitiation complex. Possible phosphorylation of NFI and DBP that are also interacting with the specific DNA sequences is consistent with the observation that majority of the potential cyclin-dependent phosphorylation sites are present outside their DNA-binding domains (Fig. 2C, D). As discussed above, phosphorylation of pTP, NFI, and DBP either by the AdPol-associated kinase or free cyclin-dependent protein kinase during DNA replication may alter their protein–protein and protein–DNA interactions. The possibility of protein phosphorylation during DNA replication is in agreement with the findings that elongation of primed DNA template in vitro was stimulated severalfold by ATP or ATP-regenerating system (Field et al. 1984; Lindenbaum et al. 1986; Leith et al. 1989), and little, if any, ATP hydrolysis occurs during DNA replication in vitro (Lindenbaum et al. 1986).

4.5 Possible Involvement of Phosphorylation in Nuclear Transport and Nuclear Matrix Attachment

As mentioned in Sect. 3.1, the virus-encoded DNA replication proteins and the replicating DNA have been shown to be associated with some discrete globular structures within the nucleus of Ad-infected cells. Ultrastructural studies support the observation that DBP is associated mainly with specific regions in the nuclei of Ad5-infected cells, which also contain large amount of viral single-stranded DNA (PUVION-DUTILLEUL and PUVION 1990a, b). PUVION-DUTILLEUL (1991) also demonstrated that concomitant detection of immunostained proteins and proteins stained using bismuth, which specifically detects substructures containing highly phosphorylated proteins, is possible. Applying this improved method to sections of Ad5-infected cells, they showed that bismuth ions and viral anti-DBP antibody bound concomitantly to intranuclear virus-induced single-stranded DNA accumulation sites, structures in which viral replication activity was intermittent, and also to the fibrillogranular peripheral replicative zones that surround the single-stranded DNA accumulation sites, in which replication of the viral genome was continuous. Though it has not been shown, it is possible that the bismuth staining of these discrete structures is due to the presence of AdPol, pTP, and cellular proteins, in addition to DBP, in their phosphorylated forms.

It should be noted that potential cyclin-dependent kinase and casein kinase II phosphorylation sites (based on consensus sequences described by KENNELLY and KREBS 1991; PEARSON and KEMP 1991) are present in the well-characterized nuclear transport signals of AdPol, pTP, and DBP (Table 1). Among these sites, in vivo phosphorylation of serine-67 in AdPol (RAMACHANDRA et al. 1993), and serines 31, 33, 35, 76, 92, and 100 in DBP have been detected (ANDERSON et al. 1986; MANN 1987; MORIN et al. 1989b). In the case of AdPol, our sequence analysis of the in vivo ^{32}P-labeled protein showed that the casein kinase II phosphorylation site, serine-73, was not phosphorylated. However, it is possible that phosphorylation at this site occurs below the level of detection or may be sensitive to phosphatase(s) present in the infected cells. The presence of cyclin-dependent kinase and casein kinase II phosphorylation sites near the NLS of Ad DNA replication proteins is very interesting, because in a large number of proteins these kinase sites occur near their NLS (DINGWALL and LASKEY 1991). Based on this observation, it was suggested that the effect of phosphorylation on nuclear proteins may be widespread. In the case of SV40 T antigen, it has been clearly demonstrated that depending upon the site of phosphorylation near NLS, nuclear import of T antigen is affected (RIHS and PETERS 1989; JANS et al. 1991). The presence of threonines-111 and -112, which are part of consensus casein kinase II sites, and their phosphorylation together with NLS greatly enhanced the rate of SV40 T antigen nuclear import (RIHS et al. 1991), whereas phosphorylation of threonine-124 mediated by p34^{cdc2} inhibited its nuclear entry (JANS et al. 1991). In addition to SV40 T antigen, many transciption regulatory proteins contain domains comprising potential casein kinase II sites, *cdc2* sites, and NLS, raising the possibility that these three elements represent a functional unit regulating nuclear protein import (JANS et al. 1991). It remains to be

Table 1. Potential and confirmed phosphorylation sites in the vicinity of nuclear localization signals (NLS)

Protein	Casein kinase II site[a]	*cdc*2 kinase site[a]	NLS
AdPol	S^{73}KDT	S^{51}PGGS S^{67}PLLD	RARRRR^{46} (BSIII)
pTP	S^{303}DPVD	S^{337}PPPT	RLPV(R_6)VP^{391}
DBP	S^{53}EDEE S^{76}AADL	S^{31}PS^{33}PS^{35}PP S^{92}PKPER S^{100}PEVI	PPKKR^{46} PKKKKK^{89}
SV40 T antigen	$S^{111}S^{112}$DDE	ST^{124}PPK	PKKKRKV^{132}

[a] The phosphorylation sites are numbered.
AdPol, adenovius DNA polymerase; pTP, preterminal protein; DBP, DNA-binding protein; SV, simian virus.

seen whether phosphorylation has any influence on the nuclear entry of Ad DNA replication proteins.

In the case of pTP, it has also been shown that pTP or TP mediates the nuclear matrix attachment of Ad DNA (Bodner et al. 1989; Schaak et al. 1990b; Fredman and Engler 1993). Mutational analysis has suggested a correlation between efficient nuclear matrix attachment and transcription of Ad DNA. Studies have implicated a role for phosphorylation in the association of other proteins with the nuclear structure. For instance, in the case of Rb, un- or underphosphorylated species are tightly associated with the nuclear structure, and such an association is essential for the growth-regulating function of Rb (Mittanacht and Weinberg 1991; Templeton 1992). Possible involvement of phosphorylation in nuclear matrix association of DNA mediated by TP or pTP remains to be determined.

Acknowledgments. We thank Dr. Jingo Kusukawa for his assistance in preparing the illustrations. M. R. was supported in part by the postdoctoral training program of the Kansas Health Foundation. Research in our laboratory was supported by an N.I.H. grant to R.P. (CA33099).

References

Akiyama T, Ohuchi T, Sumida S, Matsumoto K, Toyoshima K (1992) Phosphorylation of the retinoblastoma protein by cdk2. Proc Natl Acad Sci USA 89: 7900–7904

Alvarez E, Northwood IC, Gonzalez FA, Latour DA, Seth A, Abate C, Curran T, Davis RJ (1991) Pro-Leu-Ser/Thr-Pro is a consensus primary sequence for substrate protein phosphorylation. Characterization of the phosphorylation of c-myc and c-jun proteins by an epidermal growth factor receptor threonine 669 protein kinase. J Biol Chem 266: 15277–15285

Anderson CW, Hardy MM, Dunn JJ, Klessig DF (1983) Independent, spontaneous mutants of adenovirus type 2-simian virus 40 hybrid Ad2+ND3 that grew efficiently in monkey cells possess identical mutations in the adenovirus type 2 DNA-binding protein gene. J Virol 48: 31–39

Anderson CW, Samad A, Carroll RB (1986) Identification and characterization of the sites phosphorylated in the cellular tumor antigen p53 from SV40-transformed 3T3 cells and in the DNA-binding protein from adenovirus 2. In: Botchan M, Grodzicker T, Sharp PA (eds) DNA tumor viruses, p4. Cold Spring Harbor Press, Cold Spring Harbor, New York, pp 395–404

Anderson KP, Klessig DF (1983) Posttranscriptional block to synthesis of a human adenovirus capsid protein in abortively infected monkey cells. J Mol Appl Genet 2: 31–43
Ariga H, Klein A, Levine A, Horwitz M (1980) A cleavage product of the adenoviral DNA binding protein is active in DNA replication in vitro. Virology 101: 307–310
Arrand JR, Roberts RJ (1979) The nucleotide sequences at the termini of adenovirus 2 DNA. J Mol Biol 128: 577–594
Asselbergs FAM, Mathews MB, Smart JE (1983) Structural characterization of the proteins encoded by adenovirus early region 2A. J Mol Biol 163: 177–207
Axelrod N (1978) Phosphoproteins of adenovirus type 2. Virology 87: 366–383
Babich A, Nevins JR (1981) The stability of early adenovirus mRNA is controlled by the viral 72 kD DNA-binding protein. Cell 26: 371–379
Biedenkapp H, Borgmeyer U, Sippel AE, Klempnauer KH (1988) Viral myb oncogene encodes a sequence specific DNA-binding activity. Nature 335: 835–837
Blanco L, Bernad A, Blasco MA, Salas M (1991) A general structure for DNA-dependent DNA polymerases. Gene 100: 27–38
Bodner JW, Hanson PI, Polvino-Bodner M, Zempsky W, Ward DC (1989) The terminal regions of adenovirus and minute virus of mice DNAs are preferentially associated with the nuclear matrix in infected cells. J Virol 63: 4344–4353
Bohman D (1990) Transcription factor phosphorylation: a link between signal transduction and the regulation of gene expression. Cancer Cells 2: 337–344
Bosher J, Robinson EC, Hay RT (1990) Direct interactions between the adenovirus type 2 DNA polymerase and the DNA binding domain of nuclear factor I. New Biol 2: 1083–1090
Bosher J, Leith IR, Temperley SM, Wells M, Hay RT (1991) The DNA-binding domain of nuclear factor I is sufficient to cooperate with the adenovirus type 2 DNA-binding protein in viral DNA replication. J Gen Virol 72: 2975–2980
Bosher J, Dawson A, Hay RT (1992) Nuclear factor I is specifically targeted to discrete subnuclear sites in adenovirus type 2-infected cells. J Virol 66: 3140–3150
Boyle WJ, Smeal T, Defize LHK, Angel P, Woodgett JR, Karin M, Hunter T (1991) Activation of protein kinase C decreases phosphorylation of c-Jun at sites that negatively regulate its DNA-binding activity. Cell 64: 573–584
Branton PE, Evelegh M, Rowe DT, Graham FL, Bacchetti S (1985) Protein kinase and ATP-binding activity associated with the 72-kdalton single-stranded DNA-binding protein from early region 2A of human adenovirus type 5. Can J Biochem Cell Biol 63: 941–952
Brigati DJ, Myerson D, Leary JJ, Spalholz B, Travis SZ, Fong CK, Hsiung GD, Ward DC (1983) Detection of viral genomes in cultured cells and parafilm-embedded tissue sections using biotin-labeled hybridization probes. Virology 126: 32–50
Brough DE, Rice SA, Sell S, Klessig DF (1985) Restricted changes in the adenovirus DNA-binding protein that lead to extended host range or temperature-sensitive phenotype. J Virol 55: 206–212
Brough DE, Drouguett G, Horwitz MS, Klessig DF (1993) Multiple functions of the adenovirus DNA-binding protein are required for efficient viral DNA synthesis. Virology 196: 269–281
Cajean-Feroldi, Loeb J, Meguenni S, Girad M (1981) Protein kinase associated with the adenovirus single-stranded DNA-binding protein. Eur J Biochem 120:79–87
Cardenas ME, Walter R, Hanna D, Gasser SM (1993) Casein kinase II copurifies with yeast DNA topoisomerase II and re-activates the dephosphorylated enzyme. J Cell Sci 104: 533–543
Carter TH, Blanton RA (1978) Possible role of the 72,000-dalton in regulation of adenovirus type 5 early gene expression. J Virol 25: 664–674
Cegielska A, Virshup DM (1993) Control of simian virus 40 DNA replication by the HeLa cell nuclear kinase, casein kinase I. Mol Cell Biol 13: 1202–1211
Challberg MD, Kelly TJ (1979) Adenovirus DNA replication in vitro., Proc Natl Acad Sci USA 76: 655–659
Challberg MD, Kelly TJ (1989) Animal virus DNA replication. Annu Rev Biochem 58: 671–717
Challberg MD, Kelly TJ Jr (1981) Processing of the adenovirus terminal protein. J Virol 38: 272–277
Challberg MD, Rawlins DR (1984) Template requirement for the initiation of adenovirus DNA replication. Proc Natl Acad Sci USA 81: 100–104
Challberg MD, Desiderio SV, Kelly TJ (1980) Adenovirus DNA replication in vitro: characterization of a protein covalently linked to nascent DNA strands. Proc Natl Acad Sci USA 77: 5105–5109
Chen M, Horwitz MS (1989) Dissection of functional domains of adenovirus DNA polymerase by linker insertion mutagenesis. Proc Natl Acad Sci USA 86: 6116–6120
Chen M, Mermod N, Horwitz MS (1990) Protein-protein interactions between adenovirus DNA polymerase and nuclear factor I mediate formation of the DNA replication preinitiation complex. J Biol Chem 265: 18634–18642

Chroboczek J, Bieber F, Jacrot (1992) The sequence of the genome of adenovirus type 5 and its comparison with the genome of adenovirus type 2. Virology 186: 280–285
Clark-Lewis I, Sanghera JS, Pelech SL (1991) Definition of a consensus sequence for peptide substrate recognition by p44^{mpk,} the meiosis-activated myelin basic protein kinase. J Biol Chem 266: 15180–15184
Cleat PH, Hay RT (1989) Co-operative interactions between NFI and the adenovirus DNA binding protein at the adenovirus origin of replication. EMBO J 8: 1841–1848
Cleghon V, Klessig DF (1986) Association of the adenovirus DNA-binding protein with RNA both in vitro and in vivo. Proc Natl Acad Sci USA 83: 8947–8951
Cleghon V, Klessig DF (1992) Characterization of the adenovirus DNA binding protein's nucleic acid binding region by partial proteolysis and photochemical cross-linking. J Biol Chem 267: 7872–17881
Cleghon V, Voelkerding K, Morin N, Delsert C, Klessig DF (1989) Isolation and characterization of a viable adenovirus mutant defective in nuclear transport of the DNA-binding protein. J Virol 63: 2289–2299
De Robertis EM, Longthorne RF, Gurdon JB (1978) Intracellular migration of nuclear proteins in Xenopus oocytes. Nature 272: 254–256
Desiderio SV, Kelly TJ Jr (1981) Structure of the linkage between adenovirus DNA and the 55,000 molecular weight terminal protein. J Mol Biol 145: 319–337
De Vries E, Van Driel W, Tromp M, Van Boom J, Van der Vliet PC (1985) Adenovirus DNA replication in vitro: site-directed mutagenesis of the nuclear factor I binding site of the Ad2 origin. Nucleic Acids Res 13: 4935–4952
De Vries E, Van Driel W, Bergsma WG, Arnberg AC, Van der Vliet PC (1989) HeLa nuclear protein recognizing DNA termini and translocating on DNA forming a regular DNA-multimeric protein complex. J Mol Biol 208: 65–78
Dingwall C (1991) Transport across the nuclear envelope: enigmas and explanations. Bioessays 13: 213–218
Dingwall C, Laskey RA (1986) Protein import into the cell nucleus. Annu Rev Cell Biol 2: 367–390
Dingwall C, Laskey RA (1991) Nuclear targeting sequences—a consensus? Trends Biochem Sci 16: 478–481
Dingwall C, Laskey RA (1992) The nuclear membrane. Science 258: 942–947
Dingwall C, Sharnick SV, Laskey RA (1982) A polypeptide domain that specifies migration of nucleoplasmin to the nucleus. Cell 30: 449–458
Dobbs L, Zhao L-J, Sripad G, Padmanabhan R (1990) Mutational analysis of single-stranded DNA templates active in the in vitro initiation assay for adenovirus DNA replication. Virology 178: 43–51
Draetta G (1990) Cell cycle control in eukaryotes: molecular mechanisms of cdc2 activation. Trends Biochem Sci 15: 378–383
Draetta G, Beach D (1988) Activation of cdc2 protein kinase during mitosis in human cells: cell-cycle-dependent phosphorylation and subunit rearrangement. Cell 54: 17–26
Dutta A, Stillman B (1992) cdc2 family kinases phosphorylate a human cell DNA replication factor, RPA, and activate DNA replication. EMBO J 11: 2189–2199
Eagle PA, Klessig DF (1992) A zinc-binding motif located between amino acids 273 and 286 in the adenovirus DNA-binding protein is necessary for ssDNA binding. Virology 187: 777–787
Elroy-Stein O, Fuerst TR, Moss B (1989) Cap-independent translation of mRNA conferred by encephalomyocarditis virus 5' sequence improves the performance of the vaccinia virus/bacteriophage T7 hybrid expression system. Proc Natl Acad Sci USA 86: 6126–6130
Enomoto T, Lichy JH, Ikeda JE, Hurwitz J (1981) Adenovirus DNA replication in vitro: purification of the terminal protein in a functional form. Proc Natl Acad Sci USA 78: 6779–6783
Faha B, Harlow E, Lees E (1993) The adenovirus E1A-associated kinase consists of cyclin E-p33^{cdck2} and cyclin A-p33^{cdck2}. J Virol 67: 2456–2465
Fanning E (1992) Simian virus 40 large T antigen: the puzzle, the pieces, and the emerging picture. J Virol 66: 1289–1293
Field J, Gronostajski RM, Hurwitz J (1984) Properties of the adenovirus DNA polymerase. J Biol Chem 259: 9487–9495
Finlay DR, Meier E, Bradley P, Horecka J, Forbes DJ (1991) A complex of nuclear pore proteins required for pore function. J Cell Biol 114: 169–183
Flint SJ, Sharp PA (1976) Adenovirus transcription. V. Quantitation of viral RNA sequences in adenovirus 2-infected and transformed cells. J Mol Biol 106: 749–771
Fredman JN, Engler JA (1993) Adenovirus precursor to terminal protein interacts with the nuclear matrix in vivo and in vitro. J Virol 67: 3384–3395

Freimuth PI, Ginsberg HS (1986) Codon insertion mutants of the adenovirus terminal protein. Proc Natl Acad Sci USA 83: 7816–7820

Friefeld BR, Krevolin MD, Horwitz MS (1983) Effects of the adenovirus H5*ts*125 and H5*ts*107 DNA-binding proteins on DNA replication in vitro Virology 124: 380–389

Fuerst TR, Moss B (1989) Structure and stability of mRNA synthesized by vaccinia virus-encoded bacteriophage T7 RNA polymerase in mammalian cells: importance of the 5'-untranslated leader. J Mol Biol 206: 333–348

Garcia-Bustos J, Heitman J, Hall MN (1991) Nuclear protein localization. Biochim Biophys Acta 1071: 83–101

Georgaki A, Hubscher U (1993) DNA unwinding by replication protein A is a property of the 70 kDa subunit and is facilitated by phosphorylation of the 32 kDa subunit. Nucleic Acids Res 21: 3659–3665

Gingeras TR, Sciaky D, Gelinas RE, Bing-Dong J, Yen CE, Kelly MM, Bullock PA, Parson BL, O'Neill KE, Roberts RJ (1982) Nucleotide sequences from the adenovirus-2 genome. J Biol Chem 257: 13475–13491

Ginsberg HS, Ensinger MJ, Kauffman RS, Mayer AJ, Lundholm U (1974) Cell transformation: a study of regulation with type 5 and adenovirus temperature sensitive mutants. Cold Spring Harbor Symp Quant Biol 39: 419–426

Ginsberg HS, Lundholm V, Linne T (1977) Adenovirus DNA binding proteins in cells infected with wild-type 5 adenovirus and two DNA-minus temperature sensitive mutants H5ts125 and H5ts149. J Virol 23: 142–151

Giordano A, Lee JH, Scheppler JA, Herrmann C, Harlow E, Deuschle U, Beach D, Franza BR Jr (1991) Cell cycle regulation of histone H1 kinase activity associated with the adenoviral protein E1A. Science 253: 1271–1276

Goldfarb DS (1989) Nuclear transport. Curr Opin Cell Biol 1: 441–446

Goldfarb DS, Michaud N (1991) Pathways for the nuclear transport of proteins and RNAs. Trends Cell Biol 1: 20–24

Gounari F, De Francesco R, Schmitt J, Van der Vliet PC, Cortese R, Stunnenberg H (1990) Amino-terminal domain of NF1 binds to DNA as a dimer and activates adenovirus DNA replication. EMBO J 9: 559–566

Green M, Symington J, Brackmann KH, Cartas MA, Thornton H, Young L (1981) Immunological and chemical identification of intracellular forms of adenovirus type 2 terminal protein. J Virol 40: 541–550

Greenspan D, Palese P, Crystal M (1988) Two nuclear location signals in the influenza virus NS1 non-structural protein. J Virol 62: 3020–3026

Handa H, Kingston RE, Sharp PA (1983) Inhibition of adenovirus early region IV transcription in vitro by a purified viral DNA binding protein. Nature 302: 545–547

Hanover JA (1992) The nuclear pore: at the cross roads. FASEB J 6: 2288–2295

Hard T, Kellenbach E, Boelens R, Maler BA, Dahlman K, Kreedman LP, Carlstedt-Duke J, Yamamoto KR, Gustafsson JA, Kaptein R (1990) Solution structure of the glucocorticoid receptor DNA-binding domain. Science 249: 157–160

Hay RT, Russel WC (1989) Recognition mechanism in the synthesis of animal virus DNA. Biochem J 58: 3–16

Herrmann C, Su L-K, Harlow E (1991) Adenovirus E1a is associated with a serine/threonine protein kinase. J Virol 65: 5848–5859

Hope IA, Struhl K (1986) Functional dissection of a eukaryotic transcriptional activator protein, GCN4 of yeast. Cell 46: 885–894

Horwitz MS (1978) Temperature-sensitive replication of H5ts125 adenovirus DNA in vitro. Proc Natl Acad Sci USA 75: 4291–4295

Horwitz MS (1990) Adenoviridae and their replication. In: Fields BN, Knipe DM, Chanock RM et al. (eds) Virology, 2nd ed. Raven, New York, pp 1679–1721

Hoss A, Moarefi I, Scheidtmann K-H, Cisek LJ, Corden JL, Dornreiter I, Arthur AK, Fanning E (1990) Altered phosphorylation pattern of simian virus 40 T antigen expressed in insect cells by using a baculovirus vector. J Virol 64: 4799–4807

Hu Q, Lees JA, Buchkovich KJ, Harlow E (1992) The retinoblastoma protein physically associates with the human cdc2 kinase. Mol Cell Biol 12: 971–980

Hunter T, Karin M (1992) The regulation of transcription by phosphorylation. Cell 70: 375–387

Hurley JR, Dean AM, Sohl JL, Koshland DEJ, Stroud RM (1990) Regulation of enzyme by phosphorylation at the active site. Science 249: 1012–1016

Imamoto N, Matsuoka Y, Kurihara T, Kohno K, Miyagi M, Sakiyama F, Okada Y, Tsunaswa S, Yoneda Y (1992) Antibodies against 70-KD heat shock cognate protein inhibit mediated nuclear import of karyophilic proteins. J Cell Biol 119: 1047–1061
Jans DA, Ackermann M, Bischoff JR, Beach DH, Peters R (1991) $p34^{cdc2}$-Mediated phosphorylation at T^{124} inhibits nuclear import of SV-40 T antigen proteins. J Cell Biol 115: 1203–1212
Jeng YH, Wold WSM, Sugawara K, Gilead Z, Green M (1977) Adenovirus type 2 coded single-stranded DNA binding protein: in vivo phosphorylation and modification. J Virol 22: 402–411
Johnston JM, Anderson KP, Klessig DF (1985) Partial block to transcription of human adenovirus type 2 late genes in abortively infected monkey cells. J Virol 56: 378–385
Jones KA, Kodonaga JT, Rosenfeld PJ, Kelly TJ, Tijan R (1987) A cellular DNA-binding protein that activates eukaryotic transcription and DNA replication. Cell 48: 79–89
Joung I, Engler JA (1992) Mutations in two cysteine-histidine-rich clusters in adenovirus type 2 DNA polymerase affect DNA binding. J Virol 66: 5788–5796
Kalderon D, Richardson WD, Markham AF, Smith AE (1984a) Sequence requirements for nuclear location of simian virus 40 large-T antigen. Nature 311: 33–38
Kalderon D, Roberts BL, Richardson WD, Smith AE (1984b) A short amino acid sequence able to specify nuclear location. Cell 39:499–509
Kaplan LM, Ariga H, Hurwitz J, Horwitz MS (1979) Complementation of the temperature-sensitive defect in H5*ts*125 adenovirus DNA replication in vitro. Proc Natl Acad Sci USA 76: 5534–5538
Kawamura H, Nagata K, Masamune Y, Nakanishi Y (1993) Phosphorylation of NF-I in vitro by cdc2 kinase. Biochem Biophys Res Commun 192: 1424–1431
Kedinger C, Brison O, Perrin F, Wilhelm J (1978) Structural analysis of viral replication intermediates isolated from adenovirus type-2 infected HeLa cell nuclei. J Virol 26: 364–379
Kennelly PJ, Krebs EG (1991) Consensus sequences as substrate specificity determinants for protein kinases and protein phosphatases. J Biol Chem 266: 15555–15558
Kenny M, Hurwitz J (1988) Initiation of adenovirus DNA replication II. Structural requirements using synthetic oligonucleotide adenovirus templates. J Biol Chem 263: 9809–9817
Kitagawa M, Saitoh S, Ogino H, Okabe T, Matsumoto H, Okuyama A, Tamai K, Ohba Y, Yasuda H, Nishimura S, Taya Y (1992) cdc2-Like kinase is associated with the retinoblastoma protein. Oncogene 7: 1067–1074
Klein H, Maltzman W, Levine AJ (1979) Structure-function relationships of the adenovirus DNA-binding protein. J Biol Chem 254: 11051–11060
Kleinberger T, Shenk T (1991) A protein kinase is present in a complex with adenovirus E1A proteins. Proc Natl Acad Sci USA 88: 11143–11147
Klessig DF, Anderson CW (1975) Block of multiplication of adenovirus serotype 2 in monkey cells. J Virol 16: 1650–1668
Klessig DF, Grodzicker T (1979) Mutations that allow human Ad2 and Ad5 to express late genes on monkey cells map in the viral gene encoding the 72K DNA-binding protein. Cell 17: 957–966
Koff A, Cross F, Fisher A, Schumacher J, Leguellec K, Philipe M, Roberts JM (1991) Human cyclin E, a new cyclin that interacts with two members of the CDC2 gene family. Cell 66: 1217–1228
Kraiss S, Barnekow A, Montenarh M (1990) Protein kinase activity associated with immunopurified p53 protein. Oncogene 5: 845–855
Krevolin MD, Horwitz MS (1987) Functional interactions of the domains of the adenovirus DNA-binding protein. Virology 156: 167–170
Kusukawa J, Ramachandra M, Nakano R, Padmanabhan R (1994) Phosphorylation-dependent interaction of adenovirus preterminal protein with the viral origin of DNA replication. J Biol Chem 269: 2189–2196
Lanford RE, Butel JS (1984) Construction and characterization of an SV40 mutant defective in nuclear transport of T antigen. Cell 37: 801–813
Lanford RE, Kanda P, Kennedy RC (1986) Induction of nuclear transport with a synthetic peptide homologous to the SV40 T antigen transport signal. Cell 46: 575–582
Leegwater PAJ, Van Driel W, Van der Vliet PC (1985) Recognition site of nuclear factor I, a sequence-specific DNA-binding protein from HeLa cells that stimulates adenovirus DNA replication. EMBO J 4: 1515–1521
Lees JA, Buchkovich KJ, Marshak DR, Anderson CW, Harlow E (1992) The retinoblastoma protein is phosphorylated on multiple sites by human cdc2. EMBO J 10: 4279–4290
Leith IR, Hay RT, Russel WC (1989) Adenovirus subviral particles and cores can support limited DNA replication. J Gen Virol 70: 3235–3248
Leopald P, O'Farrell PH (1991) An evolutionary conserved cyclin homolog from Drosophila rescues yeast deficient in G1 cyclins. Cell 66: 1207–1216

Levine AJ, Van der Vliet PC, Sussenbach J (1976) The replication of papovavirus and adenovirus DNA. Curr Top Microbiol Immunol 73: 68–124
Levinson A, Levine AJ (1977) The isolation and identification of the adenovirus group C tumor antigens. Virology 76: 1–11
Levinson AD, Postel EH, Levine AJ (1977) In vivo and in vitro phosphorylation of the adenovirus type 5 single strand-specific DNA-binding protein. Virology 79: 144–159
Lew DJ, Dulic V, Reed SI (1991) Isolation of three novel cyclins by rescue of G1 cyclin (cln) function in yeast. Cell 66: 1197–1206
Lewis JB, Atkins JF, Baum PR, Solen R, Gesteland RF, Anderson CW (1976) Location and identification of the genes for adenovirus type 2 early polypeptide. Cell 7: 141–151
Lichy JH, Field J, Horwitz MS, Hurwitz J (1982) Separation of adenovirus terminal protein precursor from its associated DNA polymerase: role of both proteins in the initiation of adenovirus DNA replication. Proc Natl Acad Sci USA 79: 5225–5229
Lindenbaum JO, Field J, Hurwitz J (1986) The adenovirus DNA binding protein and adenovirus DNA polymerase interact to catalyze elongation of primed templates. J Biol Chem 261: 10218–10227
Linne T, Philipson L (1980) Further characterization of the phosphate moiety of the adenovirus type 2 DNA-binding protein. Eur J Biochem 103: 259–270
Linne T, Jornvall H, Philipson L (1977) Purification and characterization of the phosphorylated DNA-binding protein from adenovirus type 2 infected cells. Eur J Biochem 76: 481–490
Lucknow VA, Summers MD (1988) Trends in the development of baculovirus expression vectors. Biotechnology 6: 47–55
Ma J, Ptashne M (1987) Deletion analysis of GAL4 defines two transcriptional activating segments. Cell 48: 847–853
Maller JL (1990) Xenopus oocytes and the biochemistry of cell division. Biochemistry 29: 3157–3166
Mann R (1987) Identification and characterization of phosphorylation sites of adenovirus (Ad2) DNA-binding protein (DBP). Ph D thesis, New York University, New York
Matsushime H, Roussel MF, Ashmun RA, Sherr CJ (1991) Colony-stimulating factor 1 regulates novel cyclins during the G1 phase of the cell cycle. Cell 65: 701–713
Matsuura Y, Possee RD, Overton HA, Bishop DHL (1987) Baculovirus expression vectors: the requirements of high level expression of proteins including glycoproteins. J Gen Virol 69: 1233–1250
McPherson RA, Ginsberg HS, Rose JA (1982) Adeno-associated virus helper activity of adenovirus DNA binding protein. J Virol 44: 666-673
McVey D, Brizuela L, Mohr I, Marshak DR, Gluzman J, Beach D (1989) Phosphorylation of large tumor antigen by *cdc*2 kinase stimulates SV40 DNA replication. Nature 341: 503–507
McVey D, Ray S, Gluzman Y, Berger L, Wildman AG, Marshak DR, Tegtmeyer P (1993) *cdc*2 phosphorylation of threonine 124 activates the origin-unwinding functions of simian virus 40 T antigen. J Virol 67: 5206–5215
Mermod N, O'Neill EA, Kelly TJ, Tjian R (1989) The proline-rich transcriptional activator of CTF/NFI-I is distinct from the replication and DNA binding domain. Cell 58: 741–753
Meyerson M, Enders GH, Wu C, Su L, Gorka C, Nelson C, Harlow E, Tsai L (1992) A family of human *cdc*2-related protein kinases. EMBO J 11: 2909–2917
Miller LK (1988) Baculovirus as gene expression vectors. Annu Rev Microbiol 42: 177–199
Mittanacht S, Weinberg RA (1991) G1/S phosphorylation of the retinoblastoma protein is associated with an altered affinity for the nuclear compartment. Cell 65: 381–393
Moarefi IF, Small D, Gilbert I, Hopfner M, Randall SK, Schneider C, Russo AA, Ramsperger U, Arthur AK, Stahl H, Kelly TJ, Fanning E (1993) Mutation of the cyclin-dependent kinase phosphorylation site in simian virus 40 (SV 40) large T antigen specifically blocks SV 40 origin DNA unwinding. J Virol 67: 4992–5002
Mohr IJ, Stillman B, Gluzman Y (1987) Regulation of SV40 DNA replication by phosphorylation of T antigen. EMBO J 6: 153–160
Moreland RB, Langevin GL, Singer RH, Garcea RL, Hereford LM (1987) Amino acid sequences that determine the nuclear localization of yeast histone 2B. Mol Cell Biol 7: 4048–4057
Moreno S, Nurse P (1990) Substrates for p34cdc2: in vivo veritas? Cell 61: 549–551
Morin N, Delsert C, Klessig DF (1989a) Nuclear localization of the adenovirus DNA-binding protein: requirement for two signals and complementation during viral infection. Mol Cell Biol 9: 4372–4380
Morin N, Delsert C, Klessig DF (1989b) Mutations that affect phosphorylation of the adenovirus DNA-binding protein alters its ability to enhance its own synthesis. J Virol 63: 5228–5237
Moss B (1991) Vaccinia virus: a tool for research and vaccine development. Science 252: 1662–1667
Moss B, Elroy-Stein T, Mizukami T, Alexander WA, Fuerst TR (1990) New mammalian expression vectors (product review). Nature 348: 91–92

Mul YM, Van der Vliet PC (1992) Nuclear factor I enhances adenovirus DNA replication by increasing the stability of a preinitiation complex. EMBO J 11: 751–760
Mul YM, Verrijzer CP, Van der Vliet PC (1990) Transcription factors NFI and NFIII/Oct-I function independently employing different mechanisms to enhance adenovirus DNA replication. J Virol 64: 5510–5518
Mul YM, Verrijzer CP, Van der Vliet (1993) Adenovirus DNA binding protein effects the kinetics of DNA replication by a mechanism distinct from NFI or oct I. Nucleic Acids Res 21: 641–647
Murti KG, Davis DS, Kitchingman GR (1990) Localization of adenovirus-encoded DNA replication protein in the nucleus by immunogold electron microscopy. J Gen Virol 71: 2847–2857
Nagata K, Guggenheimer RA, Enomoto T, Lichy JH, Hurwitz J (1982) Adenovirus DNA replication in vitro: identification of a host factor that stimulates synthesis of the preterminal protein·dCMP complex. Proc Natl Acad Sci USA 79: 6438–6442
Nagata K, Guggenheimer RA, Hurwitz J (1983a) Adenovirus DNA replication in vitro: synthesis of full-length DNA with purified proteins. Proc Natl Acad Sci USA 80: 4266–4270
Nagata K, Guggenheimer RA, Hurwitz J (1983b) Specific binding of a cellular DNA replication protein to the origin of replication of adenovirus DNA. Proc Natl Acad Sci USA 80: 6177–6181
Nakamura H, Marita T, Sato C (1986) Structural organization of replicon domains during DNA synthetic phase in the mammalian nucleus. Exp Cell Res 165: 291–297
Nakano R, Zhao L-J, Padmanabhan R (1991) Overproduction of adenovirus DNA polymerase and preterminal protein in HeLa cells. Gene 105: 173–178
Nasheuer H-P, Moore A, Wahl AF, Wang TSF (1991) Cell cycle-dependent phosphorylation of human DNA polymerase α. J Biol Chem 266: 7893–7903
Nasheuer HP, von Winkler D, Schneider C, Dornreiter I, Gilbert I, Fanning E (1992) Purification and functional characterization of bovine RP-A in an in vitro SV40 DNA replication system. Chromosoma 102: S52–S59
Nath ST, Nayak DP (1990) Function of two discrete regions is required for nuclear localization of polymerase basic protein 1 of A/WSN/33 influenza virus (H1 N1). Mol Cell Biol 10: 4139–4145
Neale GAM, Kitchingman GR (1990) Conserved region 3 of the adenovirus type 5 DNA-binding protein is important for interaction with single-stranded DNA. J Virol 64: 630–638
Newmeyer DD, Forbes DJ (1988) Nuclear import can be separated into distinct steps in vitro: nuclear pore binding and translocation. Cell 52: 641–653
Nicolas JC, Sarnow P, Girad M, Levine AJ (1983) Host range, temperature-conditional mutants in the adenovirus DNA binding protein are defective in the assembly of infectious virus. Virology 126: 228–239
Nigg EA, Baeuerle PA, Luhrmann R (1991) Nuclear import-export: in search of signals and mechanisms. Cell 66: 15–22
O'Neill EA, Kelly TJ (1988) Purification and characterization of nuclear factor III (origin recognition protein C) a sequence-specific DNA binding protein required for efficient initiation of adenovirus DNA replication. J Biol Chem 263: 931–937
O'Neill EA, Fletcher C, Burrow CR, Heintz N, Roeder RG, Kelly TJ (1988) Transcriptional factor OTF-1 is functionally identical to the DNA replication factor NF III. Science 241: 1210–1213
Ostrove JM, Rosenfeld P, Williams J, Kelly TJ Jr (1983) In vitro complementation as an assay for purification of adenovirus DNA replication proteins. Proc Natl Acad Sci USA 80: 935–939
Pagano M, Draetta G, Jansen-Durr P (1992a) Association of cdk2 kinase with the transcription factor E2F during S phase. Science 255: 1144–1147
Pagano M, Pepperkok R, Verde F, Ansorje W, Draetta G (1992b) cyclin A is required at two points in the human cell cycle. EMBO J 11: 961–971
Paris J, Le Guellec R, Couturier A, Le Guellec K, Omilli F, Camonis J, MacNeill S, Philippe M (1991) Cloning by differential screening of a Xenopus cDNA coding for a protein highly homologous to cdc2. Proc Natl Acad Sci USA 88: 1039–1043
Pearson RB, Kemp BE (1991) Protein kinase phosphorylation site sequences and consensus specificity motifs: tabulations. Methods Enzymol 200: 62–81
Peck VM, Gerner EW, Cress AE (1993) A DNA polymerase α-associated 56 kDa protein kinase. Biochem Biophys Res Commun 190: 325–331
Pelech SL, Sanghera JS (1992) Mitogen-activated protein kinases: versatile transducers for cell signaling. Trends Biochem Sci 17: 233–238
Peters R (1986) Fluorescence microphotolysis to measure nucleocytoplasmic transport and intracellular mobility. Biochim Biophys Acta 864: 305–359
Pettit Sc, Horwitz MS, Engler JA (1988) Adenovirus preterminal protein synthesized in COS cells from cloned DNA is active in DNA replication in vitro. J Virol 62: 496–500

Picard D, Yamamoto KR (1987) Two signals mediate hormone-dependent nuclear localization of the glucocorticoid receptor. EMBO J 6: 3333–3340
Pines J, Hunter T (1989) Isolation of a human cyclin cDNA: evidence for cyclin mRNA and protein regulation in cell cycle and for interaction with $p34^{cdc2}$. Cell 58: 833–846
Postel E, Klein H, Levine AJ (1978) The fidelity of phosphorylation of the adenovirus DNA-binding protein by an in vitro nuclear protein kinase from virus-infected cells. Virology 86: 291–294
Prevelige P Jr, Fasman GD (1989) Chou-Fasman Prediction of the secondary structure of proteins: Chou-Fasman-Prevelige algorithm. In: Fasman GD (ed) Prediction of protein structure and the principles of protein conformation. Plenum, New York, pp 391
Prives C (1990) The replication functions of SV40 T antigen are regulated by phosphorylation. Cell 61: 735–738
Pruijn GJM, Van Driel W, Van der Vliet PC (1986) Nuclear factor III, a novel sequence-specific DNA-binding protein from HeLa cells stimulating adenovirus DNA replication. Nature 322: 656–659
Pruijn GJM, Van Driel W, Van Miltenburg RT, Van der Vliet PC (1987) Promoter and enhancer elements containing a conserved sequence motif are recognized by nuclear factor III, a protein stimulating adenovirus DNA replication. EMBO J 6: 3771–3778
Puvion-Dutilleul F (1991) Simultaneous detection of highly phosphorylated proteins and viral major DNA binding protein distribution in nuclei of adenovirus type 5-infected HeLa cells. J Histochem Cytochem 39: 669–680
Puvion-Dutilleul F, Puvion E (1990a) Replicating single-stranded adenovirus type 5 DNA molecules accumulate within well-delimited intranuclear areas of lytically infected HeLa cells. Eur J Cell Biol 379–388
Puvion-Dutilleul F, Puvion E (1990b) Analysis by in situ hybridization and autoradiography of sites of replication and storage of single- and double-stranded adenovirus type 5 DNA in lytically infected HeLa cells. J Struct Biol 103: 280–289
Quinlan MP, Chen LB, Knipe DM (1984) The intranuclear location of a herpes simplex virus DNA-binding protein is determined by the status of viral DNA replication. Cell 36: 857–868
Ramachandra M, Padmanabhan R (1993) Adenovirus DNA polymerase is phosphorylated by a stably associated histone H1 kinase. J Biol Chem 268: 17448–17456
Ramachandra M, Nakano R, Mohan PM, Rawitch AB, Padmanabhan R (1993) Adenovirus DNA polymerase is a phosphoptrotein. J Biol Chem 268: 442–445
Rawlins DR, Rosenfeld PJ, Wides RJ, Challberg MD, Kelly TJ Jr (1984) Structure and function of the adenovirus origin of DNA replication. Cell 37: 309–319
Rekosh DMK, Russel WC, Bellett AJD, Robinson AJ (1977) Identification of a protein linked to the ends of adenovirus DNA. Cell 11: 283–295
Rice SA, Klessig DF (1984) The function(s) provided by the adenovirus-specified, DNA-binding protein required for viral late gene expression is independent of the role of the protein in viral DNA replication. J Virol 49: 35–49
Richardson WD, Roberts BL, Smith AE (1986) Nuclear location signals in polyoma virus large-T. Cell 44: 77–85
Richardson WD, Mills AD, Dilworth SM, Laskey RA, Dingwall C (1988) Nuclear protein migration involves two steps: rapid binding at the nuclear envelope followed by slower translocation through nuclear pores. Cell 52: 655–664
Rihs HP, Peters R (1989) Nuclear transport kinetics depend on phosphorylation-site-containing sequences flanking the karyophilic signal of the simian virus 40 T antigen. EMBO J 8: 1479–1484
Rihs H-P, Jans DA, Fan H, Peters R (1991) The rate of nuclear cytoplasmic protein transport is determined by the casein kinase II site flanking the nuclear localization sequence of the SV 40 T-antigen. EMBO J 10: 633–639
Robbins J, Dilworth SM, Laskey RA, Dingwall C (1991) Two interdependent basic domains in nucleoplasmin nuclear targeting sequence: identification of a class of bipartite nuclear targeting sequence. Cell 64: 615–623
Roberts B (1989) Nuclear location signal-mediated protein transport. Biochim Biophys Acta 1008: 263–280
Roberts SB, Segil N, Heintz N (1991) Differential phosphorylation of the transcription factor oct1 during the cell cycle. Science 253: 1022–1026
Roninson I, Padmanabhan R (1980) Studies on the nature of the linkage between the terminal protein and the adenovirus DNA. Biochem Biophys Res Commun 94: 398–405
Roovers DJ, Van der Lee FM, Van der Wees J, Sussenbach JS (1993) Analysis of the adenovirus type 5 terminal protein precursor and DNA polymerase by linker insertion mutagenesis. J Virol 67: 265–276

Rosenblatt J, Gu Y, Morgan DO (1992) Human cyclin-dependent kinase 2 is activated during the S and G2 phases of the cell cycle and associates with cyclin A. Proc Natl Acad Sci USA 89: 2824–2828
Rosenwirth B, Anderson CW, Levine AJ (1976) Tryptic finger printing analysis of adenovirus 2, 5 and 12 DNA-binding proteins. Virology 69: 617–625
Russell WC, Blair GE (1977) Polypeptide phosphorylation in adenovirus-infected cells. J Gen Virol 34: 19–35
Russell WC, Webster A, Leith IR, Kemp GD (1989) Phosphorylation of adenovirus DNA-binding protein. J Gen Virol 70: 3249–3259
Saborio J, Oberg B (1976) In Vivo and in vitro synthesis of adenovirus type 2 early proteins. J Virol 17: 865–875
Sasaguri Y, Sanford T, Aguirre P, Padmanabhan R (1987) Immunological analysis of 140-kDa adenovirus-encoded DNA polymerase in adenovirus type-2 infected HeLa cells using antibodies raised against the protein expressed in E. coli. Virology 160: 389–399
Schaack J, Schedl P, Shenk T (1990a) Topoisomerase I and II cleavage of adenovirus DNA in vitro: both topoisomerase activities appear to be required for adenovirus DNA replication. J Virol 64: 78–85
Schaak J, Ho WY, Freimuth P, Shenk T (1990b) Adenovirus terminal protein mediates both nuclear matrix attachment and efficient transcription of adenovirus DNA. Genes Dev 4: 1197–1208
Schechter NM, Davies W, Anderson CW (1980) Adenovirus coded deoxyribonucleic acid binding protein: isolation physical properties and effects of proteolytic digestion. Biochemistry 19: 2802–2810
Scheidtman KH, Buck M, Schneider J, Kalderson D, Fanning E, Smith AE (1991) Biochemical characterization of phosphorylation site mutants of simian virus 40 large T antigen: evidence for interaction between amino and carboxy-terminal domains. J Virol 65: 1479–1490
Schneider J, Fanning E (1988) Mutations in the phosphorylation sites of simian virus 40 (SV40) T antigen alter its origin DNA binding specificity for sites I and II affects SV40 DNA replication activity. J Virol 62: 1598–1605
Segil N, Roberts SB, Heintz N (1991a) Mitotic phosphorylation of the Oct-1/POU homeodomain and regulation of Oct-1 DNA binding activity. Science 254: 1814–1816
Segil N, Roberts SB, Heintz N (1991b) Cell-cycle-regulated phosphorylation of the transcription factor oct-1. Cold Spring Harb Symp Quant Biol 56: 285–292
Shi Y, Thomas JO (1992) The transport of proteins into the nucleus requires the 70-kilodalton heat shock protein or its cytosolic cognate. Mol Cell Biol 12: 2186–2192
Shinagawa M, Padmanabhan R (1979) Nucleotide sequence at the inverted terminal repetition of adenovirus type 2 DNA. Biochem Biophys Res Commun 87: 671–678
Shinagawa M, Padmanabhan R (1980) Comparitive sequence analysis of the inverted terminal repetitions from different adenoviruses. Proc Natl Acad Sci USA 77: 3931–3935
Shu L, Hong JS, Wei Y-F, Engler JA (1986) Nucleotide sequence of the genes encoded in early region 2b of human adenovirus type 12. Gene 46: 187–195
Shu L, Horwitz MS, Engler JA (1987) Expression of enzymatically active adenovirus DNA polymerase from cloned DNA requires sequences upstream of the main open reading frame. Virology 161: 520–526
Shu L, Pettit SC, Engler JA (1988) The precise structure and coding capacity of mRNAs from early region 2B of human adenovirus serotype 2. Virology 165: 348–356
Silver PA (1991) How proteins enter the nucleus. Cell 64: 489–497
Simmons DT, Chou W, Rodgers K (1986) Phosphorylation downregulates the DNA-binding activity of simian virus 40 antigen. J Virol 60: 888–894
Smart JE, Stillman BW (1982) Adeovirus terminal protein precursor: partial amino acid sequence and the site of covalent linkage to virus DNA. J Biol Chem 257: 13499–13506
Sobczak-Thepot J, Harper F, Florentin Y, Zindy F, Brechot C, Puvion E (1993) Localizaion of cyclin A at the sites of cellular DNA replication. Exp Cell Res 206: 43–48
Sprang SR, Acharya KR, Goldsmith EJ, Stuart DI, Varvill K, Fletterick RJ, Madsen NB, Johnson LN (1988) Structural changes in glycogen phosphorylase induced by phosphorylation. Nature 336: 215–221
Starr CM, Hanover JA (1990) Structure and function of the nuclear pore complex: new perspectives. Bioessays 12: 323–330
Stillman BW (1981) Adenovirus DNA replication in vitro: a protein linked to the 5' end of nascent DNA strands. J Virol 37: 139–147
Stillman B (1989) Initiation of eukaryotic DNA replication in vitro. Annu Rev Cell Biol 5: 197–245
Stillman BW, Lewis JB, Chow LT, Mathews MB, Smart JE (1981) Identification of the gene and mRNA for the adenovirus terminal protein precursor. Cell 23: 497–508

Stillman BW, Topp Wc, Engler JA (1982) Conserved sequences at the origin of adenovirus DNA replication. J Virol 44: 530–537
Stuiver MH, Van der Vliet PC (1990) Adenovirus DNA binding protein forms multimeric protein complex with double-stranded DNA and enhances binding of nuclear factor I. J Virol 64: 379–386
Stunnenberg HG, Lange H, Philipson L, Van Miltenberg RT, Van der Vliet PC (1988) High expression of functional adenovirus DNA polymerase and precursor terminal protein using recombinant vaccinia virus. Nucleic Acids Res 16: 2431–2444
Sturm RA, Herr W (1988) The POU domain DNA-binding structure. Nature 336: 601–604
Sturzbecher H-W, Maimets T, Chumakov P, Brain R, Addison C, Simanis V, Rudge K, Philp R, Grimaldi M, Court W, Jenkins JR (1990) p53 interacts with $p34^{cdc2}$ in mammalian cells: implications for cell cycle control and oncogenesis. Oncogene 5: 795–801
Sugawara K, Gilead Z, Green M (1977) Purification and molecular characterization of a adenovirus type 2 DNA-binding protein. J Virol 21: 338–346
Tamura K, Kanaoka Y, Jinno S, Nagata A, Ogiso Y, Shimizu K, Hayakawa T, Nojima H, Okayama H (1993) Cyclin G: a new mammalian cyclin with homology to fission yeast Cig 1. Oncogene 8: 2113–2118
Tanaka M, Herr W (1990) Differential transcriptional activation by oct-1, and oct-2: interdependent activation domains induce oct-2 phosphorylation. Cell 60: 375–386
Temperley SM, Hay RT (1992) Recognition of the adenovirus type 2 origin of DNA replication by the virally encoded DNA polymerase and preterminal proteins. EMBO J 11: 761–768
Templeton DJ (1992) Nuclear binding of retinoblastoma gene product is determined by cell cyle-regulated phosphorylation. Mol Cell Biol 12: 435–443
Thomas G (1992) MAP kinase by any other name smells just as sweet. Cell 68: 3–6
Tokunaga O, Shinagawa M, Padmanabhan R (1982) Physical mapping of the genome and sequence analysis at the inverted terminal repetition of adenovirus type 4 DNA. Gene 18: 329–334
Tolun A, Alestrom P, Pettersson U (1979) Sequence of inverted terminal repetitions from different adenoviruses: demonstration of conserved sequences and homology between SA-7 termini and SV40 DNA. Cell 17: 705–713
Tommasino M, Adamczewski JP, Carlotti F, Barth CF, Manetti R, Contorni M, Cavalieri F, Hunt T, Crawford L (1993) HPV 16 E7 protein associates with the protein kinase $p33^{cdk2}$ and cyclin A. Oncogene 8: 195–202
Tsai LH, Harlow E, Meyerson M (1991) Isolation of the human cdk2 gene that encodes the cyclin A and adenovirus E1A-associated p33 kinase. Nature 353: 174–177
Tsernoglou D, Tsugita A, Tucker AD, Van der Vliet PC (1985) Characterization of the chymotryptic core of the adenovirus DNA-binding protein. FEBS Lett 188: 248–252
Underwood MR, Fried HM (1990) Characterization of nuclear localizing sequences derived from yeast ribosomal protein L29. EMBO J 9: 91–99
Van Bergen BGM, Van der Vliet PC (1983) Temperature sensitive initiation and elongation of adenovirus DNA replication in vitro with nuclear extracts from H5*ts*36- H5*ts*149- and H5*ts*125-infected HeLa cells. J Virol 46: 624–648
Van der Vliet PC (1990) Adenovirus DNA replication in vitro. In: Strauss PR, Wilson SH (eds) The eukaryotic nucleus, vol 1. Telford, Caldwell pp 1–29
Van der Vliet PC, Levine AJ (1973) DNA-binding proteins specific for cells infected by adenovirus. Nature 246: 170–174
Van der Vliet PC, Sussenbach JS (1975) An adenovirus type 5 gene function required for initiation of viral DNA replication.Virology 67: 415–426
Van der Vliet PC, Zandberg J, Jansz HS (1977) Evidence for a function of the adenovirus DNA binding protein in initiation of DNA synthesis as well as in elongation on nascent DNA chains. Virology 80: 98–110
Van der Vliet PC, Keegstra W, Jansz HS (1978) Complex formation between the adenovirus type 5 DNA binding protein and single-stranded DNA. Eur J Biochem 86: 389–398
Verrijzer CP, Kal AJ, Van der Vliet PC (1990) DNA binding domain (POU domain) of transcription factor oct-1 suffices for stimulation of DNA replication. EMBO J 9: 1883–1888
Voelkerding K, Klessig DF (1986) Identification of two nuclear subclasses of the adenovirus type 5 encoded DNA-binding protein. J Virol 60: 353–362
Watson CJ, Hay RT (1990) Expression of adenovirus type 2 DNA polymerase in insect cells infected with a recombinant baculovirus. Nucleic Acids Res 18: 1167–1173
Weinberg RA (1991) Tumor suppressor genes. Science 254: 1138–1146
Wides RJ, Challberg MD, Rawlins DR, Kelly TJ (1987) Adenovirus origin of DNA replication: sequence requirements for replication in vitro. Mol Cell Biol 7: 864–874

Wilcock D, Lane DP (1991) Localization of p53, retinoblastoma and host replication proteins at sites of viral DNA replication in herpes-infected cells. Nature 349: 429–431
Williams KR, Chase J (1990) Eukaryotic single-stranded nucleic acid binding proteins. In: Revzin A (ed) The biology of nonspecific DNA-protein interactions. CRC, Boca Raton, pp 197–227
Williams RT, Carbonaro-Hall DA, Hall FL (1992) Copurification of $p34^{cdc2}$/p58 cyclin A proline-directed protein kinase and the retinoblastoma tumor susceptibility gene product: interaction of an oncogenic serine/threonine protein kinase with a tumor-suppressor protein. Oncogene 7: 423–432
Xiong Y, Connolly T, Futcher B, Beach D (1991) Human D-type cyclin. Cell 65: 691–699
Xiong Y, Zhang H, Beach D (1992) D type cyclin associates with multiple protein kinases and the DNA replication and repair factor PCNA. Cell 71: 505–514
Zhao L-J (1990) Expression and nuclear transport of the adenovirus DNA polymerase and preterminal protein. PhD Thesis, University of Kansas, Kansas City, p 66
Zhao L-J, Padmanabhan R (1988) Nuclear transport of adenovirus DNA polymerase is facilitated by interaction with preterminal protein. Cell 55: 1005–1015
Zhao L-J, Padmanabhan R (1991) Three basic regions in adenovirus DNA polymerase interact differentially depending on the protein context to function as bipartite nuclear localization signals. New Biol 3: 1074–1088
Zhao L-J, Irie K, Trirawatanapong T, Nakano R, Nakashima A, Morimatsu M, Padmanabhan R (1991) Synthesis of biologically active adenovirus preterminal protein in insect cells using a baculovirus vector. Gene 100: 147–154

Homologous Recombination in the Replicative Cycle of Adenoviruses and Its Relationship to DNA Replication

C.S.H. YOUNG

1 Introductory Remarks

The observation that adenovirus genomes can undergo genetic recombination is almost a quarter of a century old (WILLIAMS and USTAÇELEBI 1971; TAKEMORI 1972; ENSINGER and GINSBERG 1972), yet despite its early discovery and its rapid exploitation as a tool for mapping mutations and for creating new genotypes (reviewed in GINSBERG and YOUNG 1977; YOUNG et al. 1984b), the mechanisms underlying it are not well understood at the molecular and biochemical levels. Such an understanding, however, can be expected to shed light not only on the specific features peculiar to adenovirus recombination itself, but also on the more general characteristics of the repair and recombinational capacities of the cell. These cellular aspects are of considerable current theoretical and practical interest,

Department of Microbiology, Columbia University, College of Physicians and Surgeons, New York, NY 10032, USA

because of recent advances both in uncovering the molecular basis of genetic defects in DNA repair in mammalian cells and in the techniques of gene targeting by homologous recombination in order to investigate mammalian development and differentiation. As with many other fundamental biological phenomena, the study of adenovirus recombination can be expected to yield valuable insights into the normal capacities of the cell.

The main purposes of this chapter are, first, to review the currently available facts and observations about adenovirus recombination, emphasizing those aspects not covered in detail in previous reviews (GINSBERG and YOUNG 1977; YOUNG et al. 1984b), and, second, to elaborate models for adenovirus recombination and to suggest experimental approaches to test their validity.

2 General Characteristics of Adenovirus Recombination

As with the bacteriophage systems upon which the experiments were based, the first attempts to define adenovirus recombination examined what might be termed the "population genetics" of the replicating mix of viral genomes. Using classical genetic approaches, with strains of virus marked with temperature-sensitive mutations and distinguishable restriction enzyme sites, it was shown that the recombinant virus fraction increased with time after infection, that recombination could take place well into the late phase of the infectious cycle, and that individual genomes probably could undergo more than one recombination event (YOUNG and SILVERSTEIN 1980; MUNZ et al. 1983). The genetic evidence also suggested that recombination was characterized by "negative interference", i.e., that recombinant genomes had undergone more than the expected number of unselected recombination events (WILLIAMS et al. 1975; YOUNG and WILLIAMS 1975; YOUNG and SILVERSTEIN 1980). The model that emerged from these studies was essentially the same as that proposed many years earlier for the T-even bacteriophages (DOERMANN 1953; VISCONTI and DELBRÜCK 1953), namely that a pool of genomes underwent random recombination events during much of the infectious cycle, until individual genomes were withdrawn from the pool by packaging into mature virus. This model left open such questions as the precise structure of the genomes entering into recombination, the nature of the recombinational intermediates, and which viral and cellular functions were involved in both the formation and the resolution of such intermediates.

3 Temporal and Functional Relationship Between Recombination and DNA Replication

The first clue as to the possible viral DNA structures involved in the formation of recombination intermediates came from the kinetics of recombination, as measured by Southern analysis of the intracellular pool of replicating molecules (Young and Silverstein 1980). In this experimental design, two parental viruses with two restriction enzyme site differences, arranged either in cis or in trans, are used to coinfect cells. Recombination between the restriction enzyme sites will yield two recombinant genomes. All four products, parental and recombinant, can be detected among the intracellular DNA by hybridizing with a specific sequence located between the two sites of restriction enzyme difference. This technique is extremely valuable in monitoring recombination, because it does not require the formation of a viable recombinant virus and thus can detect recombination in circumstances where virus is not produced, as for example in eclipse or under nonpermissive conditions. This technique, first developed in adenovirus infection, is now commonly used to monitor recombination, in a variety of systems, such as yeast meiosis (Borts et al. 1985). The kinetic analysis showed that a recombinant product could be detected only *after* the onset of DNA replication and that the proportion of product among the total pool of molecules increased with time thereafter. This could imply that some protein product, either viral or cellular and necessary for recombination, was elaborated or active only after DNA replication had commenced. Alternatively, it might mean that the process of replication itself was a prerequisite for recombination. This latter effect could be merely quantitative, for example by increasing the total number of genomes in the individual infected cell so that genomes of different genotype could find one another. It has been known for many years that adenovirus DNA replication occurs in discrete regions of the nucleus (see Puvillon-Dutilleul and Puvion 1990a,b, for recent evidence), and, if such "replication factories" initiate from single genomes, they might have to fuse before recombination could commence. Alternatively, and more interestingly, it could reflect the role of some genomic structure, produced during replication, as an active participant in recombination. The analysis of the structures produced by replication was well advanced at this time, and it seemed entirely plausible that the displaced single strands, intrinsic to adenovirus DNA replication, were involved in recombination, an idea which was first suggested by Flint et al. (1976). Support for the idea that replicating molecules could be involved in recombination was obtained by Wolgemuth and Hsu (1981), who examined the intracellular pool of adenoviral DNA using electron microscopy (EM). Prominent among the paired duplex molecules were those in which replication had occurred on one duplex and the displaced single strand had been transferred to another duplex, a viral example of the general model for recombination proposed by Meselson and Radding (1975). The EM analysis also gave evidence for classical Holliday junctions (Wolgemuth and Hsu 1980), which might suggest that recombination could also be initiated internally in the genome. This point will be considered later in the discussion of the models for adenovirus recombination. Taking these observations together, it seemed plausible that DNA replication and recombination might be intimately connected and that high levels of recombination might indeed be dependent on concurrent replication. If this latter linkage were true, then any treatment that blocked replication would also block recombination, and viral mutants with deficiencies in replication would also be recombination deficient. This was tested, again using the Southern blot analysis of intracellular DNA, and it was found that blocking DNA replication by inhibitors of Aeither replication or early protein synthesis, or the use of temperature-sensitive mutants with lesions in known DNA synthesis genes, diminished DNA recombination as well (YOUNG et al. 1984a). If the replication block was removed during the course of the experiment, replication and recombination began in concert. It should also be pointed out that the block to recombination was observed even at high multiplicities of infection, suggesting that the role of replication in promoting recombination was not merely quantitative. All of these results, using a variety of techniques, are consistent with, although they do not prove, a model of recombination in which replication produces the specific substrates necessary for the high rates of recombination observed in adenovirus infection.

4 Are Other Viral Gene Products Necessary for Adenovirus Recombination?

The results with the DNA synthesis inhibitors and with the temperature-sensitive mutants are consistent with an important role of replication in generating the substrates necessary for high rates of recombination. However, they do not address the question of whether or not viral or cellular functions are involved in the formation of recombination intermediates and for their resolution into fully recombinant products. Again, precedent from bacteriophages suggests that viral early genes might play a *primary* role in recombination i.e., they might help to form the recombination intermediates or promote their resolution. These possibilities are part of a broader question of whether or not recombination is an essential part of the viral life cycle and whether the virus encodes its own functions for this essential activity. If both suppositions are true, conditional lethal mutations in recombination could be searched for. However, the situation may be analogous to most bacteriophages, in which virally encoded recombination functions are not essential in most hosts, and thus the search for conditional lethal mutations in viral recombination functions is fruitless, unless conducted under very special host conditions. Despite these theoretical uncertainties, the accumulated efforts of the last decade using site-directed mutagenesis have generated a cornucopia of conditionally lethal host range mutations and mutations with no overt phenotype, principally in the early regions of adenovirus.

Adenovirus has an extremely varied repertoire of early gene products, many of whose functions are unknown, and thus it is possible that some of them may play a direct role in recombination. To test this, two parental viruses, both of which contained a mutation in the early region under consideration, but which could be distinguished by restriction site differences, were used to infect nonpermissive cells. The Southern hybridization technique allowed the analysis of intracellular

DNA, just as for the ts DNA synthesis mutants described above. Deletions encompassing parts or all of E1b and E4 had no effect on recombination, suggesting that none of the products encoded by these regions plays an important role in recombination (EPSTEIN and YOUNG 1991). Although these are negative results, they are worth some comment. It is known that deletions in the gene encoding the E1b 19K protein result in a failure to protect both viral and cellular DNA from the actions of a DNA endonuclease induced by E1a gene products and characteristic of the phenomenon of programmed cell death, or apoptosis (reviewed in WHITE 1993). In the recombination experiment with the E1b 19K deletion mutant, the total yield of viral DNA was reduced, as expected, and intracellular viral DNA showed the degradation characteristic of 19K mutants, suggesting that the apoptotic endonuclease was active. Despite this, the lowered amount of recombinant product was commensurate with the lowered amount of accumulated viral DNA. This suggests that the nicks, gaps, and breaks introduced into the viral DNA by the endonuclease do not increase the already high rate of recombination intrinsic to adenovirus infection. The results with the E4 deletion demonstrate not only that none of the E4 gene products are required for recombination, but also that late gene products are unlikely to be required either. The specific E4 deletions used are severely deficient in late gene product production (HALBERT et al. 1986; WEINBERG and KETNER 1986). Previous work has shown that E3 has no part to play in recombination (C.S.H. YOUNG, unpublished work), since viruses with major deletions in this area are recombination proficient. This is perhaps not surprising, since this region of the genome appears to encode gene products whose role is to modulate the response of the host organism to adenovirus infection (reviewed in WOLD and GOODING 1991). The results with an E1a deletion showed that recombination is severely delayed and is observed approximately at the time when there is an increase in DNA accumulation, which of course is delayed in nonpermissive cells. Although the timing of the onset of recombination and its dependence on DNA replication were somewhat equivocal, there can be no doubt that E1a gene products are not necessary for recombination. This suggests, but of course does not prove, that cellular genes induced by E1a are not essential for adenovirus homologous recombination.

These results with the viral early mutants suggest, therefore, that the main role of viral gene products is to produce the appropriate substrates for initiating recombination. Obviously, there are some gaps in this analysis, not the least of which is the possibility that one or more of the DNA synthesis gene products might be involved in recombination, after, and in addition to, its role in producing the substrate itself. It is possible, for example, that the single strand DNA-binding protein (DBP), known to be multifunctional, might promote annealing of displaced single strands from two different parental genotypes. It is not known whether this protein, or indeed any other of those involved in replication, has a separable domain for recombination. So far none of the temperature-sensitive mutations located in any of the DNA replication genes have been shown to be deficient in some as yet undefined function, while proficient in DNA replication. However, there are mutations located at other positions in the DBP gene which have effects on host range (Klessig and Grodzicker 1979; Anderson et al. 1983) or on assembly (Nicolas et al. 1983; Roovers et al. 1990), and these have not been tested for recombination. Another gap in the analysis is the lack of information about the possible role in recombination of such poorly characterized products as the i-leader-encoded proteins (Lewis and Anderson 1983; Symington et al. 1986; Soloway and Shenk 1990) or even RNA products such as the recently described pol III transcripts from the E2 gene (Pruzan et al. 1992). Now that the complete genome sequences of several adenoviruses are available (Chroboczek et al. 1992; Roberts et al. 1986), it is clear that there are many open reading frames (ORF) of unproven function. As more information becomes available about them, it may be necessary to test them in recombination assays. Finally, it should be pointed out that, although the known early regions have been tested in recombination assays, using large deletions in each specific region, it is always possible that such deletions have simultaneously inactivated positive and negative components in the system, leaving the mechanism essentially intact.

5 Which Cellular Functions Are Involved in Adenovirus Recombination?

Even the simple models of adenovirus recombination proposed at the end of this review include many enzymatic steps in the rearrangement of the parental genetic material. In Fig. 2, for example, these steps involve proteins to promote annealing of single strands, to recognize mismatched base pairs, to remove the mismatch, and to restore the uninterrupted duplex. Given the limited coding capacity of the adenovirus genome, it was always likely that the majority of recombination functions would be encoded by the cell and that they would be used for cellular repair and recombination purposes. Thus, although it was important to search, in a methodical fashion, for viral gene products involved in recombination, it was clearly imperative to try to discover the cellular genes involved in the process. In theory, adenovirus offers outstanding advantages in examining the processes of cellular repair and recombination. It can infect and replicate in most human cells, including primary fibroblasts, it can be recovered and examined easily, and its genetic analysis is at an advanced stage. Finally, the ability to transfect cells with DNA and recover the products, was, until the advent of polymerase chain reaction (PCR), a signal advantage. Because of these perceived advantages, there have been many investigations using adenovirus to examine DNA repair and recombination in both normal and repair-deficient human cells (see for example Rainbow 1991). However, it is probably fair to say that these investigations have not yielded any useful positive information about the mechanisms underlying adenovirus recombination. It is clear that none of the cell lines examined so far (Williams et al. 1974; Young and Fisher 1980) has had any profound effect on the frequency of recombination, and the holy grail of a human cell with a rec A-like phenotype has not been found. Cell lines which have been examined to date include those from patients with various disorders of DNA repair or chromosomal stability such as ultraviolet (UV)-sensitive xeroderma pigmentosum of complementation groups A, C, and "variant", Fanconi's anemia of unknown group, ataxia telangiectasia group D, Bloom's syndrome, and the human ligase 1 deficiency present in cell line 46BR (BARNES et al. 1992). Cell lines from a variety of other species permissive for human adenovirus replication have also been tested (WILLIAMS et al. 1974) and all show normal levels of recombination. Such negative results have been compounded, until very recently, by a virtual lack of understanding of the cellular genes and products involved in mammalian replication and repair. Fortunately this situation is changing rapidly with the genetic mapping of many of the human repair deficiency genes, isolation and sequencing of cDNA clones (reviewed in FRIEDBERG 1992), and the development of in vitro techniques for repair (see for example WOOD and COVERLEY 1991). Thus the time is ripe for a re-examination of repair-deficient cell lines' abilities to permit adenovirus recombination. This will be considered further in Sect. 8.

6 Alternative Experimental Approaches to Studying Adenovirus Recombination

6.1 Biochemical Approaches

Although the genetic and molecular descriptions of adenovirus recombination during the infectious cycle are reasonably complete, it is clear from the preceding paragraphs that much remains to be uncovered about the functions that carry out recombination. It would obviously be a great advantage to develop in vitro systems for recombination similar to those that were so successful for DNA replication (CHALLBERG and KELLY 1979). Indeed, if it is true that recombination and replication are functionally linked, a system that permits replication might permit recombination. Attempts to use the Challberg and Kelly system to do so have been unsuccessful, however (G. KETNER, personal communication, and L.H. EPSTEIN and C.S.H. YOUNG, unpublished work). In the work in my own laboratory, the assay for recombination was based on the Southern technique, used in the analysis of intracellular viral DNA, which detects the rearrangement of restriction enzyme sites. Although this is an extremely sensitive technique, which could detect the small percentage of recombinant products likely to be made in vitro, it may not be the best approach. The initial recombination intermediate will almost certainly have mismatched base pairs at the position of the restriction enzyme site difference. Many restriction enzymes fail to digest such mis-matches. Creation of a fully recombinant duplex may require the subsequent action of a mis-match repair system. Although such a system is found in mammalian cells and in vitro assays have been developed to measure it (HOLMES et al. 1990; THOMAS et al. 1991), the likelihood of both a recombinational event and mismatch repair happening to a particular genome may be very small. Nevertheless, despite such initial failures, the availability of a complete adenovirus DNA replication system, with purified products, should allow reexamination of the relationship between recombination and replication.

6.2 DNA-Mediated Transfection

One of the great advantages of using adenovirus to study the recombinational capacities of the cell is that infection can be initiated by DNA-mediated transfection (GRAHAM and VAN DER EB 1973). Different arrangements of the adenovirus genome can be presented to the cell, and the recombinant products can be recovered and analyzed in detail. The results obtained can be compared and contrasted with those from recombination occurring in virus-initiated infections. For the purposes of this review, only those aspects which pertain to homologous recombination will be covered, but it should be pointed out that adenovirus DNA fragments also undergo nonhomologous end-joining to yield viable virus (MUNZ and YOUNG 1987, 1991), and this has given insight into one of the most marked characteristics of the mammalian cell, namely the ability to join broken DNA ends at high efficiency, regardless of the precise chemical nature of the ends (reviewed in ROTH and WILSON 1988).

The investigation of homologous recombination using transfection has employed three different arrangements. The first approach, "marker rescue", was developed by FROST and WILLIAMS (1978) as a mapping tool and is formally analogous to simple gene replacement protocols in yeast and mammalian cells. In this arrangement, a full-length genome with a counter-selectable marker and a subgenomic fragment with a selectable marker are cotransfected into cells, and a recombinant product is selected. If the two parental DNA molecules differ at nonselected markers, as for example at several restriction enzyme positions, the segregation of these sites can be followed in the selected set of recombinants. Using this approach, VOLKERT et al. (1989) showed that the further away the unselected marker was from the selected one, the less likely the former was to be inherited. There was an almost perfect linear relationship between position and frequency of inheritance of unselected markers. This strongly suggests that recombination can be initiated anywhere in the parental genomes and is not confined to the ends of the subgenomic fragment. Second, if a distal marker from the fragment was inherited, all more proximal markers were also inherited. This demonstrates that genetic information is transferred as a block and is not transferred in small pieces. A second observation concerned the timing of marker rescue. The frequency of marker rescue in plaques arising from individually transfected cells varied over several orders of magnitude. This suggests that transfer of genetic information from fragment to full-length genome could occur after DNA replication had begun. The alternative formal possibility that the variability reflects the relative proportion of full-length genomes and fragments initially present in the individual transfected cell seems unlikely. The significance of these observations when compared and constrasted with viral recombination will be considered after the discussion of possible model for the latter.

The second DNA arrangement to be exploited for recombinational purposes is that of overlapping left and right terminal fragments (CHINNADURAI et al. 1979; BERKNER and SHARP 1983; VOLKERT and YOUNG 1983), so-called overlap recombination. In this arrangement, viable virus can only arise by recombination, either by employing the homology within the region of overlap or, provided the overlapping region is not too long, by end-joining. Homologous recombination within the overlap has been extensively studied (VOLKERT and YOUNG 1983; MAUTNER and MACKAY 1984). Briefly, segregation of markers is strictly determined by their position in the overlapping region, and there is very little evidence for the formation of heteroduplex DNA in the region of overlap. This latter point has been established by examining the progeny arising from individual transfected cells in which multiple markers were included in the overlapping region. Most plaques contained virus that was genetically unmixed for all of the markers in the cross. Occasionally, the marker immediately adjacent to the point of crossover was present in an equal mixture of the two parental types, suggesting that heteroduplex DNA was present close to the point of crossover. From the positions of the sites employed in the cross, which were never more than 400 bp apart, we can further suggest that heteroduplex tracts are rarely longer than a few hundred base pairs in length. The determination of the length, location, and resolution of heteroduplex DNA is one of the most important pieces of genetic evidence to be gained in elaborating models for recombination in any system. It remains to be seen whether the results on heteroduplex tract length obtained with overlap recombination can be extrapolated to viral recombination, where it is much more difficult to establish the presence of heteroduplex DNA because of the dynamic population genetics of the mixed infection.

One marked point of contrast between overlap recombination and viral recombination is in the independence of the former from viral DNA replication. Pairs of overlapping terminal fragments in which the promoter for the E2 genes is present on the right-hand member of the pair and the DNA polymerase is on the left-hand member recombine just as efficiently as pairs in which the whole E2 transcription unit is on either the left or right fragment. Expression of the adenovirus DNA polymerase, essential for viral replication, can only happen in the former case after reconstruction of the full-length viral genome. This independence of recombination from DNA replication is as expected. Gene fragments of selectable viral genes, such as that for thymidine kinase, can recombine efficiently in cotransfection experiments (FOLGER et al. 1982).

The third DNA arrangement, and the one that is most relevant to viral recombination, derives from initial observations by HAY et al. (1984) and WANG and PEARSON (1985) that minichromosomes of the adenovirus genome can be created, and will undergo multiple rounds of DNA synthesis, if adenovirus replication proteins are provided in trans. Pearson and his colleagues (AHERN et al. 1991; BENNETT and PEARSON 1993) have exploited this system to examine the possibility of hybridization of displaced single strands to form recombinant molecules and the repair of sequence mismatches in heteroduplex DNA. To examine the first possibility, cells were cotransfected with full-length adenovirus DNA, to provide the replication proteins in trans, and two linearized plasmids, one containing the right and the other the left terminal sequence of the adenovirus genome. Neither plasmid alone is capable of continuous double-strand adenovirus DNA replication, because an inverted repeat is necessary for the panhandle stage of synthesis. However, because the terminal sequences in the two plasmids were arranged so that displaced strand synthesis would generate complementary single-strands, hybridization of these strands followed by trimming of any single-stranded ends could generate a linear duplex with two inverted adenovirus termini. These duplexes would be expected to replicate to produce measurable quantities of duplex product. This was indeed the case, and the results are fully consistent with earlier observations (BODNAR and PEARSON 1980), which suggested that late in infection complementary full-length viral single strands can hybridize. This system is referred to as "replicative overlap recombination" to distinguish it from the recombination between non-replicating fragments described above.

If the hybridizing single strands contain distinguishable genetic markers, they may be recognized by mismatch repair systems and contribute to the pool of recombinant products. This will be dealt with in more detail in the consideration of models. Another arrangement of the minichromosome again involved plasmids with a single adenoviral terminal sequence, but with two nonadenoviral inverted repeats, one located immediately internal to the adenovirus terminus and the other at the distal end of the linearized plasmid. Individual single strands can form a partially duplex panhandle using the inverted repeats, and the 3' end can prime synthesis to form a completely duplex panhandle (WANG et al. 1991). If pairing occurs between single strands produced from two such plasmids, with distinguishable sites within the repeats, then the transfer of genetic information can be followed from one plasmid to the other. It was found that there was an amazingly high rate of such sequence conversion, because the "recombinant" product was as frequent as the parental types. Production of the recombinant was only found if both plasmids had an intact origin of replication, strongly suggesting that the mechanism is replication dependent. Potentially, this system can be exploited to uncover the rules governing the formation, recognition, and resolution of mismatches of various types and extent, and certain constraints on heteroduplex formation and recognition have already been determined (BENNETT and PEARSON 1993). It was already clear from the replicative overlap experiment (AHERN et al. 1991) that large mismatches are ignored by the repair system, and the authors speculated that such large, single-stranded loops might be coated by the adenovirus DBP and protected from nuclease attack. It should be pointed out that these minichromosome replication systems also hold great promise in uncovering functional aspects of repair and recombination in adenovirus-infected cells.

7 Models for Adenovirus Recombination

Results from the genetic and functional analysis of adenovirus recombination and from the minichromosome system described above strongly suggest that recombination is intimately tied in with DNA replication. Given the failure to find any viral gene products (other than those involved in replication) which play an essential or important role in recombination, it seems most fruitful to focus on the specific products of DNA replication as the key to the high rates of adenovirus recombination. Two models for adenovirus recombination will be presented, with the understanding that they are not mutually exclusive and do not exclude other more conventional mechanisms. It is hoped that they will be helpful in designing strategies to answer both genetic and functional questions about mechanism.

The first model (Fig. 1) proposes that a single strand, in the process of being displaced by ongoing DNA synthesis, is transferred to a recipient nonreplicating duplex and concurrently displaces the original strand of the same polarity, beginning at the termini of both interacting molecules. At some stage, the partially transferred strand and the recipient strand are broken and rejoined to form a partially heteroduplex molecule. The mismatched base pairs within this heteroduplex may be acted upon by the mismatch repair system of the cell, but regardless of whether or not this takes place a further round of replication will give rise to one duplex recombinant and one duplex parental genome. This model is a variant of that of MESELSON and RADDING (1975), differing in that the strand exchange begins at the termini of the two interacting molecules rather than by the formation of a D-loop. This model has certain experiment facts in its favor. First EM has shown two interacting duplexes with a single strand connecting them, albeit without the displaced recipient strand. The loss of this strand could be accounted for by endonuclease attack at the site of displacement or by a 5' or 3' exonucleolytic degradation of the displaced strand. True exonucleolytic degradation of the displaced strand from the 5'end is, however, problematic because of the protection afforded by the covalently attached terminal protein (CARUSI 1977; DUNSWORTH-BROWNE et al. 1980). The second set of observations favoring this model is that recombination with strains of adenovirus containing multiple restriction site markers shows genetic "polarity" i.e., there is a pronounced gradient of recombination from the ends of the molecule (MUNZ and YOUNG 1984). This gradient can be altered by placing regions of nonhomology close to the ends of the interacting molecules, and recombination between regions of non homology is reduced, although not blocked completely. These genetic observations make sense if recombination initiates terminally and heteroduplex DNA extends for differing lengths in different interacting pairs. This would be a viral example of a widespread genetic phenomenon in which recombination frequency falls off with distance from the initiating site, as in gradients of gene conversion in yeast (DETLOFF et al. 1992) and in recombination stimulated by Chi sites in bacteriophage lambda (CHENG and SMITH 1989).

a

b

c

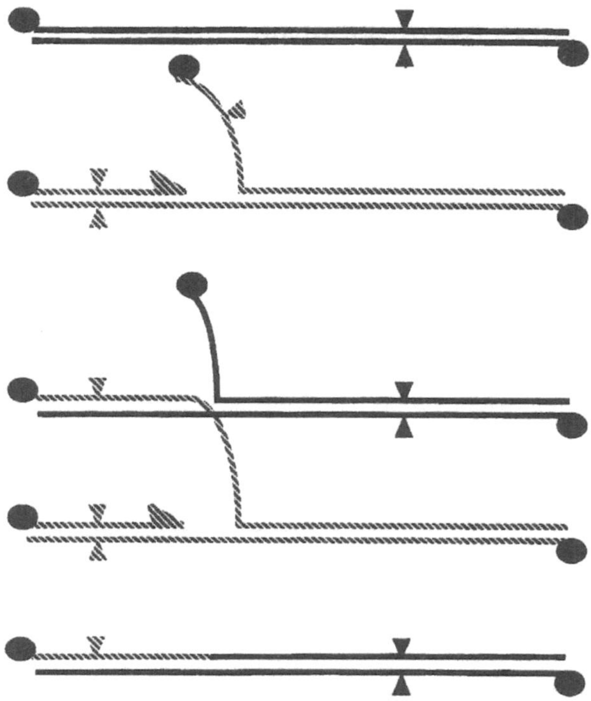

Fig. 1a–c. A modification of the Meselson and Radding strand invasion model for recombination. In this model, a single strand displaced by DNA replication of one duplex, the donor, invades a nonreplicating duplex, the recipient. The two parents are distinguished by the presence or absence of two restriction sites shown by the *triangles*, above and below the two strands of each duplex. The arrangement of sites is in trans, but could equally be in cis. The *leftward-pointing arrow* indicates the direction of replication, while the covalently attached terminal protein is shown by a solid oval at the 5' end of each stand. **a** A displaced strand is approaching a recipient duplex in the vicinity. Parent 1 (*top*) is a nonreplicating recipient and parent 2 (*bottom*) a replicating donor. **b** The single strand has invaded the recipient from the left-hand end. Invasion from the terminus could be promoted by the known protein–protein interactions between terminal proteins. The energy for strand transfer and assimilation could be derived from the ongoing replication on the donor duplex. Whether or not an independent helicase or topoisomerase type 1 activity is needed for the unwinding of the recipient, ahead of the transfer, is unknown. In a population of interacting molecules, there will be a diminishing transfer of genetic information from the donor to the recipient, with increasing distance form the end. **c** The single strands from both the donor and the recipient are cut at approximately the same genetic location; any gaps or overlapping regions are repaired by cellular nucleases and DNA polymerase and rejoined by DNA ligase. The upper strand of the joint molecule is recombinant, while the bottom strand is still parental. Fully recombinant duplexes can arise by replication or by mismatch repair of the heteroduplex. The model explains polarity (MUNZ and YOUNG 1984) and can easily accommodate the observation that multiple crossovers are usually distantly spaced (WILLIAMS et al. 1975; YOUNG and SILVERSTEIN 1980). Holliday structures, which have been seen in the electron microscope (WOLGEMUTH and HSU 1980), can also be formed if, at step b, the recipient's displaced strand invades and displaces the replicating strand of the donor. This would create the classical, single-strand crossover

The second model (Fig. 2) is based on ideas first formally presented by AHERN et al. (1991), but which were foreshadowed by the observations from pulse-chase density gradient experiments which suggested that complementary displaced

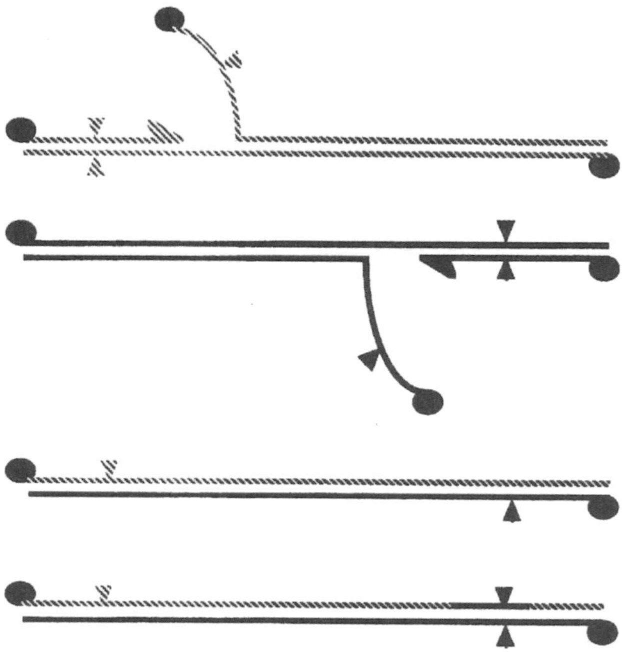

a

b

c

Fig. 2a–c. Recombination mediated by complete strand annealing and mismatch repair. **a** In this model, which is based on that of AHERN et al. (1991), complementary single strands are displaced from replication initiated on the opposite ends of two parental duplexes. **b** The single strand anneal . **c** The full-length heteroduplex is recognized and individual mismatched base pairs are corrected (only one such correction is shown). The repair tract (*solid patch*) is shown as being quite short, but longer ones might occur. All of the enzymatic activities required by this model have been demonstrated, either in adenovirus replication, the minichromosome system of AHERN et al. (1991), or in cellular extracts. However, the model does not explain polarity, nor does it account for the Holliday structures seen in the electron microscope

single strands from two molecules might reanneal (BODNAR and PEARSON 1980). In this model, recombinant products only arise if the complementary strands from the two distinguishable parents are acted upon by mismatch repair systems. There is compelling evidence for such systems both from the genetic data from the minichromosome systems discussed above and also from direct biochemical data, primarily from Modrich and collaborators (reviewed in MODRICH 1991). If the mismatch repair tracts are short, say on the order of a few hundred base pairs, the resulting recombinants would be characterized as having closely spaced double crossovers. This is illustrated in Fig. 2C. Although this mechanism could help to explain the high rates of supernumerary crossovers observed in many adenovirus recombinants (YOUNG and SILVERSTEIN 1980), the precise pattern of such crossovers is not readily explained by limited mismatch repair. Supernumerary crossovers are well spaced in individual viral recombinants, as though they arise from independent events (WILLIAMS et al. 1975; YOUNG and SILVERSTEIN 1980). The negative interference observed in most such populations of recombinants can easily be explained by a selection for a subpopulation of genomes which has undergone recombination at least once and may be in a favored environment or context for further rounds of recombination. In addition, mismatch repair of full-length heteroduplexes does not explain the overall additivity of adenovirus recombination maps (WILLIAMS et al. 1974; YOUNG and WILLIAMS 1975), nor the deviations from linearity observed in polarity (MUNZ and Young 1984). Thus, in viral crosses, this mechanism of recombination may only operate at very late times in infection, when the pool of displaced single strands may be high. The reasons for its prominence in the minichromosome experiments may be that the pool of replicating molecules may be much higher in the individual transfected cell and the products are not withdrawn into capsids. One further aspect of mammalian mismatch repair worth mentioning in passing is that, unlike the prokaryotic systems, repair is not directed at the unmethylated, newly synthesized strand. Instead, it is directed at the strand in which nicks are present (HOLMES et al. 1990). How this relates to the mechanisms observed in replicative overlap recombination is not clear, but the observed absence of methylation of adenovirus DNA (DOERFLER 1981) is not a bar to mismatch correction.

It is instructive to compare the models for viral recombination not only with the results from viral crosses, but also with those from DNA-mediated marker rescue and prereplicative overlap recombination. The segregation results from the marker rescue transfection experiments showed that, when distal unselected markers are inherited, proximal markers are almost invariably inherited too. This would not be the case if the transfer of information from the donor fragment to the full-length recipient involved a long stretch of single-stranded DNA. Mismatch repair would generate many instances of multiple crossovers in the selected recombinant. Because, as mentioned above, there is ample evidence for a mismatch system in mammalian cells, the absence of such products argues against the formation of long heteroduplexes. Similarly, in the prereplicative overlap segregation analysis, in which direct evidence for mismatched base pairs can be obtained, very little evidence for extensive heteroduplex was found. Taken together, these DNA-mediated recombination results suggest that when the cellular machinery is solely responsible for genetic exchange, short heteroduplex formation is the rule. Either DNA-mediated transfection obeys different rules to those for virus or the models outlined above for virus recombination are incorrect. In one respect, however, marker rescue is similar to viral recombination, in that transfer of genetic information can occur after DNA synthesis of the full-length molecule has commenced. However, even in this instance, it is unlikely that transfer occurs by annealing one strand of the fragment to the displaced strand of the full-length recipient, because marker rescue is just as efficient with internal fragments as with terminal ones (VOLKERT et al. 1989). Transfer of a single-strand from an internal fragment would result in an internal duplex flanked by gaps. The gap at the 5'end could not be repaired by fill-in DNA synthesis.

8 Future Directions

8.1 Tests for Heteroduplex Formation, Distribution, and Resolution

As with any genetic system, the characteristics of the heteroduplex DNA in the intermediate structures formed during adenovirus recombination are the key to understanding the mechanism. The major difference between the two models of adenovirus recombination discussed above is the absolute dependence of the second mechanism on mismatch repair to generate recombinant products. Both models predict the formation of extensive regions of heteroduplex, although the extent and location is different. Recently, techniques developed to detect the formation of heteroduplex in cells (LICHTEN et al. 1990) have been used in pox virus infections, and very high concentrations of mismatched base pairs have been detected (FISHER et al. 1991). Similar approaches could be used in adenovirus to detect the overall level of mismatch formation, although the distribution of heteroduplex along the length of the genome would probably not be revealed using such population-based techniques. The alternative genetic approach, which depends on identifying populations of mixed genotype within individual virus plaques, as was first done with plaque morphology markers in bacteriophage (HERSHEY and CHASE 1951), is extremely tedious to perform. Occasionally, there is evidence for such mixed plaques (YOUNG and SILVERSTEIN 1980), but their rarity makes it difficult to draw any conclusions. The availability of more rapid and sensitive probes for genetic heterogeneity within a DNA population in a plaque might make such screens more easy to perform. If mismatch repair does play an important role in the resolution of heteroduplex structures to yield recombinant products, it may be possible to test human cell lines from the recently described mismatch repair-deficient hereditary nonpolyposis colorectal carcinoma (FISHEL et al. 1993; PARSONS et al. 1993). If the second model (Fig. 2) is correct, viral recombination will be markedly reduced. On the other hand, if the first model (Fig. 1) is the predominant mechanism, recombination may not be much reduced, but recombinants with supernumerary crossovers, found predominantly at late times (YOUNG and SILVERSTEIN 1980) may decline.

8.2 A Continued Search for Cellular Functions Involved in Adenovirus Recombination

The uncovering of the genetic and biochemical defects behind several inherited human DNA repair disorders is proceeding rapidly, spurred both by improvements in genetic methods of gene mapping and isolation and by phylogenetic comparisons with unicellular eukaryotes, such as the yeasts (reviewed by FRIEDBERG 1992). Now that a more rational approach can be taken as to which human cell lines to examine for their recombination phenotype, it is important to go back to this approach using a number of different measures of recombination.

This is entirely feasible with adenovirus, in which both viral infection and several different DNA-mediated transfection methods can be employed. As an example, in my laboratory Patricia Munz recently examined the ligase 1-deficient cell line 46BR (BARNES et al. 1992) for its ability to undergo viral replication, recombination, prereplicative overlap recombination, and DNA end-joining. Almost all forms of recombination, both homologous and nonhomologous, require a ligation step. However, the results of all assays with this line were very similar to those obtained with any standard cell line, despite the marked reduction in the activity of the major DNA ligase of the cell. It is not surprising that viral DNA replication proceeded at close to normal rates, given that there is no ligation step involved in the mechanism of adenovirus DNA replication. However, the finding that end-joining occurred in this line (although its efficiency has not been measured yet) suggests that ligase 1 is not involved or is functionally replaceable by the other two cellular ligases. The same argument would hold for the homologous recombination mechanisms. Several of the genes involved in UV repair have been cloned recently, and that mutated in xeroderma pigmentosum group D is now known to be a helicase (SUNG et al. 1993). The model in Fig. 1 may require the action of a helicase to open up the recipient duplex, and so a rational case can be made for examining the recombinational phenotype of xeroderma pigmentosum group D cell lines. As mentioned above, it will be valuable to examine cell lines derived from hereditary nonpolyposis colorectal carcinoma, since mismatch repair may play a role in adenovirus recombination. The number of cell lines in which the precise deficiency in repair mechanism has been defined biochemically is set to increase markedly over the next few years. As information becomes available, adenovirus recombination assays can be tried in the appropriate lines. This will be a useful adjunct to plasmid- or chromosome-based recombination assays, (e.g., MEYN 1993).

8.3 Development of Biochemical Assays

As mentioned earlier, biochemical approaches to adenovirus recombination have not been vigorously pursued. In theory, there are two general biochemical approaches that could be employed. The "top-down" approach involves the establishment of a crude extract system which mimics the in vivo results, as far as is possible, followed by the gradual purification and biochemical identification of individual components. The paradigm is the development of in vitro systems for DNA replication both in adenovirus and SV40 (reviewed in CHALLBERG and KELLY 1989). Both models for adenovirus recombination described above require ongoing replication to promote recombination, so a logical place to start the search for in vitro recombination would be in the DNA replication systems already available. The most pressing issue is the development of a simple, sensitive, and quantitative assay for recombination. As mentioned previously, the Southern hybridization assay may need modification, because the initial recombination intermediate will yield restriction sites with mismatched base pairs, a chemical structure not recognized by most restriction enzymes. Testing the formation of heteroduplex DNA in the in vitro system may be possible using new techniques developed for intracellular DNA (LICHTEN et al. 1990), but this assay is cumbersome for subsequent steps of purification of the individual components. Note that the problem of assay design is intrinsic to the development of biochemical systems for recombination. The products of recombination are chemically similar, although genetically different, to the input substrates; there is no net increase in mass of the reaction products. This precludes simple assays based on incorporation of labeled substrate, such as those used for measuring DNA replication. Assuming the assay problem can be overcome, the first step would be to establish whether or not adenovirus recombination does indeed need ongoing replication. Extracts deficient in one or other of the viral replication proteins, or substrates lacking origins of replication, should be deficient in both processes. It may also be possible to define cellular components involved in adenovirus recombination using biochemical complementation. For example, if the genetic results with a particular cell line suggest that the wild-type counterpart of the deficient product is involved, it should be possible to add this product back to extracts prepared from the cell line in question.

The alternative approach to the biochemical investigation of adenovirus recombination is the "bottom-up" method, as exemplified by the extensive studies of prokaryotic recA protein. This approach is most fruitful if genetic results strongly implicate a specific gene product as being crucial for recombination. Currently, this is not the case, but might become so if, for example, the role of mismatch repair in adenovirus recombination is firmly established. The appropriate gene product would then be tested with model substrates. Note that in the absence of genetic data implicating a particular gene product, the bottom-up approach is intrinsically open to artifact, because the discovery of proteins capable of reacting with the specific substrates employed in the assay does not mean that they are involved in the reaction which takes place in the cell. The absence of convincing evidence for mammalian homologs of recA, despite considerable efforts to find and characterize them, is an example of the difficulties of this approach.

9 Concluding Remarks

Recent technical developments in the detection of specific DNA sequence and structure, in the characterization, isolation, and expression of cellular repair gene products, and in model transfection systems, can be harnessed to uncover the mechanisms underlying adenovirus recombination. Although the detailed mechanism may be strongly influenced by the mode of adenovirus replication, the ways in which displaced single strands are handled by the cell should prove to be of general interest. Repair of DNA lesions is a topic of increasing interest to both the specialist and the clinician, as evidence mounts of environmentally induced genetic defects underlying various forms of human cancer. The next few years promise to be exciting ones in the uncovering of this fundamental biological phenomenon, and adenovirus can be expected to play an important role in these discoveries.

Acknowledgments. The work from my own laboratory has been supported by grant GM31452 from the National Institute of General Medical Sciences, to which institution I am most grateful. I should also like to acknowledge the debt I owe to the late C.D.Darlington and to Brian Cox, both of the Botany School, Oxford, who stimulated an early interest in recombination, and to Jim Williams at the Institute of Virology in Glasgow and Harry Ginsberg at Columbia, New York, who taught me all I know about adenovirology. Finally, I should like to thank my technician of many years, Patricia Munz, who performed many of the experiments described from my laboratory.

References

Ahern KG, Wang K, Xu F-Y, Mathews CZ,Pearson GD (1991) Strands hybridize in postreplicative adenovirus overlap recombination. Proc Natl Acad Sci USA 88: 105–109

Anderson CW, Hardy MM, Dunn JJ, Klessig DF (1983) Independent, spontaneous mutants of adenovirus type 2-simian virus 40 hybrid Ad2$^+$ND3 that grow efficiently in monkey cells possess identical mutations in the adenovirus type 2 DNA-binding gene. J Virol 48: 31–39

Barnes DE, Tomkinson AE, Lehmann AR, Webster ADB, Lindahl T (1992) Mutations in the DNA ligase I gene of an individual with immunodeficiencies and cellular hypersensitivity to DNA-damaging agents. Cell 69: 495–503

Bennett KL, Pearson GD (1993) Sequence conversion during postreplicative adenovirus overlap recombination. Proc Natl Acad Sci USA 90: 1397–1401

Berkner KL, Sharp PA (1983) Generation of adenovirus by transfection of plasmids. Nucleic Acids Res 11: 6003–6020

Bodnar JW, Pearson GD (1980) Kinetics of adenovirus DNA replication. II Initiation of adenovirus DNA replication. Virology 105: 357–370

Borts RH, Lichten M, Haber JE (1985) Analysis of meiosis-defective mutations in yeast by physical monitoring of recombination. Genetics 113: 551–567

Carusi EA (1977) Evidence for blocked 5' termini in human adenovirus DNA. Virology 76: 380–394

Challberg MD, Kelly TJ Jr (1979) Adenovirus DNA replication in vitro. Proc Natl Acad Sci USA 76: 655–659

Challberg MD, Kelly TJ Jr (1989) Animal virus DNA replication. Annu Rev Biochem 58: 671–717

Cheng KC, Smith GR (1989) Distribution of Chi-stimulated recombinational exchanges and heteroduplex endpoints in phage lambda. Genetics 123: 5–17

Chinnadurai G, Chinnadurai S, Brusca J (1979) Physical mapping of a large plaque mutation of adenovirus type 2. J Virol 32: 623–628

Chroboczek J, Bieber F, Jacrot B (1992) The sequence of the genome of adenovirus type 5 and its comparison with the genome of adenovirus type 2. Virology 186: 280–285

Detloff P, White MA, Petes TD (1992) Analysis of a gene conversion gradient at the HIS4 locus in Saccharomyces cerevisiae. Genetics 132: 113–123

Doerfler W (1981) DNA methylation—a regulatory signal in eukaryotic gene expression. J Gen Virol 57: 1–20

Doermann AH (1953) The vegetative state in the life cycle of bacteriophage: evidence for its occurrence, and its genetic characterization. Cold Spring Harbor Symp Quant Biol 18: 3–11

Dunsworth-Browne M, Schell RE, Berk AJ (1980) Adenovirus terminal protein protects single stranded DNA from digestion by a cellular exonuclease. Nucleic Acids Res 8: 543–554

Ensinger MJ, Ginsberg HS (1972) Selection and preliminary characterization of temperature-sensitive mutants of type 5 adenovirus. J Virol 10: 328–339

Epstein LH, Young CSH (1991) Adenovirus homologous recombination does not require expression of the immediate-early E1a gene. J Virol 65: 4475–4479

Fishel R, Lescoe MK, Rao MRS, Copeland NG, Jenkins NA, Garber J, Kane M, Kolodner R (1993) The human mutator gene homolog MSH2 and its association with hereditary nonpolyposis colon cancer. Cell 75: 1027–1038

Fisher C, Parks RJ, Lauzon ML, Evans DH (1991) Heteroduplex DNA formation is associated with replication and recombination in poxvirus-infected cells. Genetics 129: 7–18

Flint SJ, Berget SM, Sharp PA (1976) Characterization of single-stranded viral DNA sequences present during replication of adenovirus types 2 and 5. Cell 9: 559–571

Folger KR, Wong EA, Wahl G, Capecchi MR (1982) Patterns of integration of DNA microinjected into cultured mammalian cells: evidence for homologous recombination between injected plasmid DNA molecules. Mol Cell Biol 2: 1372–1387

Friedberg EC (1992) Xeroderma pigmentosum, Cockayne's syndrome, helicases, and DNA repair: what's the relationship? Cell 71: 887–889

Frost E, Williams J (1978) Mapping temperature-sensitive and host range mutations of adenovirus type 5 by marker rescue. Virology 91: 39–50

Ginsberg HS, Young CSH (1977) The genetics of adenovirus. In: Fraenkel-Conrat H, Wagner RR (eds) Comprehensive virology, vol 9. Plenum, New York, pp 27–88

Graham FL, Van der Eb AJ (1973) A new technique for the assay of infectivity of human adenovirus 5 DNA. Virology 52: 456–467

Halbert DN, Cutt JR, Shenk T (1986) Adenovirus early region 4 encodes functions required for efficient DNA replication, late gene expression, and host cell shutoff. J Virol 56: 250–257

Hay RT, Stow ND, McDougall IM (1984) Replication of adenovirus mini-chromosomes. J Mol Biol 175: 493–510

Hershey AD, Chase M (1951) Genetic recombination and heterozygosis in bacteriophage. Cold Spring Harbor Symp Quant Biol 16: 471–479

Holmes J, Clark S, Modrich P (1990) Strand-specific mismatch correction in nuclear extracts of human and *Drosophila melanogaster* cell lines. Proc Natl Acad Sci USA 87: 5837–5841

Klessig DF, Grodzicker T (1979) Mutations that allow human Ad2 and Ad5 to express late genes in monkey cells map in the viral gene encoding the 72K DNA binding protein. Cell 17: 957–966

Lewis JB, Anderson CW (1983) Proteins encoded near the adenovirus late messenger RNA leader segments. Virology 127: 112–123

Lichten M, Goyon C, Schultes NP, Treco D, Szostak JW, Haber JE, Nicolas A (1990) Detection of heteroduplex DNA molecules among the products of Saccharomyces cerevisiae meiosis. Proc Natl Acad Sci USA 87: 7653–7657

Mautner V, Mackay N (1984) Recombination in adenovirus: analysis of crossover sites in intertypic overlap recombinants. Virology 139: 43–52

Meselson MS, Radding CM (1975) A general model for recombination. Proc Natl Acad Sci USA 72: 358–361

Meyn MS (1993) High spontaneous intrachromosomal recombination rates in ataxia-telangiectasia. Science 260: 1327–1330

Modrich P (1991) Mechanisms and biological effects of mismatch repair. Annu Rev Genet 25: 229–253

Munz PL, Young CSH (1984) Polarity in adenovirus recombination. Virology 135: 503–514

Munz PL, Young CSH (1987) The creation of adenovirus genomes with viable, stable, internal redundancies centered about the E2b region. Virology 158: 52–60

Munz PL, Young CSH (1991) End-joining of DNA fragments in adenovirus transfection of human cells. Virology 183: 160–169

Munz Pl, Young C, Young CSH (1983) The genetic analysis of adenovirus recombination in triparental and superinfection crosses. Virology 126: 576–586

Nicolas JC, Sarnow P, Girard M, Levine AJ (1983) Host range temperature-conditional mutants in the adenovirus DNA binding protein are defective in the assembly of infectious virus. Virology 126: 228–239

Parsons R, Li G-M, Longley MJ, Fang W-H, Papadopoulos N, Jen J, de la Chapelle A, Kinzler KW, Vogelstein B, Modrich P (1993) Hypermutability and mismatch repair deficiency in RER$^+$tumor cells. Cell 75: 1227-1236

Pruzan R, Chatterjee PK, Flint SJ (1992) Specific transcription from the adenovirus E2E promoter by RNA polymerase III requires a subpopulation of TFIID. Nucleic Acids Res. 20: 5705–5712

Puvion-Dutilleul F, Puvion E (1990a) Analysis by in situ hybridization and autoradiography of sites of replication and storage of single- and double-stranded adenovirus type 5 DNA in lytically infected HeLa cells. J Struct Biol 103: 280–289

Puvion-Dutilleul F, Puvion E (1990b) Replicating single-stranded adenovirus type 5 DNA molecules accumulate within well-delimited intranuclear areas of lytically infected HeLa cells. Eur J Cell Biol 52: 379–388

Rainbow AJ (1991) Host cell reactivation of sunlamp-exposed adenovirus in fibroblasts from patients with Bloom's syndrome, ataxia telangiectasia, and Huntington's disease. Environ Mol Mutagen 17: 98–103
Roberts RJ, Akusjärvi G, Aleström P, Gelinas RE, Gingeras TR, Sciaky D, Pettersson U (1986) A consensus sequence for the adenovirus-2 genome In: Doerfler W (ed) Adenovirus DNA. Martinus Nijhoff, Boston, pp 1–51
Roovers DJ, Young CSH, Vos HL, Sussenbach JS (1990) Physical mapping of two temperature-sensitive adenovirus mutants affected in the DNA polymerase and DNA binding protein. Virus Genes 4: 53–61
Roth D, Wilson J (1988) Illegitimate recombination in mammalian cells. In: Kucherlapati R, Smith GR (eds) Genetic recombination. American Society for Microbiology, Washington DC, pp 621–653
Soloway PD, Shenk T (1990) The adenovirus type 5 i-leader open reading frame functions in cis to reduce the half-life of L1 mRNAs. J Virol 64: 551–558
Sung P, Bailly V, Weber C, Thompson LH, Prakash L, Prakash S (1993) Human xeroderma pigmentosum group D gene encodes a DNA helicase. Nature 365: 852–855
Symington JS, Lucher LA, Brackmann KH, Virtanen A, Pettersson U, Green M (1986) Biosynthesis of adenovirus type 2 i-leader protein. J Virol 57: 848–856
Takemori N (1972) Genetic studies with tumorigenic adenoviruses. III. Recombination in adenovirus type 12. Virology 47: 157–167
Thomas DC, Roberts JD, Kunkel TA (1991) Heteroduplex repair in extracts of human HeLa cells. J Biol Chem 266: 3744–3751
Visconti N, Delbrück M (1953) The mechanism of genetic recombination in phage. Genetics 38: 5–33
Volkert FC, Young CSH (1983) The genetic analysis of recombination using adenovirus overlapping terminal DNA fragments. Virology 125: 175–193
Volkert FC, Munz PL, Young CSH (1989) A genetic investigation of the mechanism of adenovirus marker rescue. Virology 173: 77–88
Wang K, Pearson GD, (1985) Adenovirus sequences required for replication in vivo. Nucleic Acids Res 13: 5173–5187
Wang K, Xu F-Y, Ahern KG, Pearson GD (1991) Inverted repeats direct repair of adenovirus minichromosome ends. Virology 183: 44–51
Weinberg DH, Ketner G (1986) Adenoviral early region 4 is required for efficient viral DNA replication and for late gene expression. J Virol 57: 833–838
White E (1993) Regulation of apoptosis by the transforming genes of the DNA tumor virus adenovirus. Proc Soc Exp Biol Med 204: 30–39
Williams JF, Ustaçelebi S (1971) Complementation and recombination with temperature-sensitive mutants of adenovirus type 5. J Gen virol 13: 345–348
Williams JF, Young CSH, Austin PE (1974) Genetic analysis of human adenovirus type 5 in permissive and nonpermissive cells. Cold Spring Harbor Symp Quant Biol 39: 427–437
Williams J, Grodzicker T, Sharp P, Sambrook S (1975) Adenovirus recombination: physical mapping of crossover events. Cell 4: 113–119
Wold WSM, Gooding LR (1991) Region E3 of adenovirus: a cassette of genes involved in host immunosurveillance and virus–cell interactions. Virology 184: 1–8
Wolgemuth DJ, Hsu M-T (1980) Visualization of genetic recombination intermediates of human adenovirus type 2 DNA from infected HeLa cells. Nature 287: 168–171
Wolgemuth DJ, Hsu M-T (1981) DNA replication-mediated recombination of molecules of adenovirus 2 DNA. Proc Natl Acad Sci USA 78: 5076–5080
Wood RD, Coverley D (1991) DNA excision repair in mammalian cell extracts. Bioessays 13: 447–453
Young CSH, Fisher PB (1980) Adenovirus recombination in normal and repair-deficient human fibroblasts. Virology 100: 197–184
Young CSH, Silverstein SJ (1980) The kinetics of adenovirus recombination in homotypic and heterotypic genetic crosses. Virology 101: 503–515
Young CSH, Williams JF (1975) Heat-stable variant of human adenovirus type 5: characterization and use in three-factor crosses. J virol 15: 1168–1175
Young CSH, Cachianes G, Munz P, Silverstein S (1984a) Replication and recombination in adenovirus-infected cells are temporally and functionally related. J Virol 51: 571–577
Young CSH, Shenk T, Ginsberg HS (1984b) The genetic system In. Ginsberg HS (ed) The adenoviruses. Plenum, New York, pp 125–172

The Mechanism of Adenovirus DNA Integration: Studies in a Cell-Free System

K. FECHTELER[1], J. TATZELT[1,2], S. HUPPERTZ[1], P. WILGENBUS[1,3], and W. DOERFLER[1]

1 Scope of Review

The understanding of recombinative interactions between foreign (viral) DNA and mammalian DNA is of interest not only in tumor virology, gene therapy, and the generation of transgenic animals, but also for models about the possible evolutionary role of foreign DNA integration into established genomes. Foreign DNA can recombine with the host DNA in mammalian cells and thus become integrated into the cellular genome by what has been termed "nonhomologous recombination." Nonhomologous recombination plays a central role in the biology

[1] Institut für Genetik, Cologne University, Weyertal 121, 50931 Cologne, Germany
[2] Present Address: Department of Neurology, University of California San Francisco, CA 94143-0518, USA
[3] Present Address: Deutsches Krebsforschungszentrum, Im Neuenheimer Feld 280, 69120 Heidelberg, Germany

of mammalian systems and is thought to be more common in mammalian cells than homologous recombination.

The various DNA rearrangements known to occur in mammalian cells include homologous recombination, depending on extensive sequence homologies, and nonhomologous recombination, which requires little or no sequence identity. Examples for homologous recombination events include genetic recombination during meiosis (STAHL 1979), sister chromatid exchange (LATT 1981), interchromosomal recombination during mitosis (KUNZ and HAYNES 1981; WASMUTH and HALL 1984), and nonallelic gene conversion (SCHERER and DAVIS 1980; WEISS et al. 1983; LISKAY and STACHELEK, 1984). Examples of nonhomologous recombination include chromosome translocations (GERONDAKIS et al. 1984), immunoglobulin and T cell receptor gene rearrangements (ALT and BALTIMORE 1982; HONJO 1983; HEDRICK et al. 1984; MALISSEN et al. 1984; YANAGI et al. 1984) and movements of retroviruses and transposable elements (contributions in SHAPIRO 1983). Last but not least, the induction of tumors by adenovirus type 12 (Ad12) in hamsters (TRENTIN et al. 1962) is associated with the integration of Ad12 DNA into the hamster cell genome by a nonhomologous recombination mechanism (DOERFLER 1968, 1970; DOERFLER et al. 1983). Ad12 DNA in surprisingly large amounts can become associated with hamster chromosomes early after the infection of hamster cells (SCHRÖER and DOERFLER, submitted for publication).

DNA rearrangements classified as nonhomologous events are rare in bacteria and yeast, but occur frequently in mammalian cells. Targeted integration of exogenous DNA at its homologous chromosomal location in mammalian cells is masked by a 1000-fold higher frequency of integration at random genomic sites (SMITH and BERG 1984; LIN et al. 1985; SMITHIES et al. 1985; THOMAS et al. 1986). In contrast, nontargeted integration events are difficult to detect in bacteria and yeast (HICKS et al. 1978; HINNEN et al. 1978; ORR-WEAVER et al. 1981; ORR-WEAVER and SZOSTAK 1983). Sequence analyses of recombinant junction sites created by a variety of DNA rearrangements revealed only minimal homology requirements around these sites. For this reason, these recombination events were defined as "illegitimate" or nonhomologous recombination (FRANKLIN 1971).

The molecular and biochemical mechanisms underlying nonhomologous recombination are largely unknown. Upon foreign DNA integration, new genetic potential is introduced into the cell's repertoire that can either reduce or stimulate the "fitness" of its host or can lead to alterations in the overall structure of the affected genomes. Since insertions can lead to the disruption of genes (insertional mutagenesis), damage to cellular genes caused by integration of foreign DNA has to be kept at a minimum. Furthermore, it is probably essential for host survival that the expression of foreign genes potentially toxic to the cell can be shut off. Only those insertion events will eventually be survived by the cell which do not place the recipient host at a selective disadvantage.

Among the functional consequences of foreign DNA integration into established mammalian genomes are: (a) the de novo methylation of foreign DNA, and (b) alterations of DNA methylation patterns in the preexisting cellular genome. Integrated adenovirus genomes have been valuable models to study different aspects of changes in viral DNA methylation, since the DNA in the infecting virus particle is not detectably methylated (GÜNTHERT et al. 1976). De novo methylation can inactivate foreign genes, and the host cell can thus be protected from possibly detrimental effects of foreign gene products. This modification reaction has been interpreted as a cellular defense mechanism against the activity of foreign genes integrated into established eukaryotic genomes (DOERFLER 1991a). Additionally, increases or decreases in the methylation of host genes can cause the shutoff or activation of cellular target genes, respectively (DOERFLER 1983a). Based on the observation that patterns of cellular DNA methylation (HELLER et al. 1995) as well as expression levels of certain cellular genes (ROSAHL and DOERFLER 1992) can be substantially altered upon adenovirus or bacteriophage lambda DNA integration at chromosomal locations remote from adenovirus insertion, we formulated an extended version of the classical concept of insertional mutagenesis. Integration of adenovirus (foreign) DNA into a mammalian genome does not only alter cellular host DNA sequences and functions in the immediate vicinity of the insertion site via changes in methylation patterns, but can also affect methylation and transcriptional patterns at cellular sites at a considerable distance from the integration site and on different chromosomes.

There are several reasons to investigate the interdependency between foreign DNA integration, alterations in methylation patterns of the host DNA, and changes in cellular gene expression. A wide field of research will have to be addressed here:

1. It is important to analyze whether and how the integration of viral or more generally of foreign DNA into mammalian genomes can be involved in oncogenesis.
2. Some of the current concepts in gene therapy are based on the integration of foreign genes into established human genomes. The constant, regulated expression of these artificially introduced genes would be a prerequisite for successful gene therapy. De novo methylation and inactivation of the therapeutic genes would jeopardize any beneficial effects.
3. In mammalian cells, gene targeting by site-specific homologous recombination has become a routine method of generating mutant genes in transgenic animals in order to analyze the function of mammalian genes. However, the mechanisms of mitotic and meiotic recombination in mammalian cells are still unclear, and we have only limited knowledge about possible alterations of host DNA sequences caused by foreign DNA integration.

Thus, for many reasons it would be desirable to understand the mechanism and consequences of integrative recombination of foreign DNA sequences with established mammalian genomes. Cell-free systems have proven to be useful in studying genetic recombination mechanisms (KUCHERLAPATI et al. 1984; BROWN et al. 1987; JESSBERGER et al. 1989b; JESSBERGER and Berg 1991; SYMINGTON 1991; TATZELT et al. 1992, 1993). We have continued our efforts to develop and apply a cell-free system to elucidate the mechanism and enzymatic functions of this type of genetic recombination in mammalian cells.

2 Characteristics of Integration Patterns in Adenovirus-Transformed Cells and in Adenovirus Type 12-Induced Tumors

2.1 Adenovirus as a Model System

The discovery of adenovirus oncogenesis in newborn hamsters (TRENTIN et al. 1962; HUEBNER et al. 1962) and the ability of these viruses to transform hamster cells in culture (POPE and ROWE 1964; FREEMAN et al. 1967) have highlighted the human adenoviruses as valuable models to investigate DNA viral oncology and cell transformation. The molecular biology and the biochemistry of this virus have been studied in considerable detail (recent reviews in DOERFLER and BOEHM 1993 and this volume). Our laboratory has been interested in the fate of adenovirus DNA in infected and transformed hamster cells and has concentrated on studies of the mechanism of adenovirus DNA integration. In addition, the differential transcription of viral and cellular genes and alterations in DNA methylation patterns in the integrated and host genomes have been subjects of our research (for reviews, see DOERFLER 1991b, 1992, 1993).

Depending on a number of factors, including characteristics of the virus and the susceptibility of the cell, viral infections can lead to an abortive or to a productive infection or to the malignant transformation of the host cell. It is thought that one of the prerequisites for the transformation of nonpermissive cells by DNA or RNA tumor viruses is the persistence of at least parts of the viral genome in the affected cell, which ensures the continued expression of viral functions. In adenovirus-transformed or Ad12-induced hamster tumor cells, the viral genome is found exclusively in a chromosomally integrated state. Free episomal forms of adenovirus genomes have not been detected in transformed or Ad12-induced tumor cells. It still remains to be investigated whether viral DNA insertion into the cellular host genome is involved in oncogenesis in a direct or indirect manner. In our investigations, major emphasis has therefore been placed on the analysis of adenoviral DNA integration. It has also been shown that in some cases even after the loss of all viral genomes, as determined by Southern blot hybridization, the oncogenic phenotype of the cells can be maintained (KUHLMANN et al. 1982). It cannot be ruled out that minute amounts of fragments of Ad12 DNA are still present in these cells.

The infection of hamster cells with Ad12 leads to an abortive infection during which Ad12 DNA can become integrated into the hamster cell genome (DOERFLER 1968, 1970; SCHRÖER and DOERFLER, submitted for publication). Thereby, the whole viral genome or fragments of adenovirus DNA are covalently linked to the host genome (DOERFLER 1982; GRONEBERG et al. 1977; SUTTER et al. 1978; STABEL et al. 1980; DOERFLER et al. 1983). This nonproductive infection is characterized by the complete failure of Ad12 virion production. Only the expression of early Ad12 functions can be observed. Neither viral DNA replication nor transcription of late viral functions can be detected. The major late promotor (MLP) of Ad12 DNA is unable to function in hamster cells. A mitigator element in the downstream sequence of this promotor is, at least in part, responsible for the inactivity of the MLP of Ad12 DNA in hamster cells (ZOCK and DOERFLER 1990; ZOCK et al. 1993a, b). The factors in virus-host interactions that determine the outcome of a viral infection are complex (for reviews, see DOERFLER 1991b; ZOCK et al. 1993b). Based on the absolutely nonpermissive nature of the Ad12–hamster cell interaction, natural selection does not work against the persistence of the entire Ad12 genome in hamster cells. This lack of selective pressure facilitates the persistence and integration of multiple copies of the Ad12 genome that are nearly intact and colinearly arranged as compared to the arrangement and orientation of the viral genes in virion DNA. In contrast, natural selection in a productive infection of hamster cells with Ad2 seems to select against the persistence of an intact viral genome free in the cell or integrated into the host genome. Persistence of intact Ad2 genomes would lead to cell death, since intact Ad2 genomes are able to replicate in hamster cells and thus Ad2-transformed cell lines would not get a chance to arise. In contrast to the situation in Ad12-transformed hamster cells, we have found mainly fragments of integrated Ad2 sequences or Ad2 genomes with varying lengths of internal deletions in Ad2-transformed hamster cells (DOERFLER et al. 1983). Therefore, it seems likely that the type of virus–host interaction can influence patterns of persistence and integration of viral DNA into the cellular host genome.

Ad12 DNA becomes integrated early after viral infection (DOERFLER 1968, 1970). Most of our studies on viral DNA integration have been performed in adenovirus-transformed cell lines or in Ad12-induced tumors or tumor cell lines (for reviews, see DOERFLER 1982, 1992, 1993; DOERFLER et al. 1983). These data pertain to analyses of the state of viral DNA in cell lines established and maintained in culture for a long period of time. In all these transformation or tumorigenesis experiments, virions and not viral DNA or DNA fragments have been used. Distinct differences in integration patterns exist between cells after the initial integration event following virus infection and the clonal cell lines. One possible explanation for these differences could be rearrangements and losses of integrated viral genomes in the time period between the initial virus infection and the malignant transformation of the cell. We have so far restricted our studies on the parameters of Ad12 DNA integration mainly to cells obtained under cell culture conditions. Detailed studies on Ad12 DNA integration early after the infection of hamster cells have just been reinitiated (SCHRÖER and DOERFLER, submitted for publication). Although there may be limitations with respect to the generality of data adduced with the adenovirus system, we have argued that this virus offers obvious advantages, since its molecular biology has been studied in considerable detail (DOERFLER 1983b, 1984; GINSBERG 1985; DOERFLER et al. 1993; DOERFLER and BOEHM 1993 and this volume).

2.2 Basic Features of Adenovirus DNA Integration

The integration of foreign (viral) DNA into an established genome of mammalian cells immediately raises the question of how viral DNA insertion proceeds. Does foreign DNA integrate randomly at many different, totally unrelated sites or at highly specific sites in the genome of the affected host cell? These two possibilities are, of course, at the opposite ends of a scale of many conceivable intermediate possibilities.

Three different analytical approaches have been chosen to prove and investigate the integrated state of adenovirus genomes in the cellular host genome:

1. We have analyzed the localization of internal and terminal fragments of integrated adenovirus DNA with respect to cellular sequences in the host genome by using the technique of Southern blot hybridization (SOUTHERN 1975; SUTTER et al. 1978; STABEL et al. 1980). With this protocol, the DNA from transformed or tumor cells has been cleaved with restriction endonucleases, the fragments have been separated by gel eletrophoresis, and the adenovirus-specific DNA fragments have been identified by DNA–DNA hybridization to adenovirus DNA or to the cloned terminal fragments of virion DNA.
2. In order to prove the direct covalent linkage of the two DNA recombination partners, junction sites between adenoviral and cellular DNA sequences have been cloned, and their nucleotide sequences have been determined from the Ad12-transformed hamster cell lines T637 (OREND et al. 1995a) and HA12/7 (JESSBERGER et al. 1989a), the Ad2-transformed cell line HE5 (GAHLMANN et al. 1982; GAHLMANN and DOERFLER 1983), the Ad12-induced hamster tumor cell lines CLAC1 (STABEL and DOERFLER 1982), CLAC3 (DEURING et al. 1981b), T1111/2 (LICHTENBERG et al. 1987), and H191 (OREND et al. 1995c), the Ad12-induced mouse tumor cell line CBA12/T1 (SCHULZ and DOERFLER 1984), and the symmetric recombinant (SYREC2) of Ad12 virus (DEURING et al. 1981a; DEURING and DOERFLER 1983).
3. Recently, we have applied the fluorescent in situ hybridization technique (FISH) to determine the chromosomal locations of the integrated viral genomes (HELLER et al., 1995; OREND et al. 1995b; SCHRÖER and DOERFLER, submitted for publication).

We have discovered morphological revertants of the adenovirus-transformed cell line T637 that have lost all or most of the formerly integrated viral genomes after cell passages, but still retained the tumorigenic phenotype (GRONEBERG et al. 1978; EICK et al. 1980). Similar revertants have been isolated from the Ad12- induced tumor cell line T1111(1) (KUHLMANN et al. 1982). With the sensitive method of Southern blot hybridization we could not detect Ad12 DNA sequences in some of the revertant cell lines. We are now reexamining these cell lines with the highly sensitive technique of the polymerase chain reaction (PCR; SAIKI et al. 1988) for the presence of minute amounts of Ad12 DNA sequences. For this purpose, however, these cell lines have to be extensively recloned cellularly.

In summary, we have studied the topology of integrated adenovirus genomes in rodent cells transformed by adenoviruses, in Ad12-induced tumors, or in cells cultured from these Ad12-induced tumors. Though it is not certain whether these different cell types can be considered comparable with respect to their tumorigenic phenotype, the results of the analyses on integrated viral DNA can be summarized as follows:

a

b

Fig. 1a, b. Adenovirus type 12 (Ad12) DNA integration in the Ad12-transformed hamster cell line T637 as demonstrated by fluorescent in situ hybridization. **a** Interphase nuclei. **b** Chromosomal location

1. Ad12 DNA integrates chromosomally. As exemplified by the approximately 20 copies of Ad12 DNA integrated in the Ad12-transformed hamster cell line T637, the viral DNA copies can be localized by fluorescent in situ hybridization to one chromosomal site (Fig.1b). The Ad12 genomes can be detected by the same technique in interphase nuclei as well (Fig.1a), some of which might be partly diploid.
2. In adenovirus-transformed cells or in Ad12-induced tumor cells, adenovirus DNA has never been found in an episomal free form, but always in an integrated state. Ad12 DNA molecules are covalently integrated intact or nearly intact via sites at or close to the termini of the virion DNA molecule. Frequently, the orientation of viral DNA integration is colinear with that of virion DNA at least for some of the integrants. In contrast, Ad2 DNA is inserted in a fragmented form or with internal deletions (DOERFLER 1982; DOERFLER et al. 1983). Some of the integrated viral DNA molecules can be rearranged in their sequence array (EICK and DOERFLER 1982; OREND et al. 1995a).
3. So far, results from analyses of several junction sites in transformed or tumor cells from different species have provided no evidence for the notion that adenovirus DNA had integrated in a highly specific cellular DNA sequence (DOERFLER et al. 1983). Adenoviral DNA can integrate into unique as well as repetitive cellular DNA sequences. In the productive system of human cells, Ad12 DNA was preferentially found in association with human chromosome 1, as described for early and late stages after infection (McDOUGALL et al. 1972; ROSAHL and DOERFLER 1988).
4. Most frequently, multiple copies of viral genomes per cell can become integrated into the host genome; copy numbers range between 1 and 30 or more (DOERFLER 1982; DOERFLER et al. 1983).
5. Apparently, multiple copies of viral DNA are often not integrated in a truly tandem fashion. Either cellular or rearranged viral sequences separate individual viral DNA molecules from each other. In some of the integrated adenoviral DNA sequences, viral DNA termini have been found to be rearranged, amplified, or partly inverted (SUTTER et al. 1978; STABEL et al. 1980; KUHLMANN and DOERFLER 1982; KUHLMANN et al. 1982; OREND et al. 1991, 1994, 1995a). These rearrangements can be quite complicated in individual instances. We cannot rule out the possibility that rearrangements as well as transpositions and other alterations occurred in the period between the original integration event after virus infection or development of a tumor and the establishment and maintenance of cells in culture.
6. Adenovirus DNA integration can entail the deletion of cellular as well as terminal viral DNA sequences. However, integration can also proceed without the deletion of a single nucleotide in the viral or the cellular recombination partner. At the junction sites between viral and cellular DNA sequences, up to 174 terminal viral nucleotides have been found deleted in different recombination events. Occasionally, the terminal viral nucleotides are completely preserved, e.g., the left end of integrated Ad12 DNA in the Ad12-transformed hamster cell line HA12/7 (JESSBERGER et al. 1989a). Apparently, the process of viral DNA

insertion can be associated with deletions of larger segments of host DNA sequences (SCHULZ and DOERFLER 1984) or could proceed without the loss of a single nucleotide at the sites of linkage (GAHLMANN and DOERFLER 1983).

7. Patchy homologies, short DNA stretches of sequence homology, between the two recombination partners at the sites of linkage are frequently observed and seem to play a role in the selection of sites for recombination. These homologies are found either between the linked viral and cellular DNA sequences or between the two recombination partners replacing each other after adenovirus DNA integration. In only a few instances, junctions devoid of such stretches of sequence identities have been observed. Nevertheless, patchy homologies occur too frequently to be dismissed as random events.
8. The mechanism of insertional recombination in mammalian cells is characterized by an additional feature. Cellular sequences at the junction sites that have served as recombination targets for adenovirus DNA insertion, the so-called preinsertion sequences, have been shown to be transcriptionally active (GAHLMANN et al. 1984; SCHULZ et al. 1987; JESSBERGER et al. 1989a). This transcriptional activity can be documented both in the original host cells prior to contact with adenoviral DNA as well as in adenovirus-transformed cells and Ad12-induced tumor cells. Analyses of RNA molecules reveal differences in length and quality of the transcription products; they represent either short RNA without an intact open reading frame (ORF) or large products that contain ORFs. Actively transcribed cellular sequences might have assumed a chromatin configuration that facilitates foreign (viral) DNA integration.
9. In many instances, Ad12 genomes are stably integrated into the host genome and the integration patterns remain constant for decades over many cell passages (SUTTER et al. 1978; OREND et al. 1994). Morphological revertant cell lines have originated from the Ad12-transformed hamster cell line T637. In these revertants, the integrated viral DNA molecules are partly or completely lost (GRONEBERG et al. 1978; GRONEBERG and DOERFLER 1979; EICK et al. 1980; EICK and DOERFLER 1982; KUHLMANN et al. 1982). In the Ad12-induced tumor cell line T1111(1), this loss of Ad12 DNA sequences does not affect the tumorigenic phenotype of these revertants (KUHLMANN et al. 1982). Obviously, the persistence of viral DNA or an intact E1 region is not an absolute prerequisite for the maintenance of the tumorigenic phenotype. This finding would argue against the importance of viral functions encoded in the E1 region of the adenovirus genome to maintain the oncogenic phenotype. Nevertheless, integration and continued expression of Ad12 genes in the host genome are somehow involved in the transformation event, but are not an absolute requirement in all instances. It has not been rigorously ruled out that the revertants might carry minute amounts of Ad12 DNA which cannot be detected by Southern blotting.

For our understanding of the molecular mechanism of viral transformation, it seems conceivable that the integration of Ad12 DNA into the mammalian genome subsequent to viral infection and the secondary loss of formerly integrated viral DNA can entail an overall change either in the organization of the host genome or in the expression patterns of cellular genes.

2.3 Towards a Cell-Free Recombination System

The data accumulated from in vivo studies of the junction sites between viral DNA and cellular DNA sequences support our concept that the mechanism of integrative recombination of foreign DNA into the mammalian genome is apparently highly flexible. Therefore, it seems of little value to categorize the mechanism according to simple parameters, e.g., sequence specificity or homology requirements. Studies on the requirements for sequence homologies between the two recombination partners demonstrate that it is difficult to classify Ad12 DNA insertion as homologous or nonhomologous recombination (see also Sect 4.3). Studies on the sequence specificity of this integrative recombination reaction reveal that, at least in established cell lines of Ad12-induced tumor cells or adenovirus-transformed cells, a common or specific nucleotide sequence does not exist. However, under certain conditions and in cell lines cultivated for longer periods, Ad12 DNA has been found integrated at selective cellular sequences (see also Sect. 4.2).

Another major question concerning foreign DNA integration into mammalian genomes in general is whether cellular proteins can catalyze the insertion by themselves or are dependent on the presence and/or participation of viral gene products. It is conceivable that adenoviral functions, though perhaps themselves not essential components of the recombination machinery, play a modifying or scaffolding role in the process. The fact that adenovirus DNA transfected into mammalian cells can become integrated in the absence of viral infection favors the existence of a general recombination machinery supplied by the affected host cell that is responsible for foreign DNA integration in general.

Such a wide scale of obligatory and facultative requirements for the course of the insertion reaction may render this mechanism quite flexible and therefore successful and efficient. Questions concerning the interplay of several different parameters still remain unanswered. Despite the sophistication of the junction site analyses, a complete understanding of the recombination process requires a detailed biochemical analysis of the reaction in a cell-free system. In this way, we hope to shed light on the mechanism of integrative recombination of foreign (viral) DNA into the mammalian genome.

3 Cell-Free Recombination System

We have explored the mechanism and the requirements for enzymatic functions of Ad12 DNA integration in more detail by developing and applying a cell-free system using nuclear extracts of BHK21 hamster cells. High-salt extracts from uninfected BHK21 cells are capable of catalyzing the in vitro recombination reaction, when a hamster preinsertion sequence and fragments of Ad12 DNA are used as recombination targets (Jessberger et al. 1989b). It has been reasoned that preinsertion sequences which had previously served as targets for Ad12 DNA integration in vivo might carry elements essential in eliciting recombination and hence might be recognized again even under cell-free reaction conditions. The preinsertion sequence p7 from BHK21 hamster cells was cloned into the plasmid pBR322. This sequence represents the cellular integration site of Ad12 DNA in the Ad12-induced hamster tumor cell line CLAC1 (STABEL and DOERFLER 1982). Possibly because of apparent size limitations, we have so far not been able to use the intact Ad12 DNA molecule as the recombination partner for this preinsertion sequence. In previous work, a fragment of Ad12 DNA comprising nucleotides 20 885–24 053 of the full-length Ad12 DNA molecule (SPRENGEL et al. 1994) has proved to recombine at an increased frequency with the hamster p7 sequence as compared to other segments of the Ad12 genome (JESSBERGER et al. 1989b).

With the help of this cell-free recombination system we have started to gain insight into the mechanism of integrative recombination between the hamster preinsertion sequence p7 and Ad12 DNA. In the course of this work, we have purified the nuclear extracts to only a limited number of protein bands which have retained activity in cell-free recombination.

3.1 Assays for the Identification of Cell-Free Recombinants

To monitor integrative recombination reactions in mammalian cells under cell-free conditions, we have to be prepared to detect relatively rare recombination products. Assay systems for the identification of these infrequently arising products have to fulfil several preconditions to be useful in monitoring enzyme purification and in surveying the reaction mechanism, its requirements, and kinetics. The assay has to be reliable and yield reproducible results; it has to be as direct and rapid as possible. We have used two methods based on different principles to assess the generation of cell-free recombinants in our cell-free system.

One test system relies on the transfection of possible recombination products into *Escherichia coli* followed by colony hybridization to Ad12 DNA. Ad12 DNA–hamster DNA recombinants generated upon incubation with the BHK21 hamster cell nuclear extract or with fractions purified from it were identified via transfection of the reextracted DNA into the recA$^-$ strain HB101/LM1035 of *E. coli* and followed by hybridization to ^{32}P-labeled Ad12 DNA. Subse-quently, the presumptive recombinants thus isolated were verified by restriction analyses and in part by sequence determinations across the sites of linkage between the two recombination partners. Numerous control experiments have validated the reliability of this identification assay (JESSBERGER et al. 1989b; TATZELT et al. 1992,1993). In some of these controls, the recombination partners were separately incubated with extracts and then reextracted, mixed, and transfected into *E. coli*. recombinants were never found in these or other control experiments.

Although recombination-deficient bacteria were used as test organisms, it was still difficult to stringently rule out the possibility that *E. coli* could somehow have made a contribution to the recombination process. Events initiated in vitro in the cell-free system could have been completed by bacterial enzymes in vivo. Furthermore, the transfection assay could detect only recombinants infectious for *E. coli* under transfection conditions.

We have, therefore, started to apply the polymerase chain reaction (PCR) technique for the identification of cell-free-generated recombinants (TATZELT et al. 1993). These in vitro products were analyzed with synthetic oligodeoxyribonucleotide primers, whose locations were either close to the termini of the viral DNA fragment or within the circular p7 plasmid, close to but outside the hamster preinsertion sequence, since it could not be predicted where inside the p7 preinsertion segment recombination might have occurred. After the identification of the amplified recombination products by subsequent Southern blot hybridization to ^{32}P-labeled Ad12 DNA, the recombination products were further amplified by using primers located slightly internally to those employed in the first round of amplification. The identity of at least some of the thus documented recombinants was confirmed by nucleotide sequence analyses across the sites of junction. The negative results of many control experiments attested to the specifity of the amplification reaction and argued against PCR artifacts (TATZELT et al. 1993). The results obtained by the *E. coli* transfection assay system with the recombinationally active protein fractions and the structural characteristics of the cell-free-generated recombinants were thus confirmed by the PCR-based test system.

Both assay systems have now been working reproducibly and reliably for several years and for several different investigators in this laboratory and, therefore, seem to be well suited for the purification of the recombinationally active protein(s) and for the characterization of the cell-free-generated recombination products. This optimistic evaluation does not preclude endeavors to develop a simpler and even more direct assay system (see Sect. 4.1).

3.2 Purification of Recombinationally Active Proteins

By using standard chromatographic procedures we have been able to purify enzymatic activities catalyzing cell-free recombination between the 20 885- to 24 053-nucleotide fragment of Ad12 DNA and the hamster preinsertion sequence p7. Figure 2 summarizes our present purification scheme for extracts from BHK21 cell nuclei. Crude nuclear extracts were prepared from uninfected BHK21 hamster cells to yield crude nuclear fraction I (JESSBERGER et al. 1989b). The crude extract was then applied to a Sephacryl S-300 gel filtration column. The recombinationally active fractions eluted in the shoulder of the absorbancy profile at 280 nm and were pooled as fraction II (TATZELT et al. 1992). In subsequent purification steps on MonoS and MonoQ ion-exchanger columns, the activity promoting cell-free recombination resided in the flowthrough of the cation exchanger (MonoS) as fraction III and could be eluted from the anion exchanger (MonoQ) with a salt concentration of about 500 m*M* NaCl as fraction IV (Tatzelt et al. 1993). After dialyzing fraction IV against the standard buffer (see the legend to Fig. 2), it was loaded onto a Heparin-Sepharose column. Proteins that still retained the activity eluted in the flowthrough as recombinationally active fraction V. Figure 3 shows the protein composition of the most highly purified fraction V as determined by sodium dodecyl sulfate polyacrylamide gel electrophoresis (SDS-PAGE) followed by silver staining. Only about five major bands are left in this preparation. Table 1 summarizes quantitative aspects of the purification procedure. The results demonstrate the purification towards a limited set of recombinationally active protein(s) in the cell-free system.

In successive chromatographic steps an about 470-fold purification was achieved. This degree of purification was calculated relative to the activity in the crude nuclear extract. The determination of the recombination frequency of protein fractions I–V was based on results obtained with the *E. coli* transfection assay (Jessberger et al. 1989b; Tatzelt et al. 1992, 1993). The frequency of recombination was standardized as Ad12-positive colonies per ampicillin-resistant (amp^r) colonies. The specific activity was given in arbitrary units. One unit was equivalent to 1% Ad12-positive colonies per μg protein. Fraction V was obtained by the consecutive fractionation of nuclear extracts over individual chromatographic steps as described or, alternatively, by purifying fraction II directly over a MonoQ column followed by Heparin-Sepharose purification (see Fig. 2). There was no obvious difference in protein composition or in activity between fraction V purified by either protocol.

We are now working toward the further purification of the active fraction V by size fractionation on a Superdex 75 column. Since we do not expect a single protein to catalyze recombination, but more likely a recombination complex consisting of several proteins, we have also initiated extensive functional characterizations of the components of fraction V. Recent results demonstrate that this procedure yields fraction VI which is still active in cell-free recombination (Fechteler et al. 1995b).

Results from in vitro studies have validated the reliability of the purified cell-free recombination system. Apparently, the system works without a contribution of viral gene products. Hence, we postulate the existence of a functional recombination machinery to be supplied exclusively by the host cell. We will now concentrate on the enzymatic purification and characterization of its components. Nevertheless, an auxiliary or modifying role of viral proteins cannot be excluded based on data presently available from the cell-free system. We have, therefore, initiated the purification of nuclear extracts from BHK21 hamster cells, prepared at 36 h after Ad12 infection.

3.3 Characteristics of the In Vitro-Generated Ad12 DNA–Hamster DNA Recombinants

The quality of an in vitro recombination system is dependent on the cell-free reaction being as close as possible to the integration event in the living cell and on

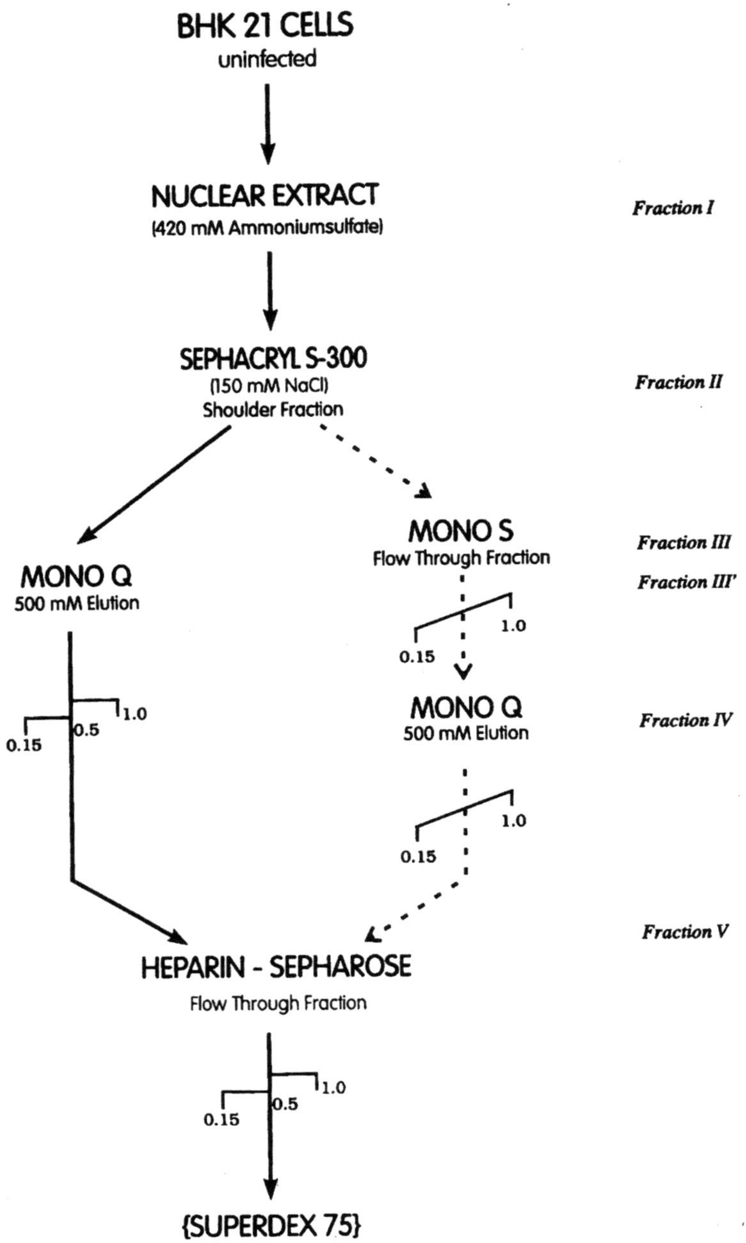

Fig. 2. Fractionation of nuclear extracts from BHK21 hamster cells. Crude nuclear extracts were prepared from uninfected BHK21 hamster cells grown in suspension culture. Fraction I was applied to a Sephacryl S-300 column equilibrated and developed with buffer A (20 m*M* Hepes, pH 7.9, 20% glycerol, 150 m*M* NaCl, 0.2 m*M* each ethylenediaminetetra-acetic acid, EDTA, and ethylene glycol tetra-acetic acid, EGTA, 2 mM dithiothreitol, 1 m*M* phenylmethylsulfonyl fluoride). The active fractions from the shoulder of the $OD_{280\ nm}$ absorbancy profile were pooled. This fraction II was loaded onto a Mono S column equilibrated with buffer A. Fractions catalyzing cell-free recombination eluted in the flowthrough of the absorbancy profile as fraction III, which was adsorbed onto a Mono Q column equilibrated with

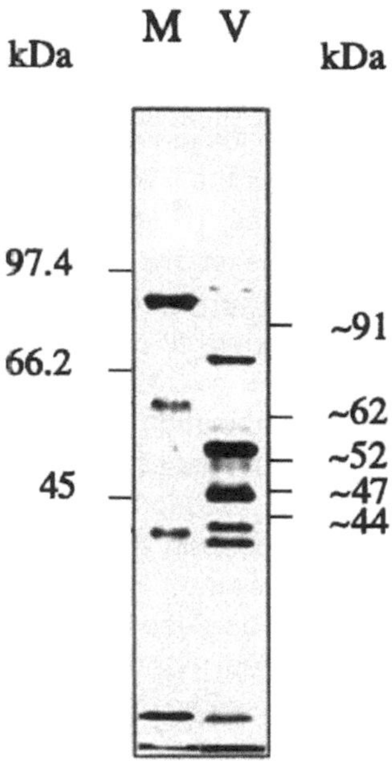

Fig. 3. Protein composition of fraction V purified from nuclear extracts from BHK21 hamster cells. The proteins in fraction V were analyzed by electrophoresis in sodium dodecyl sulfate (SDS)-polyacrylamide gels followed by silver staining. These proteins ranged in size between about 91 and about 44 kDa. In *lane M*, marker proteins of known molecular masses (kDa) were coelectrophoresed

Table 1. Purification of proteins catalyzing recombination between adenovirus type 12 (Ad12) DNA and the p7 hamster preinsertion sequence

Fraction	Procedure	Total (mg)	Recovery of protein (%)	Recombination frequency (%)[a]	Specific activity (unit)[b]	Purification fold
I	Nuclear extract	30	100	0.2	4×10^{-3}	–
II	Sephacryl S-300	5.5	18.5	0.1	0.02	5
III	Mono S	3.5	11.6	0.1	0.02	5
IV	Mono Q	1.2	4	0.5	0.2	50
V	Heparin Sepharose	0.5	1.7	4.25	1.89	472

[a] Recombination frequencies of protein fractions I–V were determined by the *Escherichia* coli transfection assay. Frequency is measured as Ad12-positive colonies per ampicillin-resistant (Amp) colonies. Values given are approximate values.

[b] Specific activity is given in arbitrary units. One unit is equivalent to 1% Ad12-positive colonies per µg protein.

buffer A. Recombinationally active proteins were eluted with buffer A containing 500 m*M* NaCl, pooled and dialyzed against buffer A (fraction IV). This fraction was applied onto a Heparin-Sepharose column equilibrated with buffer A. The recombinationally active fractions eluted in the flowthrough and were pooled as fraction V. Alternatively, the purification scheme was performed by loading fraction II directly on a Mono Q column (fraction III') followed by fractionation of the proteins over a Heparin-Sepharose column. There is no apparent difference in protein composition or in the activity of the two protein fractions. Quantitative aspects of the purification of the nuclear extracts are shown in Table 1

the recombination products exhibiting the same characteristics as in vivo-produced recombinants.

In order to determine the sequence characteristics of the cell-free-generated recombinants in more detail, we have analyzed numerous junction sites across viral and cellular host DNA sequences. As described for the linkage sites in adenovirus-transformed and Ad12-induced tumor cell lines, patchy homologies between the two recombination partner sequences at the sites of integration are present in most, though not all, of the cell-free recombinants (TATZELT et al. 1992). Therefore, patches of short homology seem to play a role in the selection of sites for recombination, in recombination events in living cells, and in the cell-free system. Cell-free-generated recombination products also bear resemblance to in vivo-isolated recombinants with respect to their target preference. Though data from analyses of several junction sites in adenovirus-transformed and Ad12-induced tumor cell lines support the notion that a specific or common cellular insertion sequence does not exist, it is striking that the sequence motif 5'-CCTCTCCG-3' or sequences close to it have repeatedly served as preferred target sequences in cell-free recombination events (TATZELT et al. 1992, 1993). These analyses have revealed an interesting clustering of recombination sites in certain regions of the preinsertion DNA near or inside the sequence motif or at very closely related sequences. Possibly, proteins in all recombinationally active fractions were targeted to very similar sites, thereby facilitating the integration reaction and eliciting independent cell-free recombi-nation events. Furthermore, the recombination site of Ad12 DNA integration in the original tumor cell line CLAC1, from which the p7 pre-insertion sequence has been derived, also falls directly inside a CCTCTCCG element (STABEL and DOERFLER 1982). We must however, emphazise that we have also observed integration sites in vitro which were different from those just described.

Preinsertion sequences have been used as recombination targets in vivo and in the cell-free system. Randomly selected hamster sequences of the unique or repetitive sequence type do not give rise to in vitro recombinants, at least not at a frequency comparable to that of recombination with the p7 sequence (JESSBERGER et al. 1989b; TATZELT et al. 1992). However, in cell-free recombination reactions with another cloned preinsertion sequence p16 (LICHTENBERG et al. 1987), we have been able to detect cell-free recombination products (JESSBERGER et al. 1989b). This apparent preference for preinsertion sequences is probably not absolute, since we have identified a few integration sites outside the preinsertion segment in the immediately adjacent pBR322 DNA sequences of the same p7 construct.

Taken together, the documented characteristics of this cell-free recombination system are consistent with the interpretation that this system mimics the integrative recombination reaction observed in living cells sufficiently closely to pursue the further development of this nuclear extract system.

3.4 Proteins in the Recombinationally Active Fractions Interact with DNA

It was reasoned that proteins catalyzing cell-free recombination between viral and cellular host DNA sequences might specifically bind to sequence motifs inside the preinsertion sequence which were identified as recombination targets in vitro. Two synthetic double-stranded, 50-bp oligodeoxyribonucleotides were used in electrophoretic mobility shift assays to probe the recombinationally active proteins for their ability to interact specifically with this target DNA sequence. Both sequences derived from the preinsertion sequence could serve as acceptor sites for Ad12 DNA integration in independent cell-free recombination events, and both elements contained the target motif 5'-CCTCTCCG-3'.

Interestingly, proteins in all recombinationally active fractions bound to double-stranded oligodeoxyribonucleotides carrying multiple sites for cell-free recombination and formed specific protein–DNA complexes (TATZELT et al. 1992, 1993; Fig. 4a, b). This complex formation could be inhibited by the same unlabeled oligodeoxyribonucleotides, attesting to the specifity of this protein–DNA interaction (Fig. 4a,b). Moreover, a double-stranded, 40-bp oligodeoxyribonucleotide comprising the original Ad12 DNA integration site in the Ad12-induced tumor cell line CLAC1 could also compete for this protein binding. There was no competition detectable with nonspecific poly-dA:dT or with oligodeoxyribonucleotides with unrelated sequences, e.g., sequences from the late E2A promotor of Ad2 DNA (TATZELT et al. 1993).

We also tested the capability of proteins in the recombinationally active fractions to bind to single-stranded DNA. We used the same oligodeoxyribonucleotide sequences in the single-stranded form and also observed protein–DNA interactions (Fig. 4c). However, in contrast to the interactions with the double-stranded oligodeoxyribonucleotides, the newly identified interactions with single-stranded DNA could be competed not only by the same unlabeled oligodeoxyribonucleotide, but also with totally unrelated sequences, suggesting unspecific interactions of the recombinationally active proteins with single-stranded DNA (Fig. 4c).

4 Integrative Recombination

4.1 Does the Model Reaction Resemble the Reaction in Cells?

Cell-free systems have been developed to investigate the integrative recombination of viral genomes (BROWN et al. 1987; JESSBERGER et al. 1989b; TATZELT et al. 1992, 1993). Working with cell-free abstractions immediately raises the question to what extent the model reaction analyzed in vitro resembles the naturally

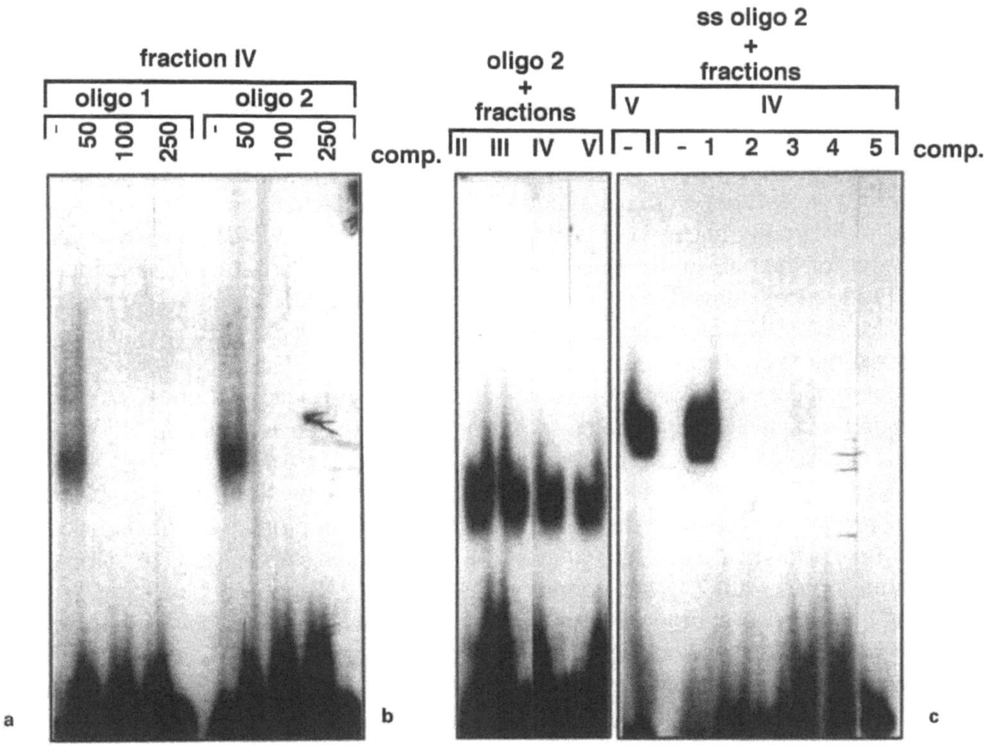

Fig. 4. Binding characteristics of recombinationally active proteins. Electrophoretic mobility shift assays of proteins from fraction II–V with the synthetic oligodeoxyribonucleotides 1 and 2. The sequences are indicated below (the preferred target motif 5'-CCTCTCCG-3' is underlined): (1) 5'-CACTCCACCGACGCGG*CCTCTCCG*CACGCTTGCACAAGCAGCAACCAGCT-3';(2) 5'-CCCGCCCGCACTCCACCGACGCGG*CCTCTCCG*CACGCTTGCACAAGCAGC-3';. **a** Mobility shift assay of active proteins from fraction IV with the double-stranded, 50-bp oligodeoxyribonucleotide 1 (lane -, without competitor, *comp*), which represents a segment of the hamster preinsertion sequence and has served as acceptor target for adenovirus type 12 (Ad12) DNA in cell-free recombination. This delayed migration can be inhibited by the same unlabeled oligodeoxyribonucleotide, attesting to the specifity of this protein–DNA interaction. Different amounts of competitor have been used (50-, 100-, and 250-fold). Unrelated oligodeoxyribonucleotides have not been able to compete for this protein binding. There is no competition with nonspecific poly-dA: dT (data not shown; TATZELT et al. 1992, 1993). **b** Mobility shift assays of recombinationally active proteins from fraction II–V with the double-stranded, 50-bp oligodeoxyribonucleotide 2. This sequence also encompasses the preferred target sequence for Ad12 DNA. Proteins in all recombinationally active fractions form the same protein–DNA complex. These interactions seem to be identical to those observed between proteins in fraction IV and with oligodeoxyribonucleotide 1. **c** Mobility shift assays of proteins from fractions IV and V with the single-stranded oligodeoxyribonucleotide 2 (lane-, without comp). The observed protein–DNA interactions can be inhibited by the same unlabeled oligodeoxyribonucleotide (*lane 1*) as well as by a single-stranded (*ss*), 40-bp oligodeoxyribonucleotide comprising the original site of Ad12 DNA integration in the tumor CLAC1 (*lanes 2* and *3*; both strands). Both strands of an unrelated single-stranded oligodeoxyribonucleotide (*lanes 4* and *5*) derived from frog virus 3 DNA are also able to compete for protein binding, attesting to an unspecific single-stranded binding of the recombinationally active proteins. The competitors are listed below (the preferred target sequence is underlined in 1–3). There is no target sequence in unrelated DNA (items 4 and 5): 1) 5'CCCGCCCGCACTCCACCGACGCGG*CCTCTCCG*CACGCTTGCACAAGCAGC-3';, (2) 5'-ACTTGAAGGAGACGCCGC*CCTCTCCG*GGGGTGCGAGTGCC-3;, (3) 5'-GGCACTCGCACCCC*CGGAGAGG*GCGGCGTCTCCTTCAAGT-3';,(4) 5'-GGATAGAACGGCCACATCTACCTCTGTGGCCGTCGGGTTCGAGGCGCAGG-3';,(5) 5'-CCTGCGCCTCGAACCCGACGGCCACAGAGGTAGATGTGGCCGTTCTATCC-3'

occurring event in vivo. In order to monitor enzyme purification, to survey the reaction mechanism and its requirements, and the kinetics of the process, the analyzed reaction need be "simplified" to accommodate the system to work in vitro. Data obtained with the cell-free system demonstrate that the model recombination reaction, at least in certain aspects, resembles the natural Ad12 DNA integration event in mammalian cells. In particular, details of the reaction parameters and characteristics of the recombination products are similar to the situation in the living cell and indicate that the cell-free system will be well suited for mechanistic and enzymatic studies of this recombination reaction. The system has enabled us to purify the crude extracts of BHK21 cell nuclei to five major protein bands still retaining recombinational activity.

At that point, further work will be directed, on the one hand, towards a dissection of the complete reaction in order to study enzymatic functions of the components in the recombination machinery in more detail. We further wish to investigate the possibility of an adenovirus-specific integrative recombination mechanism.

4.2 Specific Versus Nonspecific Adenovirus Type 12 DNA Integration

One of the major questions concerning viral DNA integration is whether this type of recombination reaction occurs in a random or nonrandom manner. Does a common or specific target sequence exist at which viral DNA integrates into the mammalian genome? Though data derived from the junction site analyses of adenovirus-transformed cell lines and of Ad12-induced tumor cell lines do not support the notion of a site-specific type of recombination, it has to be analyzed whether, under certain conditions, adenovirus DNA may integrate at selective sequences.

Cell-free recombination experiments with randomly selected hamster sequences of repetitive or unique origin did not give rise to in vitro recombinants (Jessberger et al. 1989b; Tatzelt et al. 1992). These negative controls validated a model of target preference. Preinsertion sequences might be endowed with features, primary sequence or chromatin structure, that render a region of the host genome a preferred target sequence that is recognized by the catalyzing enzymatic activities for foreign DNA insertion. Furthermore, since we have not found frequent integration events for Ad12 DNA in the backbone of the pBR322 vector sequences, integrative recombination seems to be directed somehow by sequence characteristics of the hamster preinsertion sequence. Integration, therefore, does not exhibit features of a completely random event. The finding that, in a few instances, Ad12 DNA recombination sites were located in the immediately adjacent pBR322 DNA sequences suggests a possible contribution of preinsertion sequences on neighboring DNA sequences, in spite of their bacterial origin. Moreover, the observation that proteins in all recombinationally active fractions bind specifically to double-stranded, synthetic oligodeoxyribonucleotides comprising certain regions of the preinsertion sequence that carry the target motif CCTCTCCG (TATZELT et al. 1992, 1993) supports our interpretation of a target preference of Ad12 DNA integration in vivo and in vitro. It seems conceivable that sequence preferences may exist for adenoviral (foreign) DNA insertion under certain conditions, but insertion is not strictly dependent on these requirements.

Recombination reactions, though based on similar requirements, may be diverse with respect to their specificity and detailed mechanisms. This diversity can be illustrated by various integration patterns of retroelements. The scale reaches from site-specific to nonrandom integration reactions. It is beyond the scope of this review to discuss all these different reactions in various systems. Although retroviruses can insert at many different sites (HUGHES et al. 1978), much evidence suggests that integration of retroviruses is not completely random in vivo. Experiments in the mid-1980s with MoMLV infection of murine cells suggested that integration was nonrandom. These analyses also revealed that DNA integrated by retroviral infection, compared to DNA introduced by transfection, tended to be in transcribed regions of the genome. Based on different approaches several investigators verified that insertions are associated with transcribed regions of the host genome (HWANG and GILBOA 1984; KING et al. 1985; SCHERDIN et al. 1990).

Based on this diversity of integration patterns, one could imagine different ways in which genomic DNA was identified for preferential targeting. These possibilities ranged from specific to general and included parameters such as sequence content, chromatin structure, nuclear localization, accessibility to DNA by removal of chromatin proteins, and association with other non-nucleosome proteins, e.g., nuclear matrix proteins, topoisomerases, transcription, or replication factors. DNA or protein motifs could constitute a specific target for recombination. The targeting mechanism by itself could act directly or indirectly to influence integration site preferences. Transcription factors associated with DNA in nucleosome-free regions could target integration directly by association with the recombination machinery or indirectly by inducing changes in the DNA that increased the accessibility of recombination active protein(s) to target DNA.

4.3 Homologous Versus Nonhomologous Recombination

The observation that the recombination machinery is not dependent on the presence of long stretches of perfect sequence identity, but may take advantage of patchy homologies, raises the question of whether this type of integration belongs to the class of homologous or nonhomologous recombination events.

The genetic integrity of all organisms depends on the ability to repair damage to DNA caused by endogenous or exogenous agents or mechanisms. The gene products involved in DNA repair probably operate in multiple and perhaps functionally redundant pathways. Different modes of in vivo repair of double-strand breaks (DSB) have been described for various organisms: the recombinational DSB repair (DSBR) mode (SZOSTAK et al. 1983), the single-strand annealing (SSA) mode (LIN et al. 1984), and end-to-end joining related to illegitimate recombination since it can join nonhomologous DNA ends (VIEIRA and MESSING 1987; THODE et al. 1990). Although most DSB repair in *Saccharomyces cerevisiae* probably occurs via recombination involving homologous DNA sequ-ences (DSB repair or SSA modes), nonhomologous repair events can occur utilizing very weak or no sequence homologies, suggesting the existence of a recombination mechanism which is nearly homology independent. Such events are usually classified as illegitimate recombination.

Interestingly, the recombinant junctions of the rare products created during nonhomologous repair in *S. cerevisiae* are very similar to those observed in the repair of DSB in mammalian cells (ROTH and WILSON 1986, 1988). However, DSB repair by illegitimate recombination is a minor pathway in *S. cerevisiae*. The frequency of illegitimate events either intra- or intermolecularly is at least 100-fold lower than in the presence of a cointroduced homologous sequence. In contrast, targeted integration of exogenous DNA at its homologous chromosomal location in mammalian cells is masked by a 1000-fold higher frequency of random integration events (SMITH and BERG 1984; LIN et al. 1985; SMITHIES et al. 1985; THOMAS et al. 1986). The presence of short, one- to five-nucleotide homologies at many of nonhomologous junctions has led several investigators to propose that these homologies play a role in nonhomologous recombination (EFSTRATIADIS et al. 1980; ALBERTINI et al. 1982; ALT and BALTIMORE 1982; STRINGER 1982; GAHLMANN et al. 1982; MARVO et al. 1983; RULEY and FRIED 1983; HOGAN and FAUST 1984; BULLOCK et al. 1985; ROTH et al. 1985). However, because these homologies are so short, it is difficult to distinguish between mechanistic relevance and chance occurrence.

Patchy homologies between the hamster preinsertion DNA and adenoviral DNA sequences at the recombinant junction sites are not an absolute prerequisite for the mechanism of integrative recombination to take place in the mammalian cell. Nevertheless, the small but significant number of nucleotide identities may facilitate the Ad12 DNA integration event in vivo and in vitro. The stretches of homology vary in size ranging from one up to eight nucleotides. They might direct recombination reactions in mammalian cells, as discussed for adenovirus DNA integration (GAHLMANN et al. 1982) and also suggested by ROTH et al. (1985) and others. Possible explanations for the recombination machinery to take advantage of patchy homologies are end-to-end ligation of DNA ends in lower eukaryotes (SCHIESTL et al. 1993) or in mammalian cells (ROTH et al. 1985; ROTH and WILSON 1986) or a stabilization of the two recombination partner molecules through an annealing event. We also suggest that patchy homologies may offer advantages over longer perfect homologies in partner recognition, since structural features can then be more distinctly recognized by the enzymatic activities of the host cell (TATZELT et al. 1993). The results from in vivo and in vitro studies indicate the existence of a recombination mechanism that is sensitive to very small regions of homology for the integration of adenoviral DNA into the mammalian genome. The terms "homologous" and "nonhomologous" recombination are not very precisely defined in the literature. Hence, we prefer to classify foreign (viral) DNA integration into mammalian genomes as integrative recombination.

4.4 Possible Links Between Transcription and Integrative Recombination of Adenovirus Type 12 DNA

Many cellular preinsertion sequences of adenovirus DNA integration analyzed so far have proven to be transcriptionally active (GAHLMANN et al. 1984; LICHTENBERG et al. 1987; SCHULZ et al. 1987; JESSBERGER et al. 1989a). Transcriptional activity has been documented in the original host cell from different species (hamster, mouse, human) prior to any contact with viral DNA during infection, integration, and transformation. Furthermore, transcriptional activity at the junction sites between viral and cellular DNA has also been demonstrated in Ad12-induced tumor cells and in adenovirus-transformed cells. The hypothesis of a possible stimulatory effect of transcription on adenovirus DNA integration is also supported by the fact that in other viral and cellular systems integrative recombination frequently takes place at transcriptionally active cellular sites (MOOSLEHNER et al. 1990; SCHERDIN et al. 1990).

The observation that transcription might be somehow involved in the control of homologous recombination is not restricted to bacteria or lower eukaryotes. It has been also suggested, that transcription levels may influence homologous recombination in mammalian cells (MANSOUR et al. 1988; JOHNSON et al. 1989). Transcription has been implicated in the control of recombinational events in the course of the development of mammalian immune systems. During the development of immunoglobulin genes in B and T cell receptor genes, specific recombination events mediate the assembly of mature genes from component gene segments (TONEGAWA 1983; ALT et al. 1986). ALT et al. (1986) and BLACKWELL et al. (1986) presented evidence for the notion that transcriptionally active immunoglobulin gene segments recombine at high frequencies. Furthermore, NICKOLOFF and REYNOLDS (1990) have demonstrated that transcription stimulates homologous recombination between transfecting plasmids in mammalian cells.

If integration of adenoviral DNA into the mammalian genome was somehow influenced by transcription, then the question immediately arises of how transcription can influence or stimulate nonhomologous recombination. Assuming these proteins of the transcriptional machinery are present at the preferred cellular target sequences, it seems possible that these proteins are involved in the mechanism of integrative recombination. This functional role is quite speculative, but at least it seems plausible that these cellular proteins can render the preinsertion sequences accessible for the recombination machinery (see also Sect. 4.2). Two possible explanations for a stimulating role of transcription are discussed: (1) accessibility of a chromosomal region may control whether recombination-active proteins (recombinases) are able to catalyze the reaction at the target DNA, and this accessibility may be controlled by transcription; (2) alternatively, transcription might be controlled by factors that influence target accessibility to RNA polymerase and recombination-active protein(s). Considering the wide distribution of transcription-induced recombination in many animal systems, it seems conceivable that a variety of mechanisms are involved, not only in different organisms, but also in the different processes of homologous and nonhomologous recombination as well as foreign (viral) DNA integration.

4.5 Do Viral Gene Products Make a Contribution?

Assuming that the cell-free recombination system described does indeed resemble the natural integration event, the postulated recombination machinery could be supplied by the affected host cell. This interpretation is supported by the fact that cell-free recombination between the adenoviral 20 885- to 24 053-nucleotide fragment and the hamster preinsertion sequence p7 is catalyzed by crude nuclear extracts or purified protein fractions from uninfected BHK21 hamster cell nuclear extracts. This finding does not necessarily imply that viral-encoded functions are not also involved in the recombination process in vivo, either by contributing an obligatory enzymatic or structural component of the recombination machinery or as a modifier of the reaction in the living cell. Insertion of adenovirus DNA is very different from the prototype retroviral integration during which integrases encoded and introduced into the cell by the retroviruses facilitate the integration event as catalysts. In our cell-free system, viral counterparts could modify a general cellular recombination reaction for the virus' aims. Apparently, the system works well without viral proteins under the chosen conditions, but we had to "simplify" the reaction in that we did not use the entire Ad12 genome as one of the DNA recombination substrates. Although adenovirus DNA was also found integrated in a fragmented form, the insertion of the intact Ad12 molecules proceeded mainly via the viral terminal sequences in vivo. The virion DNA termini are bound to the terminal protein at their 5' ends. It has not yet been investigated whether the terminal viral protein participates in integrative recombination of Ad12 DNA.

Hence, one of our goals for the future remains to determine whether viral functions participate in the course of integrative recombination between foreign (viral) DNA and the mammalian genome. We have initiated different approaches to investigate several possibilities: (a) adenovirus DNA integration is exclusively catalyzed by the host cell recombination machinery; (b) viral gene products contribute to the recombination process enzymatically and/or structurally. A combination of items a and b also seems possible.

We have cloned a 4-kbp fragment of BHK21 DNA into a cosmid vector that contains the hamster preinsertion sequence p7. This approach may help to use the entire Ad12 genome in the cell-free recombination reaction. Furthermore, instead of using protein-free, "naked" Ad12 DNA as the recombination partner, we wish to test the recombinational activity of subviral Ad12 particles. It is known that subviral particles reach the nucleus early after infection (LONBERG-

HOLM and PHILIPSON 1969; MORGAN et al. 1969). Alternatively, we have initiated experiments to test nuclear extracts or purified fractions from Ad12-infected BHK21 hamster cells for recombinational activity. Crude nuclear extracts were prepared from BHK21 cells at 36 h after Ad12 infection and will be tested in the conventional cell-free system as well as in the cosmid-based recombination system. Our more recent data indicate that this cell-free system may work more efficiently than that from uninfected BHK21 cells (FECHTELER et al. 1995b).

5 Outlook

Results described in the preceding sections demonstrate that the integrative recombination reaction between Ad12 DNA and a hamster preinsertion sequence can be imitated in a cell-free system derived and purified from extracts of hamster BHK21 cell nuclei. This system will enable us to identify proteins active in catalyzing this type of nonhomologous recombination in mammalian cells and to survey mechanism, requirements, and kinetics of the reaction. Furthermore, we have adduced evidence for the general importance of the nonhomologous integration reaction in eukaryotic cells. We have been able to document that at least certain aspects of nonhomologous recombination between the E1 fragment of Ad2 DNA and the *Eco*RI-O fragment of *Autographa californica* nuclear polyhedrosis virus (AcNPV) DNA (XIONG et al. 1991) can be reproduced in a cell-free system derived from partly purified nuclear extracts of *Spodoptera frugiperda* insect cells (SCHORR and DOERFLER 1992). With respect to reaction parameters and characteristics of the in vitro recombinants, there seem to be no major differences between the mammalian cell and insect cell systems, arguing for the existence of a widely distributed mechanism that operates generally in mammalian as well as in insect cells.

The possible existence of common recombination mechanisms exceeding species limitations is supported by the isolation and cloning of recombination genes in higher organisms that share extensive homologies with RAD51 and recA (SHINOHARA et al. 1993; MORITA et al.1993). In *E. coli* and many other prokaryotes, the recA protein or a recA-like protein plays an essential role in homologous recombination and in a variety of repair recombination processes elicited by DNA damage (ROCA and COX 1990; KOWALCZYKOWSKI 1991). In yeast, the RAD51 gene has recently been cloned, and its gene product shows structural similarity to the *E. coli* recA protein with adenosine triphosphate (ATP)-dependent DNA-binding activity (SHINOHARA et al. 1992; ABOUSSEKHARA et al. 1992; BASILE et al. 1992). The wide distribution of structural homologs of the yeast RAD51 protein among eukaryotes and prokaryotes, including *E. coli*, suggests a common fundamental mechanism of recombination through evolution.

For the immediate future, further work in our own research will be directed towards a detailed analysis of the enzymatic functions of the cellular proteins in the recombination machinery. Currently, we are also exploring the possible participation of viral gene products. It will be mandatory to investigate whether the integration proceeds more efficiently in adenovirus-infected cells. Furthermore, we plan to use subviral Ad12 particles instead of purified Ad12 DNA as a substrate to investigate a putative influence of viral structural proteins on the integration event. We will also try to gain insight into the actual integration reaction of the entire Ad12 genome. Moreover, we will extend analyses of protein–DNA interactions between the recombinationally active proteins and regions of the hamster preinsertion sequences. It should also be interesting to evaluate sequence or structural requirements of recombination targets as they participate in the recombination reaction.

Acknowledgments. K.F. was the recipient of a grant from the Boehringer-Ingelheim Stiftung; S.H. received a grant through the Graduiertenprogramm of the Deutsche Forschungsgemeinschaft (DFG). Research in the authors' laboratory was supported by the Bundesministerium für Forschung und Technologie through Genzentrum Köln, TP 2.03, and by the DFG through Sonderforschungsbereich 274–A1.

References

Aboussekhara A, Chanet R, Adjiri A, Fabre F (1992) Semidominant suppressors of Srs2 helicase mutations of Saccharomyces cerevisiae map in the RAD51 gene, whose sequence predicts a protein with similarities to prokaryotic RecA proteins. Mol Cell Biol 12: 3224–3234

Albertini AM, Hofer M, Calos MP, Miller JH (1982) On the formation of spontaneous deletions: the importance of short sequence homologies in the generation of large deletions. Cell 29: 319–328

Alt FW, Baltimore D (1982) Joining of immunoglobulin heavy chain gene segments: implications from a chromosome with evidence of three D-J_H fusions. Proc Natl Acad Sci USA 79: 4118–4122

Alt FW, Blackwell TK, Depinho RA, Reth MG, Yancopoulos GD (1986) Regulation of genome rearrangement events during lymphocyte differentiation. Immunol Rev 89: 5–30

Basile G, Aker M, Mortimer RK (1992) Nucleotide sequence and transcriptional regulation of the yeast recombinational repair gene RAD51. Mol Cell Biol 12: 3235–3246

Blackwell TK, Moore MW, Yancopoulos GD, Suh H, Lutzker S, Selsing E, Alt FW (1986) Recombination between immunoglobulin variable region gene segments is enhanced by transcription. Nature 324: 585–589

Brown PO, Bowermann B, Varmus HE, Bishop JM (1987) Correct integration of retroviral DNA in vitro. Cell 49: 347–356

Bullock P, Champoux JJ, Botchan MR (1985) Association of crossover points with topoisomerase I cleavage sites: a model for nonhomologous recombination. Science 230: 954–958

Deuring R, Doerfler W (1983) Proof of recombination between viral and cellular genomes in human KB cells productively infected by adenovirus type 12: structure of the junction site in a symmetric recombinant (SYREC). Gene 26: 283–289

Deuring R, Klotz G, Doerfler W (1981a) An unusual symmetric recombinant between adenovirus type 12 DNA and human cell DNA. Proc Natl Acad Sci USA 78: 3142–3146

Deuring R, Winterhoff U, Tamanoi F, Stabel S, Doerfler W (1981b) Site of linkage between adenovirus type 12 and cell DNAs in hamster tumour line CLAC3. Nature 293: 81–84

Doerfler W (1968) The fate of the DNA of adenovirus type 12 in baby hamster kidney cells. Proc Natl Acad Sci USA 60: 636–643

Doerfler W (1970) Integration of deoxyribonucleic acid of adenovirus type 12 into the deoxyribonucleic acid of baby hamster kidney cells. J Virol 6: 652–666

Doerfler W (1982) Uptake, fixation, and expression of foreign DNA in mammalian cells: the organization of integrated adenovirus DNA sequences. Curr Top Microbiol Immunol 101: 127–194

Doerfler W (1983a) DNA methylation and gene activity. Annu Rev Biochem 52: 93–124
Doerfler W (ed) (1983b) The molecular biology of adenoviruses, vols 1, 2. Springer, Berlin Heidelberg NewYork (Current topics in microbiology and immunology, vols 109, 110)
Doerfler W (ed) (1984) The molecular biology of adenoviruses, vol 3. Springer, Berlin Heidelberg NewYork (Current topics in microbiology and immunology, vol 111)
Doerfler, W (1991a) Patterns of DNA methylation—evolutionary vestiges of foreign DNA inactivation as a host defense mechanism: a proposal. Biol Chem Hoppe Seyler 372: 557–564
Doerfler W (1991b) Abortive infection and malignant transformation by adenoviruses: integration of viral DNA and control of viral gene expression by specific patterns of DNA methylation. Adv Virus Res 39: 89–128
Doerfler W (1992) Transformation of cells by adenoviruses: less frequently discussed mechanisms. In: Doerfler W, Boehm P (eds) Malignant transformation by DNA viruses: molecular mechanisms. Verlag Chemie, Weinheim, pp 141–169
Doerfler W (1993) Adenoviral DNA integration and changes in DNA methylation patterns: a different view of insertional mutagenesis. Prog Nucleic Acids Res Mol Biol 46: 1–36
Doerfler W, Boehm P (eds) (1993) Virus strategies. Verlag Chemie, Weinheim
Doerfler W, Gahlmann R, Stabel S, Deuring R, Lichtenberg U, Schulz M, Eick D, Leisten R (1983) On the mechanism of recombination between adenoviral and cellular DNAs: the structure of junction sites. Curr Top Microbiol Immunol 109: 193–228
Doerfler W, Tatzelt J, Orend G, Schorr J, Rosahl T, Fechteler K, Zock C, Lichtenberg U (1993) Viral and cellular strategies on integrated adenovirus genomes. In: Doerfler W, Boehm P (eds) Virus strategies. Verlag Chemie, Weinheim, pp 369–400
Efstratiadis A, Posakony JW, Maniatis T et al. (1980) The structure and evolution of the human β-globin gene family. Cell 21: 653–668
Eick D, Doerfler W (1982) Integrated adenovirus type 12 DNA in the transformed hamster cell line T637: Sequence arrangements at the termini of viral DNA and mode of amplification. J Virol 42: 317-321
Eick D, Stabel S, Doerfler W (1980) Revertants of adenovirus type 12-transformed hamster cell line T637 as tools in the analysis of integration patterns. J Virol 36: 41–49
Fechteler K, Herbertz S, Doerfler W (1995b) Integrative recombination of adenovirus type12 DNA with cellular DNA: Studied with purified nuclear proteins from uninfected or Ad12-infected hamster cells. (submitted)
Franklin N (1971) Illegitimate recombination. In: Hershey AD (ed) The bacteriophage lambda. Cold Spring Harbor Lab, Cold Spring Harbor NY, pp 175–194
Freeman AE, Black PH, Vanderpool EA, Henry PH, Austin JB, Huebner RJ (1967) Transformation of primary rat embryo cells by adenovirus type 2: Proc Natl Acad Sci USA 58: 1205–1212
Gahlmann R, Doerfler W (1983) Integration of viral DNA into the genome of the adenovirus type 2-transformed hamster cell line HE5 without loss or alteration of cellular nucleotides. Nucleic Acids Res 11: 7347–7361
Gahlmann R, Leisten R, Vardimon L, Doerfler W (1982) Patch homologies and the integration of adenovirus DNA in mammalian cells. EMBO J 1: 1101–1104
Gahlmann R, Schulz M, Doerfler W (1984) Low molecular weight RNAs with homologies to cellular DNA at sites of adenovirus DNA insertion in hamster or mouse cells. EMBO J 3: 3263–3269
Gerondakis S, Cory S, Adams J (1984) Translocation of the myc cellular oncogene to the immunoglobulin heavy chain locus in the murine plasmacytomas is an imprecise reciprocal exchange. Cell 36: 973–982
Ginsberg H (ed) (1985) Adenoviruses. Plenum, New York
Groneberg J, Doerfler W (1979) Revertants of adenovirus type-12-transformed hamster cells have lost part of the viral genomes. Int J Cancer 24: 67–74
Gronenberg J, Chardonnet Y, Doerfler W (1977) Integrated viral sequences in adenovirus type 12-transformed cells. Cell 10: 101–111
Groneberg J, Sutter D, Soboll H, Doerfler W (1978) Morphological revertants of adenovirus type 12-transformed hamster cells. J Gen Virol 40: 635–645
Günthert U, Schweiger M, Stupp M, Doerfler W (1976) DNA methylation in adenovirus, adenovirus-transformed cells, and host cells. Proc Natl Acad Sci USA 73: 3923–3927
Hedrick SM, Cohen DI, Nielsen EA, Davis MM (1984) Isolation of cDNA clones encoding T cell-specific membrane-associated proteins. Nature 308: 149–153
Heller H, Kämmer C, Wilgenbus P, Doerfler W (1995) The chromosomal insertion of foreign (adenovirus type 12 or bacteriophage lambda) DNA is associated with enhanced methylation of cellular DNA segments. Proc Natl Acad Sci USA 92 (in press)

Hicks JB, Hinnen A, Fink GR (1978) Properties of yeast transformation. Cold Spring Harbor Symp Quant Biol 43: 1305–1313
Hinnen A, Hicks JB, Fink GR (1978) Transformation of yeast. Proc Natl Acad Sci USA 75: 1929–1933
Hogan A, Faust EA (1984) Short direct repeats mediate spontaneous high-frequency deletions in DNA of minute virus of mice. Mol Cell Biol 4: 2239–2242
Honjo T (1983) Immunoglobulin genes. Annu Rev Immunol 1: 499–528
Huebner RJ, Rowe WP, Lane WT (1962) Oncogenic effects in hamsters of human adenovirus types 12 and 18. Proc Natl Acad Sci USA 48: 2051–2058
Hughes SH, Shank PR, Spector DH, Kung H-J, Bishop JM et al. (1978) Proviruses of avian sarcoma virus are terminally redundant, co-extensive with unintegrated linear DNA and integrated at many sites. Cell 15: 1397–1410
Hwang L-HS, Gilboa E (1984) Expression of genes introduced into cells by retroviral infection is more efficient than that of genes introduced into cells by DNA transfection. J Virol 50: 417–424
Jessberger R, Berg P (1991) Repair of deletions and double-strand gaps by homologous recombination in a mammalian in vitro system. Mol Cell Biol 11: 445–457
Jessberger R, Weisshaar B, Stabel S, Doerfler W (1989a) Arrangement and expression of integrated adenovirus type 12 DNA in the transformed hamster cell line HA12/7: amplification of Ad12 and c-myc DNAs and evidence for hybrid viral-cellular transcripts. Virus Res 13: 113–128
Jessberger R, Heuss D, Doerfler W (1989b) Recombination in hamster cell nuclear extracts between adenovirus type 12 DNA and two hamster preinsertion sequences. EMBO J 8: 869–878
Johnson RS, Sheng M, Greenberg ME, Kolodner RD, Papaioannou VE, Spiegelman BM (1989) Targeting of nonexpressed genes in embryonic stem cells via homologous recombination. Science 245: 1234–1236
King W, Patel MD, Lobel LI, Goff SP, Nguyen-Huu MC (1985) Insertion mutagenesis of embryonal carcinoma cells by retroviruses. Science 228: 554–558
Kowalczykowski SC (1991) Biochemistry of genetic recombination: energetics and mechanism of DNA strand exchange. Annu Rev Biophys Biophys Chem 20: 539–575
Kucherlapati RS, Ayares D, Hanneken A, Noonan K, Rauth S, Spencer JM, Wallace L, Moore PD (1984) Homologous recombination in monkey cells and human cell-free extracts. Cold Spring Harbor Symp Quant Biol 49: 191–197
Kuhlmann I, Doerfler W (1982) Shifts in the extent and patterns of DNA methylation upon explantation and subcultivation of adenovirus type 12-induced hamster tumor cells. Virology 118: 169–180
Kuhlmann I, Achten S, Rudolph R, Doerfler W (1982) Tumor induction by human adenovirus type 12 in hamsters: loss of the viral genome from adenovirus type 12-induced tumor cells is compatible with tumor formation. EMBO J 1: 79–86
Kunz BA, Haynes RH (1981) Phenomenology and genetic control of mitotic recombination in yeast. Annu Rev Genet 15: 57–89
Latt SA (1981) Sister chromatid exchange formation. Annu Rev Genet 15: 11–55
Lichtenberg U, Zock C, Doerfler W (1987) Insertion of adenovirus type 12 DNA in the vicinity of an intracisternal A particle genome in Syrian hamster tumor cells. J Virol 61: 2719–2726
Lin F-LM, Sperle K, Sternberg N (1984) Model for homologous recombination during transfer of DNA into mouse L cells: role for DNA ends in the recombination process. Mol Cell Biol 4: 1020–1034
Lin F-LM, Sperle K, Sternberg N (1985) Recombination in mouse L cells between DNA introduced into cells and homologous chromosomal sequences. Proc Natl Acad Sci USA 82: 1391–1395
Liskay RM, Stachelek JL (1984) Evidence for intrachromosomal gene conversion in cultured mouse cells. Cell 35: 157–165
Lonberg-Holm K, Philipson L (1969) Early events of virus-cell interaction in an adenovirus system. J Virol 4: 323–338
Malissen M, Minard K, Mjolsness S et al. (1984) Mouse T cell antigen receptor: structure and organization of constant and joining gene segments encoding the β polypeptide. Cell 37: 1101–1110
Mansour SL, Thomas KR, Capecchi MR (1988) Disruption of the proto-oncogene int-2 in mouse embryo-derived stem cells: a general strategy for targeting mutations to non-selectable genes. Nature 336: 348–352
Marvo SL, King SR, Jaskunas SR (1983) Role of short regions of homology in intermolecular illegitimate recombination events. Proc Natl Acad Sci USA 80: 2452–2456
McDougall JK, Dunn AR, Jones KW (1972) In situ hybridization of adenovirus RNA and DNA. Nature 236: 346–348
Mooslehner K, Karls U, Harbers K (1990) Retroviral integration sites in transgenic Mov mice frequently map in the vicinity of transcribed DNA regions. J Virol 64: 3056–3058

Morgan C, Rosenkranz HS, Mednis B (1969) Structure and development of viruses as observed in the electron microscope. V. Entry and uncoating of adenovirus. J Virol 4: 777–796
Morita T, Yoshimura Y, Yamamoto A, Murata K, Mori M, Yamamoto H, Matsushiro A (1993) A mouse homolog of the *Escherichia coli* recA and Saccharomyces cerevisiae RAD51 genes. Proc Natl Acad Sci USA 90: 6577–6580
Nickoloff JA, Reynolds RJ (1990) Transcription stimulates homologous recombination in mammalian cells. Mol Cell Biol 10: 4837–4845
Orend G, Kuhlmann I, Doerfler W (1991) Spreading of DNA methylation across integrated foreign (adenovirus type 12) genomes in mammalian cells. J Virol 65: 4301–4308
Orend G, Linkwitz A, Doerfler W (1994) Selective sites of adenovirus (foreign) DNA integration into the hamster genome: changes in integration patterns. J Virol 68:187–194
Orend G, Knoblauch M, Kämmer C, Tjia S, Schmitz B, Linkwitz A, Meyer zu Altenschildesche G, Maas J, Doerfler, W (1995a) The initiation of de novo methylation of foreign DNA integrated into a mammalian genome is not exclusively target by nucleotide sequence. J Virol 69: 1226–1242
Orend G, Knoblauch M, Doerfler W (1995b) Selective loss of unmethylated segments of integrated Ad12 genomes in revertants of the adenovirus type 12-transformed cell line T637. (submitted)
Orr-Weaver TL, Szostak JW (1983) Yeast recombination: the association between double-strand gap repair and crossing-over. Proc Natl Acad Sci USA 80: 4417–4421
Orr-Weaver TL, Szostak JW, Rothstein RJ (1981) Yeast transformation: a model system for the study of recombination. Proc Natl Acad Sci USA 78: 6354–6358
Pope JH, Rowe WP (1964) Immunofluorescent studies of adenovirus 12 tumors and of cells transformed or infected by adenoviruses. J Exp Med 120: 577–588
Roca AT, Cox MM (1990) The recA protein: structure and function. Crit Rev Biochem Mol Biol 25: 415–456
Rosahl T, Doerfler W (1988) Predominant association of adenovirus type 12 DNA with human chromosome 1 early in productive infection. Virology 162: 494–497
Rosahl T, Doerfler W (1992) Alterations in the levels of expression of specific cellular genes in adenovirus-infected and -transformed cells. Virus Res 26: 71–90
Roth D, Wilson J (1986) Nonhomologous recombination in mammalian cells: role for short sequence homologies in the joining reaction. Mol Cell Biol 6: 4295–4304
Roth D, Wilson J (1988) Illegitimate recombination in mammalian cells. In: Kucherlapati RS, Smith GR (eds) Genetic recombination. ASM, Washington, pp 621–653
Roth D, Porter TM, Wilson J (1985) Mechanisms of nonhomologous recombination in mammalian cells. Mol Cell Biol 5: 2599–2607
Ruley HE, Fried M (1983) Clustered illegitimate recombination events in mammalian cells involving very short sequence homologies. Nature 304: 181–184
Saiki RK, Gelfand DH, Stoffel S, Scharf SJ, Higuchi R, Horn GT, Mullis KB, Erlich HA (1988) Primer-directed enzymatic amplification of DNA with a thermostable DNA polymerase. Science 239: 487–491
Scherdin U, Rhodes K, Breindl M (1990) Transcriptionally active genome regions are preferred targets for retrovirus integration. J Virol 64: 907–912
Scherer S, Davis RW (1980) Recombination of dispersed repeated DNA sequences in yeast. Science 209: 1380–1384
Schiestl RH, Dominska M, Petes TD (1993) Transformation of Saccharomyces cerevisiae with nonhomologous DNA: illegitimate integration of transforming DNA into yeast chromosomes and in vivo ligation of transforming DNA to mitochondrial DNA sequences. Mol Cell Biol 13: 2697–2705
Schorr J, Doerfler W (1992) Non-homologous recombination between adenovirus and AcNPV DNA fragments in cell-free extracts from insect Spodoptera frugiperda nuclei. Virus Res 28: 153–170
Schröer J, Doerfler W (1995) Extensive association of adenovirus type 12 DNA with chromosomes early after the infection of hamster or human cells. (submitted)
Schulz M, Doerfler W (1984) Deletion of cellular DNA at site of viral DNA insertion in the adenovirus type 12-induced mouse tumor CBA–12-1-T. Nucleic Acids Res 12: 4959–4976
Schulz M, Freisem-Rabien U, Jessberger R, Doerfler W (1987) Transcriptional activities of mammalian genomes at sites of recombination with foreign DNA. J Virol 61: 344–353
Shapiro JA (ed) (1983) Mobile genetic elements. Academic, New York
Shinohara A, Ogawa, H, Ogawa, T (1992) Rad51 protein involved in repair and recombination in S. cerevisiae is a recA-like protein. Cell 69: 457–470
Shinohara A, Ogawa H, Matsuda Y, Ushio N, Ikeo K, Ogawa T (1993) Cloning of human, mouse and fission yeast recombination genes homologous to *RAD51* and *recA*. Nature Genet 4: 239–243

Smith AJH, Berg P (1984) Homologous recombination between defective neo genes in mouse 3T6 cells. Cold Spring Harbor Symp Quant Biol 49: 171–181
Smithies O, Gregg RG, Boggs SS, Koralewski MA, Kucherlapati RS (1985) Insertion of DNA sequences into the human chromosomal β-globin locus by homologous recombination. Nature 317: 230–234
Southern EM (1975) Detection of specific sequences among DNA fragments separated by gel electrophoresis. J Mol Biol 98: 503–517
Sprengel J, Schmitz B, Heuss-Neitzel D, Zock C, Doerfler W (1994) Nucleotide sequence of human adenovirus type 12 DNA: comparative functional analysis. J Virol 68: 379–389
Stabel S, Doerfler W (1982) Nucleotide sequence at the site of junction between adenovirus type 12 DNA and repetitive hamster cell DNA in transformed cell line CLAC1. Nucl Acids Res 10: 8007–8023
Stabel S, Doerfler W, Friis RR (1980) Integration sites of adenovirus type 12 DNA in transformed hamster cells and hamster tumor cells. J Virol 36: 22-40
Stahl, FW (1979) Genetic recombination: thinking about it in phage and fungi. Freeman, San Francisco
Stringer JR (1982) DNA sequence homology and chromosomal deletion at a site of SV 40 DNA integration. Nature 296: 363–366
Sutter D, Westphal M, Doerfler W (1978) Patterns of integration of viral DNA sequences in the genomes of adenovirus type 12-transformed hamster cells. Cell 14: 569–585
Symington LS (1991) Double-strand-break repair and recombination catalyzed by a nuclear extract of Saccharomyces cerevisiae. EMBO J 10: 987–996
Szostak JW, Orr-Weaver TL, Rothstein RJ, Stahl FW (1983) The-double-strand-break repair model for recombination. Cell 33: 25–35
Tatzelt J, Scholz B, Fechteler K, Jessberger R, Doerfler W (1992) Recombination between adenovirus type 12 DNA and a hamster preinsertion sequence in a cell-free system. J Mol Biol 226: 117–126
Tatzelt J, Fechteler K, Langenbach P, Doerfler W (1993) Fractionated nuclear extracts from hamster cells catalyze cell-free recombination at selective sequences between adenovirus DNA and a hamster preinsertion site. Proc Natl Acad Sci USA 90: 7356–7360
Thode S, Schäfer A, Pfeiffer P, Vielmetter W (1990) A novel pathway of DNA end-to-end joining. Cell 60: 921–928
Thomas KR, Folger KR, Capecchi MR (1986) High frequency targeting of genes to specific sites in the mammalian genome. Cell 44: 419–428
Tonegawa S (1983) Somatic generation of antibody diversity. Nature 302: 575–581
Trentin JJ, Yabe Y, Taylor G (1962) The quest for human cancer viruses. Science 137: 835–841
Vieira J, Messing J (1987) Production of single-stranded plasmid DNA. Methods Enzymol 153: 3–11
Wasmuth, JJ, Hall LV (1984) Genetic demonstration of mitotic recombination in cultured Chinese hamster cell hybrids. Cell 36: 697–707
Weiss EH, Mellor A, Golden L, Fahrner K, Simpson E, Hurst J, Flavell RA (1983) The structure of a mutant H-2 gene suggests that the generation of polymorphism in H-2 genes may occur by gene conversion-like events. Nature 301: 671–674
Xiong G, Schorr J, Tjia ST, Doerfler W (1991) Heterologous recombination between Autographa californica nuclear polyhedrosis virus DNA and foreign DNA in non-polyhedrin segments of the viral genome. Virus Res 21: 65–85
Yanagi Y, Yoshikai Y, Leggett K, Clark SP, Aleksander I, Mak TW (1984) A human T cell-specific cDNA clone encodes a protein having extensive homology to immunoglobulin chains. Nature 308: 145–149
Zock C, Doerfler W (1990) A mitigator sequence in the downstream region of the major late promotor of adenovirus type 12 DNA. EMBO J 9: 1615–1623
Zock C, Iselt A, Doerfler W (1993a) A unique mitigator sequence determines the species specificity of the major late promotor in adenovirus type 12 DNA. J Virol 67: 682–693
Zock C, Iselt A, Doerfler W (1993b) Role of the mitigator sequence in the species-specific expression of the major late promotor in adenovirus type 12 DNA. In: Doerfler W, Boehm, P (eds) Virus strategies. Verlag Chemie, Weinheim, pp 305–317

Post-transcriptional Control of Adenovirus Gene Expression

M.J. Imperiale[1], G. Akusjärvi[2], and K.N. Leppard[3]

1 Introduction

Adenovirus genes are expressed in a defined, temporally controlled manner during the course of a lytic infection. The mechanisms responsible for this control

[1]Department of Microbiology and Immunology, University of Michigan Medical School, Ann Arbor, MI 48109-0620, USA
[2]Department of Cell and Molecular Biology, Karolinska Institute, 171 77 Stockholm, Sweden
[3]Department of Biological Sciences, University of Warwick, Coventry CV4 7AL, UK

have been the subject of intense study. These studies have shown that, although control of transcription initiation is a major determinant of the observed pattern of viral gene expression, post-transcriptional control is also crucial to a successful outcome of infection. It is these latter mechanisms which are the subject of this chapter.

Of the many human adenovirus serotypes isolated, the group C adenoviruses types 2 (Ad2) and 5 (Ad5) have been studied in greatest detail. The genomes of these viruses have been completely sequenced (ROBERTS et al. 1984; CHROBOCZEK et al. 1992), and they show a very high level of sequence identity over most of their length and virtually identical gene organization. The limited sequence variation between these serotypes is concentrated in genome segments encoding parts of the hexon and fiber proteins and in the E3 gene (KINLOCH et al. 1984; CLADARAS and WOLD 1985; CHROBOCZEK and JACROT 1987). These sequence differences do not affect materially the processes of post-transcriptional control discussed below; data deriving from one or other of these viruses can therefore be taken to apply equally to both serotypes and, in the following discussion, these serotypes generally are not distinguished.

2 Regulation of Adenovirus Gene Expression at the Level of Alternative RNA Splicing

2.1 Exons and Introns: General Considerations

Soon after the revolutionary discovery of split genes in the adenovirus system in 1977 (BERGET et al. 1977; CHOW et al. 1977; DUNN and HASSELL 1977; KLESSIG 1977; LEWIS et al. 1977) it was shown that most viral transcription units encode not one but multiple alternatively spliced mRNAs (Fig. 1); many of these are translated into proteins that have unique biological activities (reviewed in AKUSJÄRVI et al. 1986). Furthermore, it was rapidly demonstrated that accumulation of alternatively spliced mRNAs in this system was a regulated process, with specific mRNA species accumulating at different time points during the infectious cycle. This provided the first evidence that eukaryotic gene expression may be regulated at the level of alternative RNA processing.

Much of our knowledge about the biochemistry of RNA splicing comes from studies of the major late first intron. The conclusions from these and other studies have been extensively reviewed elsewhere (M.J. MOORE et al. 1993). Only a brief summary of the sequence elements and *trans*-acting factors necessary for splicing will be given here. Pre-mRNA splicing takes place in a large ribonucleoprotein particle, the spliceosome. In addition to the pre-mRNA, spliceosomes contain the major small nuclear ribonucleoprotein particles (snRNPs) U1, U2, U4/U6, and U5 and many non-snRNP protein factors. The exon–intron boundaries, which are loosely conserved in metazoans, are defined in part by direct RNA–RNA

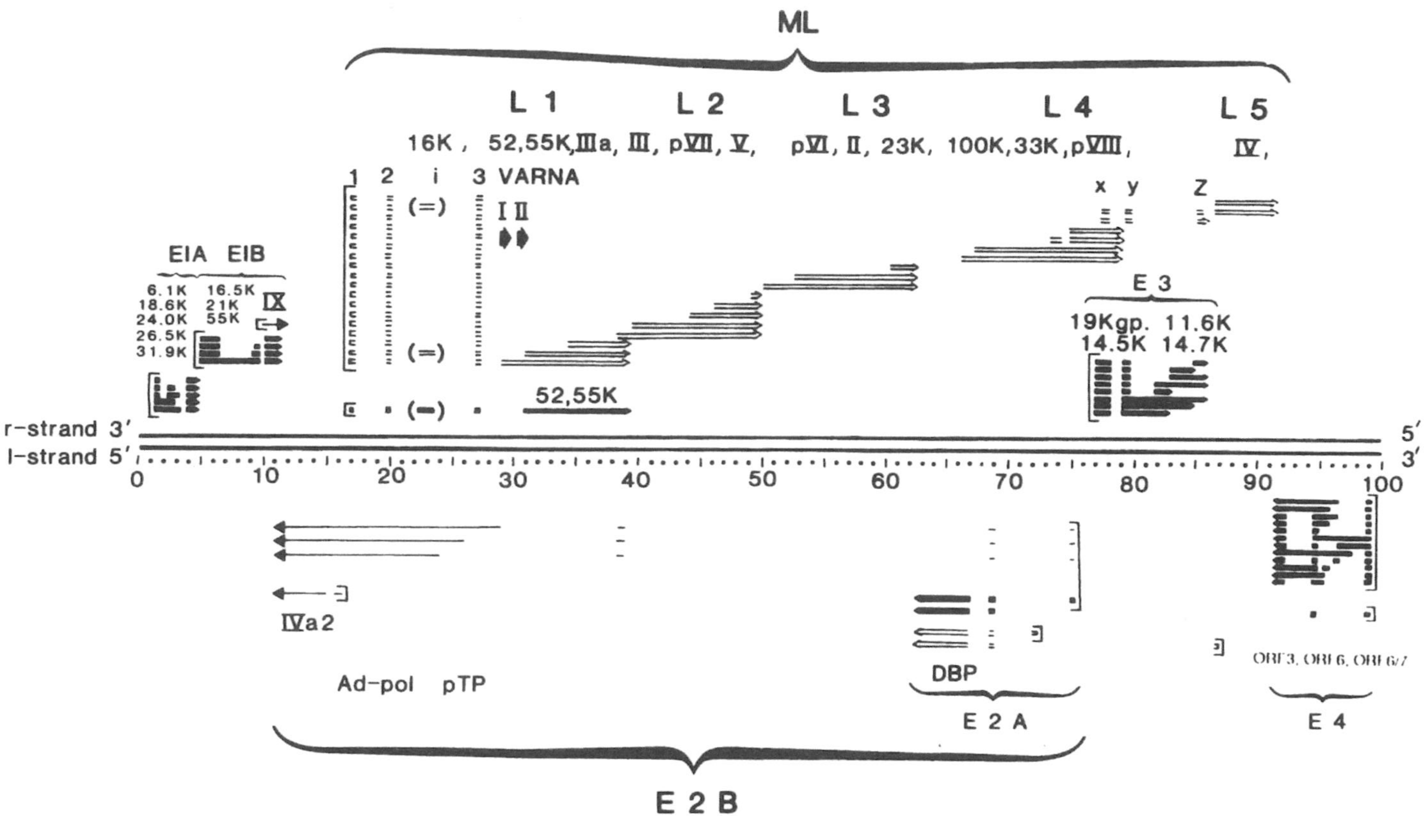

Fig. 1. Schematic map of genes and transcription units encoded by the adenovirus type 2 (Ad2)/Ad5 genomes. The linear double-stranded viral DNA is depicted in the *center* of the figure, with *r* and *l* refering to rightward and leftward transcription. *Thick lines* indicate early mRNAs and *thin lines* indicate mRNAs expressed at intermediate times of infection. *Open arrows* show mRNAs expressed late after infection. *Gaps* in the *arrows* denote the position of introns. *Arrowheads* show the locations of 3' ends of the mRNAs, and the promoter sites are indicated by *brackets*. Selected polypeptides that have been assigned to different regions are indicated. *ML*, major late transcription unit; *ORF*, open reading frame; *VARNA*, Virus-associated RNA; *gP*, glycoprotein; *DBP*, DNA-binding protein; *Ad-pol*, adenovirus DNA polymerase; *pTP*, preterminal protein. (Adapted from CHOW et al. 1979 and AKUSJÄRVI et al. 1986)

base pairing. The 5' splice site is recognized through base pairing with U1 snRNA and the 3' splice site is similarly defined by base pairing between U2 snRNA and the branch point. Assembly of the spliceosome has been shown to proceed through two stable intermediate stages, the commitment complex (or E complex) and the A complex (prespliceosome). Formation of the commitment complex, which is the earliest detectable stable precursor to the spliceosome, requires the 5' splice site and the polypyrimidine tract at the 3' splice site of the pre-mRNA. The recognition and functional interaction of the 5' splice site and 3' splice site in the commitment complex appear to require U1 snRNP, which recognizes the 5' splice site, U2AF, which binds to the polypyrimidine tract at the 3' splice site, and non-snRNP protein factors such as SC35, ASF/SF2, and/or other SR proteins. The commitment complex appears to be converted to the A complex by incorporation of U2 snRNP in a reaction that requires adenosine triphosphate (ATP) and other protein factors. The mature spliceosome is then formed by incorporation of the U4/U6-U5 triple snRNP. It is believed that formation of the commitment complex irreversibly defines the exon–intron boundaries in the pre-mRNA. Thus, formation of the commitment complex is likely to be a key step at which alternative splicing is regulated.

The adenovirus system provides an excellent opportunity to study regulatory events that are manifested at the level of RNA splicing, since most of the viral transcription units encode two or more alternatively spliced mRNA. There are numerous ways in which multi-intronic pre-mRNA can be alternatively spliced. These include multiple 5' splice sites that can be paired with a single 3' splice site, a single 5' splice site that can be paired with multiple 3' splice sites, or both ends of an exon that can be bypassed, leading to exon skipping.

Adenovirus genes contain few introns compared with cellular genes. Most viral mRNAs are matured by removal of one to three introns. The extreme examples are the polypeptide IX (pIX) mRNA, which contains no introns (Aleström et al. 1980), and the variant fiber mRNA containing the auxiliary x, y, and z leaders (Chow and Broker 1978), which requires the removal of six introns. By contrast, splicing of cellular mRNAs is often more complex, involving removal of many introns. An extreme example is the gene encoding dystrophin, which contains more than 70 introns (Mandel 1989). A valid argument is, of course, that a virus has fewer introns since it has to compress much genetic information into the viral capsid.

A second noticeable difference between the exon–intron arrangement in adenovirus genes compared to cellular genes is that the viral introns usually do not interrupt the protein-coding portion of the gene (the noticeable exception is E1A). Most often the introns are positioned in the 5' or the 3' noncoding portion of the precursor RNA. This is in striking contrast to cellular genes, where introns often interrupt the coding portion of the gene. This fragmentation of the protein-coding portion of eukaryotic genes into modules is believed to be important by allowing for assembly of new gene combinations through exon shuffling. In the virus such evolutionary considerations are probably not as important, since the virus can compensate for a lack of sophistication in gene organization by having a short replication cycle.

2.2 Temporal Regulation of Adenovirus Alternative Splicing

The potential to produce alternatively spliced mRNAs from a pre-mRNA is of great importance, since it allows for a novel mechanism to regulate adenovirus gene expression. It is clear that the temporal control of alternative splicing is of vital importance for lytic virus growth by controlling the synthesis of proteins that are needed at certain stages of the life cycle, e.g., activation of the synthesis of the mRNAs encoding the structural proteins of the viral capsid late during infection (Fig. 2). The general tendency is that shorter mRNAs produced by splicing out larger introns, accumulate at later time points of infection. For example, the E1A 13S and the E1B 22S mRNAs are the most abundant E1 mRNAs expressed at early times of infection, whereas the E1A 9S and E1B 13S mRNAs predominate at late times of infection. This shift in RNA production is a regulated process.

Although mRNA accumulation from most viral transcription units is subjected to a temporal control, current research efforts have mostly been concentrated on the mechanisms involved in the control of alternative splicing of the E1A, E3, and major late transcription units (MLTU). The conclusions reached from these studies will be briefly summarized here.

2.3 Temporal Regulation of Tripartite Leader Assembly

A number of studies have shown that the temporal shift in major late mRNA accumulation requires late viral protein synthesis. This conclusion has been reached by comparing the early and late pattern of L1 mRNA acccumulation in

Fig. 2. Physical map of the major late transcription unit. The linear viral genome is shown at the *bottom* in map units (*m.u.*). The position of the major late promoter (*MLP*) at 16.4 map units is indicated below the line. Above the line are the major exons encompassing, from left to right: the tripartite leader (*1*, *2*, and *3*) and *i* exons; *L1* (with the 52/55K and IIIa mRNA bodies indicated) *L2*, *L3*, *L4*, the *x*, *y*, and *z* leaders, and *L5*. (Modified from DEZAZZO 1990)

various adenovirus mutant-infected cells (JOHNSTON et al. 1985; PILDER et al. 1986b; SVENSSON and AKUSJÄRVI 1986; LARSSON et al. 1991) and wild type-infected cells treated with inhibitors of DNA or protein synthesis (SHAW and ZIFF 1980; THOMAS and MATHEWS 1980; AKUSJÄRVI and PERSSON 1981; NEVINS and WILSON 1981; LARSSON et al. 1991, 1992). However, it is only recently that two virus-encoded proteins with the properties expected of RNA splicing factors have been identified (NORDQVIST et al. 1994).

The MLTU generates a primary transcript of approximately 28, 000 nucleotides that can be processed into a minimum of 20 cytoplasmic mRNAs. These are grouped into five families (L1–L5, Fig. 2), where each family consists of multiple, alternatively spliced species with coterminal 3' ends (see Sect. 3). An important consequence of the processing pathway is that all mRNAs expressed from the MLTU have a common 201-nucleotide tripartite leader sequence at their 5' end. A variant form of this leader contains the 440-nucleotide i-leader exon (Fig. 2). Splicing of the i-leader is temporally regulated during infection. Thus, splicing of major late mRNA at early times of infection usually leads to the inclusion of the i-leader exon (1,2, i, 3), while the large majority of mRNAs expressed late after infection contain the classical tripartite leader (1, 2, 3). The biological significance of the i-leader exon inclusion/skipping reaction is far from clear. However, it should be noted that the i-leader encodes a stable and abundant 16-kDa protein that is expressed at intermediate and late times of infection (SYMINGTON et al. 1986). It appears to be translated predominantly from the early-transcribed mRNA, since large quantities of the 16-kDa protein accumulate in the absence of viral DNA replication (SYMINGTON et al. 1986). Although this protein is dispensable for virus growth in tissue culture cells (SOLOWAY and SHENK 1990) it may have an important function connected with virus growth in humans.

2.4 E4 Open Reading Frames 3 and 6 Are Virus-Encoded RNA Splicing Factors that Regulate Major Late Tripartite Leader Assembly

Adenovirus mutants lacking early region 4 (E4) show a very complex phenotype, which includes defects in viral DNA accumulation, late viral mRNA accumulation, and protein synthesis, as well as a failure to shut off host cell protein synthesis (HALBERT et al. 1985; WEINBERG and KETNER 1986; see also Sect. 4.6). The primary lesion responsible for these defects could be the inability of E4 deletion mutants to efficiently accumulate nuclear and cytoplasmic RNA derived from the MLTU. This deficiency is not due to reduced transcriptional activity of the major late promoter in E4 mutant-infected cells (SANDLER and KETNER 1989). More likely, E4 gene products are required for stable nuclear accumulation of the major late pre-mRNA. To this end it has been shown that E4 post-transcriptionally stimulates tripartite leader assembly both during virus growth and in a transient expression assay (NORDQVIST and AKUSJÄRVI 1990; OHMAN et al. 1993; NORDQVIST et al. 1994). Expression of a transcription unit containing either the first, second, or both tripartite leader introns is enhanced by region E4, whereas expression of a cDNA of the tripartite leader is not. The E4 enhancement works post-transcriptionally and requires efficient splicing signals. Enhancement is not strictly dependent on virus-specific sequences, although it works best with constructs containing tripartite leader introns. A surprising positional effect of the intron was observed. Thus, the intron has to be located within approximately 250 nucleotides of the 5' end of the pre-mRNA to be E4 responsive. The significance of this positional effect is still not clear, although some speculative models have been proposed (NORDQVIST and AKUSJÄRVI 1990).

The E4 transcription unit is very complex. DNA sequence analysis in combination with mRNA mapping studies have suggested that E4 encodes a minimum of seven proteins (Fig. 3; FREYER et al. 1984; TIGGES and RASKAS 1984; VIRTANEN et al. 1984). Phenotypic analysis of E4 mutants has shown that the E4 open reading frame (ORF) 3 and the E4-ORF6 proteins appear to have redundant activities during infection, and expression of either one seems to be sufficient to establish an essentially wild-type virus infection (HEMSTRÖM et al. 1988; BRIDGE and KETNER

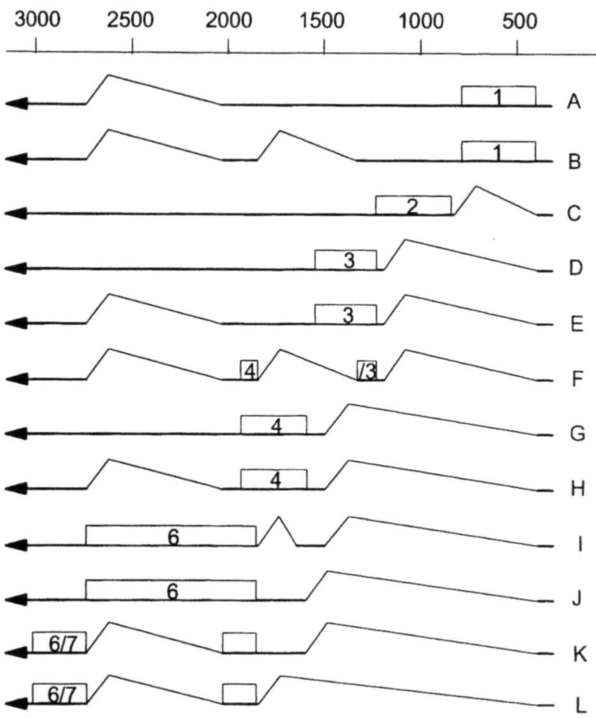

Fig. 3. Physical map of the E4 transcription unit. The linear genome is shown at the *top* with numbering in base pairs from the right end of the genome (conventional orientation). A single primary transcript gives rise to multiple mRNA species, which are indicated as *bold lines* and lettered *A–L* (other minor mRNAs formed by use of alternative splice acceptor sites in the 3′ end of the gene have also been detected, but are not shown here). The protein coding regions ORF1 to ORF6/7 are indicated as *boxes*, with those referred to in the text being *shaded*. (Based on a figure in VIRTANEN et al. 1984)

1989; HUANG and HEARING 1989; KETNER et al. 1989). It is interesting to note that the same two E4 proteins are also the E4 products that regulate major late tripartite leader assembly in a transient assay (OHMAN et al. 1993; NORDQVIST et al. 1994). The E4-ORF3 protein functions as an exon inclusion factor, facilitating i-leader exon inclusion, and the E4-ORF6 protein functions as an exon skipping activity, preferentially stimulating tripartite leader assembly (1,2,3). Preliminary experiments suggest that E4-ORF3 and E4-ORF6 also regulate alternative splicing in an in vitro system (K. OHMAN and G. AKUSJÄRVI, unpublished work).

The exon skipping activity of E4-ORF6 is not universal. For example, leader 2 is never skipped under conditions where the i-leader exon is efficiently excluded. Available data suggest that the exon skipping activity of E4-ORF6 is restricted to exons that are unusual in some respect, for example those that are unusually long (e.g., the i-leader exon: internal exons in eukaryotes are normally less than 350 nucleotides) or internal exons that have weak splice signals at their borders (NORDQVIST et al. 1994). Collectively, the data are compatible with the hypothesis that E4-ORF3 and E4-ORF6 are virus-encoded alternative splicing factors that either facilitate splice site communication across an exon (E4-ORF3) or cause exon skipping by disrupting weak 5' and 3' splice site interactions (E4-ORF6). It remains to be determined whether the E4-ORF3 and E4-ORF6 proteins are of global importance for alternative splicing of virus-specific mRNA. However, it seems unlikely that the virus would devote two proteins to regulate only i-leader splicing, especially since the i-leader protein appears to be dispensable for lytic virus growth (SOLOWAY and SHENK 1990).

E4-ORF3 and E4-ORF6 do not show any extensive primary sequence homology with other cloned metazoan RNA splicing factors. However the biological activities of E4-ORF3 and E4-ORF6 appear to be very similar to two cellular splicing activities, ASF/SF2 and hnRNP A1/DSF (HARPER and MANLEY 1991; MAYEDA and KRAINER 1992; MAYEDA et al. 1993; NORDQVIST et al. 1994). ASF/SF2 has been shown to function as an exon inclusion factor and prevents exon skipping in complex pre-mRNAs (MAYEDA et al. 1993). This activity resembles the exon inclusion activity of E4-ORF3. By contrast, hnRNP A1 has been shown to cause exon skipping in transcripts with improper exon length or in transcripts with a weak 3' splice site preceding an internal exon (MAYEDA et al. 1993). This is reminiscent of the activity of E4-ORF6. Similar to the E4-ORF6 protein, hnRNP A1 does not appear to cause inappropriate exon skipping on natural, consititutively spliced pre-mRNAs containing multiple exons (MAYEDA et al. 1993). The balanced expression of ASF/SF2 and hnRNP A1 has been suggested to play an important role in regulation of alternative splicing in eukaryotic cells. Similarly, the level of E4-ORF3 and E4-ORF6 expression during adenovirus infection is likely to have important consequences for alternative splicing in the infected cell.

2.5 Temporal Control of L1 Alternative Splicing

Expression of the L1 mRNA family represents an example of alternative splicing in which the last intron is spliced using a common 5' splice site and two alternative

3' splice sites, generating two cytoplasmic mRNAs: the 52/55K (proximal 3' splice site) and the IIIa (distal 3' splice site) mRNAs (Fig. 4). The pattern of splicing of the L1 pre-mRNA is regulated during the infectious cycle (CHOW et al. 1979; AKUSJÄRVI and PERSSON 1981; NEVINS and WILSON 1981). The 52/55K mRNA is produced both early and late after infection, whereas the IIIa mRNA is produced exclusively late (Fig. 4). Although IIIa splicing is a late-specific event, recent studies have shown that the IIIa 3' splice site can be recognized in uninfected cells both in vivo (DELSERT et al. 1989) and in vitro (KREIVI et al. 1991). The efficiency of IIIa splicing, however, is only around 10% of that of 52/55K splicing (KREIVI et al. 1991). This difference in splice site activity correlates with a reduced affinity of the IIIa 3' splice site for cellular polypyrimidine-binding proteins such as U2AF (J.P. KREIVI, unpublished data); this protein binds with a strong preference for long, pyrimidine-rich sequences, the longer the better (ZAMORE et al. 1992). It appears common in nature that one of the alternative splice sites in a regulated system has a weaker relative strength when compared to the others (SMITH et al. 1989). The 52/55K 3' splice site region is characterized by a very long polypyrimidine tract (18 pyrimidines out of 19 nucleotides), whereas the IIIa 3' splice site lacks such an extended polypyrimidine tract (KREIVI et al. 1991). Selective nuclear extraction procedures have suggested that IIIa splicing may require a distinct cellular *trans*-acting factor(s) that does not have a significant effect on splicing of transcripts which conform to the RNA 3' splice site consensus sequence (ZERIVITZ et al. 1992). However, the identity of this factor has not been determined.

The order of RNA splice site presentation is very important in L1 alternative RNA splicing. Although the IIIa 3' splice site is much weaker than the 52/55K 3' splice site, it competes efficiently with the 52/55K 3' splice site when it is presented as the proximal site in a tandem construct both in vivo (SCHMITT et al. 1987; DELSERT et al. 1989) and in vitro (KREIVI et al. 1991). Thus, the lack of IIIa splicing in early-infected cells can in part be explained by the distal location of the IIIa 3' splice site on the L1 pre-mRNA.

Efficient IIIa splicing requires late viral protein synthesis. The shift from the early pattern of exclusive 52/55K splicing to the late pattern of enhanced IIIa splicing can be reproduced in vitro, using splicing extracts prepared from the late virus-infected cells (KREIVI and AKUSJÄRVI 1994). The activation of IIIa splicing under

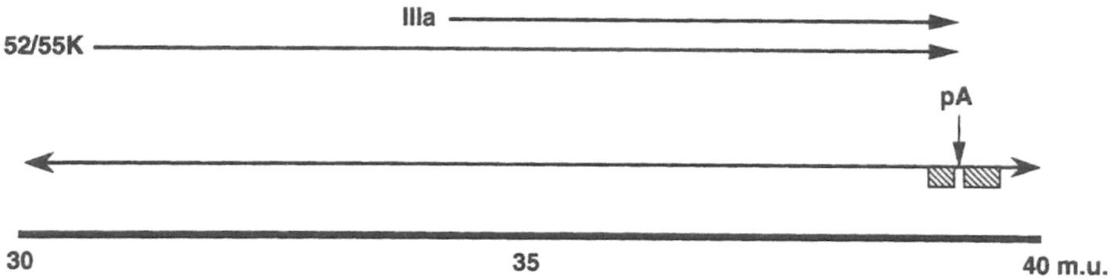

Fig. 4. Physical map of the L1 region. The linear genome is shown at the *bottom* with map units (m.u.) for reference. The *top two lines* represent the IIIa and 52/55K mRNA bodies. The *third line* shows the primary RNA transcript transversing this region, with the L1 poly (A) site regulatory regions as discussed in the text, –50 to –113 and +33 to +170 with respect to the cleavage site (*pA*), shown as *hatched boxes*

such conditions results from an enhanced efficiency of IIIa splicing in late extracts combined with a virus-induced repression of 52/55K splicing. Three important findings have emerged to explain the temporal shift in L1 3' splice site usage. First, the shift to increased IIIa splicing observed in late-infected extracts results from an increase in commitment complex formation using the distal (IIIa) 3' splite site. Second, splicing of the proximal (52/55K) mRNA is repressed in extracts prepared from late extracts. Third, the shift in L1 3' splice site usage does not require *cis* competition between splice sites, since the upregulation of IIIa splicing and inhibition of 52/55K splicing can be observed on transcripts encoding the individual 3' splice sites. Interestingly, splicing of other pre-mRNA that contain long polypyrimidine tracts are also repressed in late virus-infected extracts (. MÜHLEMANN, J.-P. KREIVI, and G. AKUSJÄRVI, submitted for publication). Thus, the shift in specificity of RNA splice site choice from pyrimidine-rich to more purine-rich 3' splice sites may be a general feature in the control of alternative splicing in late virus-infected cells.

The IIIa 3' splice site is intrinsically weak and has a short, and frequently interrupted, polypyrimidine tract (KREIVI et al. 1991) which is poorly recognized by cellular polypyrimidine-binding proteins. The drastic increase in IIIa 3' splice site recognition in late extracts may be accomplished by the presence of a virus-encoded splicing factor(s) that shows specificity for nonconsensus 3' splice sites. Alternatively, virus infection may lead to an increased synthesis or post-translational modification of a cellular 3' splice site factor. Such a hypothetical virus-induced modulation would probably not involve the general splicing factor U2AF, since the steady state concentration of U2AF is not altered in virus-infected cells (KREIVI and AKUSJÄRVI 1994). Also, the activity of U2AF in late extracts does not appear to be changed since prespliceosome formation on the 52/55K 3' splice site, which efficiently binds U2AF, appears to be unchanged in virus-infected compared to uninfected extracts (KREIVI and AKUSJÄRVI 1994). Thus, regulation of IIIa splicing through U2AF would require a modification of that protein such that its specificity is extended to include short polypyrimidine tracts. It is also possible that adenovirus induces the synthesis of, or post-transcriptionally modifies, a second cellular 3' splice site factor which stimulates the activity of nonconsensus 3' splice sites such as the IIIa 3' splice site. The existence of such a hypothetical HeLa cell 3' splice site factor is supported by the observation that uninfected HeLa cell nuclear extracts contain a distinct *trans*-acting factor that stimulates IIIa splicing but does not appear to have a significant effect on 52/55K or β-globin splicing (ZERIVITZ et al. 1992).

2.6 Regulation of E1A Alternative Splicing

Considerable efforts have been spent in attempts to unravel the mechanism of adenovirus E1A alternative splicing. The E1A pre-mRNA gives rise to three major mRNAs, the 13S, 12S, and 9S mRNAs (Fig. 5), by alternative use of three 5' splice sites (BERK and SHARP 1978; CHOW et al. 1979; PERRICAUDET et al. 1979), and two

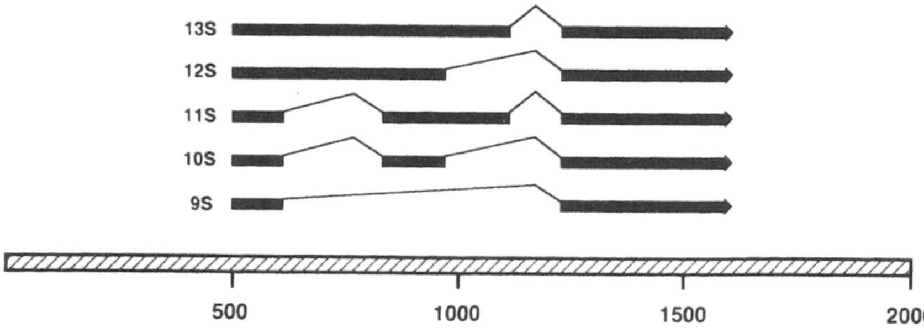

Fig. 5. Physical map of the E1A transcription unit. *Black arrows* represent the exons and *thin lines* represent the introns. For reference, the relevant part of the linear viral DNA is shown at the *bottom* with numbering in base pairs from the left end of the genome. (Adapted from ULFENDAHL et al. 1987; STEPHENS and HARLOW 1987)

minor mRNAs, the 10S and 11S mRNAs, that were first detected in vitro (SCHMITT et al. 1987) and later shown to be predominantly expressed late after infection (STEPHENS and HARLOW 1987; ULFENDAHL et al. 1987). They differ from the 13S and 12S mRNAs in that they are formed by the removal of an additional intron; the 9S 5' splice site is joined to a novel 3' splice site located upstream of the 12S 5' splice site. The inefficiency of 10S and 11S mRNA production appears to result from the unusually long distance between the branch point sequence and the 3' splice site in the first intron (more than 50 nucleotides; GATTONI et al. 1988; CHEBLI et al. 1989); in eukaryotes the distance between the branch site and the splice acceptor is usually between 18 and 40 nucleotides (M.J. MOORE et al. 1993). Evidence has also been presented that the 13S 5' splice site (POPIELARZ et al. 1994) and a hairpin structure, which shortens the operational distance between the branch site and the 3' splice site (CHEBLI et al. 1989), are important *cis*-acting elements for first intron splicing. Mutations that destroy or weaken this hairpin result in a reduction of 10S and 11S splicing, whereas mutations that strengthen this hairpin improve the splicing efficiency (CHEBLI et al. 1989).

The 9S mRNA is predominantly expressed late after infection and is not detected under standard in vitro splicing conditions in uninfected extracts (SCHMITT et al. 1987). A shift to 9S mRNA production can be induced in several ways. For example, reduction of the ionic strength in uninfected extracts has been shown to partially activate 9S splicing (SCHMITT et al. 1987). The significance of this for regulation of E1A splicing in vivo is not clear. Since 5' splice sites are recognized in part by RNA–RNA interactions, such an effect may result from differential destabilization of hairpins or U snRNA/pre-mRNA interactions. Also, the relative concentration of two pairs of cellular splicing factors, hnRNP-A1 and ASF/SF2 (MAYEDA and KRAINER 1992) or DSF and ASF/SF2 (HARPER and MANLEY 1991), have been shown to modulate 9S splicing in vitro. A high concentration of ASF/SF2 favors 13S 5' splice site use, whereas a high concentration of hnRNP A1 or DSF activates 9S 5' splice site usage. Changes in the relative concentration of general splicing factors which have antagonistic effects on splice site choice are a potentially important way to regulate E1A alternative splicing in vivo. Since E4-ORF3 and E4-ORF6 have analogous biological activities to ASF/SF2 and hnRNP A1 (NORDQVIST et al. 1994), they are candidate factors for regulating E1A alternative splicing during infection. Most interestingly, extracts prepared from late adenovirus-infected cells have been shown to have an increased capacity to splice the 9S mRNA (GATTONI et al. 1991). This effect did not require late viral proteins and could be reproduced by addition of the nuclear RNA fraction prepared from the late extracts. This modulation of alternative E1A splicing has been explained by a sequestering of general splicing factor(s) by high molecular weight nuclear RNA expressed from the MLTU (GATTONI et al. 1991). This RNA fraction is rich in RNA splice sites and may efficiently compete for binding of cellular splicing factors. The same conclusion concerning regulation of E1A alternative splicing was reached based on in vivo experiments using adenovirus infections (LARSSON et al. 1991).

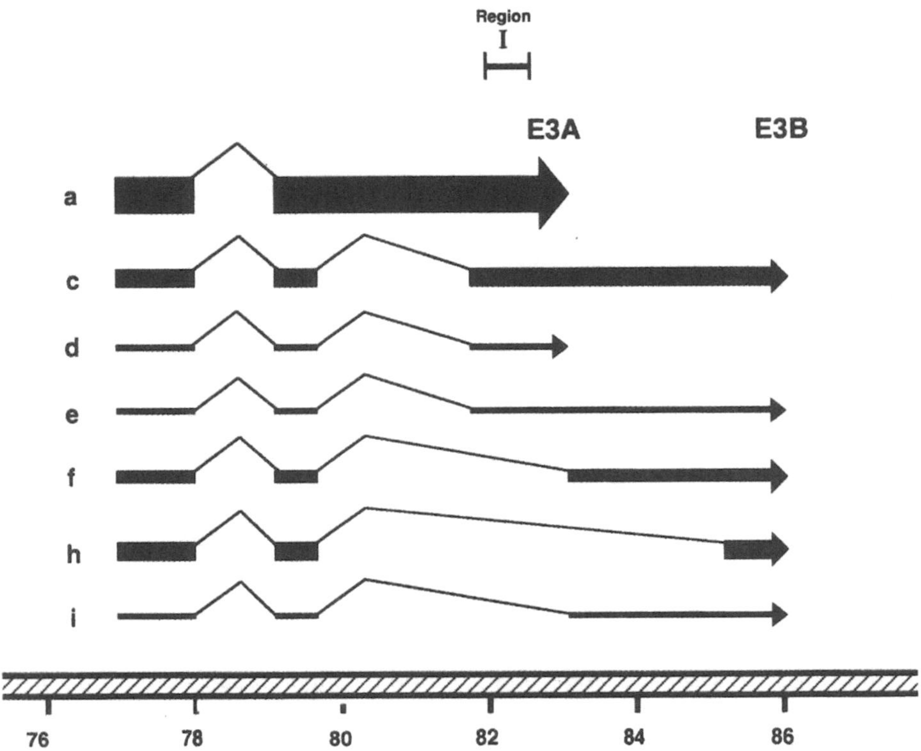

Fig. 6. Physical map of the E3 transcription unit. *Black arrows* represent the exons of mRNA a–i, with the *thickness* of the arrow indicating the relative abundance of the mRNA. *Thin lines* indicate the introns removed during mRNA maturation. *E3A* and *E3B* denote two alternative polyadenylation sites used during E3 mRNA formation. For reference, the relevant part of the linear viral DNA is shown at the *bottom* with numbering in map units. *Region I* designates the position of the splice suppressor sequence discussed in the text. (Adapted from WOLD and GOODING 1991)

2.7 Regulation of E3 Splicing

Region E3 is embedded within the MLTU (reviewed in AKUSJÄRVI et al. 1986). It expresses approximately nine alternatively spliced mRNA which are polyadenylated at one of two sites, thus generating the E3A and the E3B families of mRNAs (Fig. 6). There are nine predicted E3 proteins in group C adenoviruses, most of which have been detected in infected cells by immunoprecipitation (reviewed in WOLD and GOODING 1991). Most appear to have a function in virus–host cell interactions or in combating immunoserveillance (reviewed by W.S. WOLD et al., this volume). Accumulation of E3 mRNAs is subject to complex regulation. At early times after infection mRNAs a, c, f, and h are the most abundant E3 mRNAs. Late after infection transcription from the E3 promoter is drastically reduced (NEVINS et al. 1979; BHAT and WOLD 1986) and E3 mRNA synthesis is directed by the major late promoter. At this stage the mRNA profile changes such that mRNAs d and e (containing the tripartite leader) are the most abundant E3 mRNAs (TOLLEFSON et al. 1992).

E3 mRNA expression has been analyzed both in vivo (summarized below) and in vitro (DOMENJOUD et al. 1991). An impressive number of mutant viruses have been constructed to study the control of E3 mRNA accumulation. These studies have shown that multiple *cis* sequences around the E3A polyadenylation site have a crucial effect on E3 mRNA processing (SCARIA and WOLD 1994). The major conclusions are summarized here. A 129-nucleotide sequence element, designated region I (Fig. 6), that increases synthesis of mRNAs a and c has been structurally characterized (BRADY et al. 1992; SCARIA and WOLD 1994). This element is believed to function as a splice suppressor sequence, since region I deletion mutants show a dramatic increase in synthesis of mRNAs f and h at the expense of mRNAs a and c. The easiest interpretation of these studies would be that the outcome of E3 processing represents a simple competition between splicing (production of mRNAs f and h) and polyadenylation (production of mRNA a) at the E3A poly(A) (polyadenylation) site. However, this simple interpretation does not seem to be true, since region I mutants produce mRNA d with wild-type efficiency (SCARIA and WOLD 1994), suggesting that the E3A polyadenylation site is indeed functional. Collectively, the mutant studies have been interpreted to indicate that region I may have a specific function as a splice suppressor sequence favoring mRNAs a and c production and as a consequence efficient E3 19K production. Since this protein works stoichiometrically in reducing cell surface expression of major histocompatibility complex class I molecules, its production has to be maintained at a high level during infection. Similar splice suppressor sequences have been shown to increase the synthesis of unspliced mRNA in other systems, such as the genomic gag-pol RNA in Rous sarcoma virus (MCNALLY and BEEMON 1992). This conclusion does not exclude the possibility that competition between splicing and polyadenylation plays a role in E3 mRNA production. The 3' splice site for mRNA f is positioned four nucleotides upstream of the E3A AAUAAA hexanucleotide, and thus competition between binding of splicing and polyadenylation factors at this site is likely to regulate mRNA f production. Indeed, a mutation in which the AAUAAA sequence was changed to ATCGAT completely abolished E3A poly(A) site usage and resulted in almost exclusive mRNA f production (BRADY and WOLD 1988). However, in the natural context of the E3 promoter this *cis* competition does not appear to be sufficient to explain the phenotype of region I mutants (SCARIA and WOLD, 1994).

The mechanism by which region I controls E3 mRNA processing is not yet known. It is noteworthy that region I is not located near a conventional splice signal or polyadenylation signal. The speculation is that region I binds a cellular or viral *trans*-acting factor that regulates E3 RNA splice site choice (SCARIA and WOLD, 1994). However, so far no obvious sequence homology between region I and binding sites for cellular or viral *trans*-acting factors has been found.

2.8 Effect of Viral DNA Replication on Alternative RNA Splice Site Choice

Although the evidence indicates that the temporal shift in viral mRNA profile is dependent on viral protein synthesis, recent data have shown that viral DNA replication per se plays an important role in the shift from the early to the late pattern of E1A, E1B, and L1 mRNA accumulation in vivo (ADAMI and BABISS 1991; LARSSON et al. 1991). However, in the absence of late viral protein synthesis this shift was not complete, suggesting that viral proteins play a major regulatory role. The requirement for late viral protein synthesis differs among E1A, E1B, and L1 alternative splicing, with L1 showing the most pronounced dependence on viral protein translation. Without efficient late translation the L1 mRNAs were incompletely spliced, with a large fraction of the cytoplasmic mRNAs retaining the i-leader exon (LARSSON et al. 1991).

The effect of viral DNA replication on the control of late mRNA structure has been explained by a sequestration model in which synthesis of large quantities of viral transcripts results in titration of one or more limiting cellular splicing factors (LARSSON et al. 1991). High molecular weight nuclear RNA expressed from the viral genome is rich in RNA splice sites and is proposed to preempt the cellular splicing factors that are necessary to maintain the early splice pattern.

3 Regulation of mRNA 3' End Formation

3.1 Cleavage and Polyadenylation of 3' Ends: General Considerations

The regulation of gene expression by the alternative use of poly(A) sites is relatively rare compared to regulation by alternative splicing. In adenovirus, such regulation is most evident in the MLTU. We use the term "3' end formation" to refer to the two-step process of endonucleolytic cleavage of the pre-mRNA followed by addition of the poly(A) tail. Much of our understanding of the biochemistry of 3' end formation has come from studies on the adenovirus L3 poly(A) site. This has been reviewed in depth elsewhere (MANLEY 1988; WICKENS 1990; WAHLE and KELLER 1992), but a brief description of the sequences and factors involved is in order here. The pre-mRNA molecule usually contains three sequence elements that make up the core poly(A) site. These are the cleavage site itself, the highly conserved hexanucleotide AAUAAA, found just upstream of the cleavage site, and a G+U- (or U-) rich sequence located just downstream of the cleavage site. At least four biochemically definable factors are involved in 3' end formation. The first factor, CPSF (cleavage and polyadenylation specificity factor), binds to the poly(A) site through recognition of the AAUAAA. This interaction is unstable, however. Binding of a second factor, CStF (cleavage stimulatory factor), is mediated through the G+U-rich element and results in stabilization of the processing complex and commitment to proceed with the reaction, which also requires the poly(A) polymerase and at least one other accessory factor. Cleavage occurs, and the poly(A) tail of approximately 200 residues is added to the newly formed 3' end. It is interesting to note that, unlike splicing, 3' end processing does not appear to require snRNP or other RNA components except for the substrate itself.

3.2 Expression from the Major Late Transcription Unit Late in Infection

The MLTU encodes five poly(A) sites, one for each family of mRNA, L1–L5 (LE MOULLEC et al. 1983; see Fig. 2). During the late phase of the infection, RNA polymerase transcribes the entire MLTU each time it initiates (with the exception of some transcripts that are prematurely terminated very close to the promoter; EVANS et al. 1979), terminating somewhere just short of the right end of the genome (FRASER et al. 1979; DRESSLER and FRASER 1987). Pulse-labeling experiments performed in the 1970s showed that each of the five poly(A) sites is used at roughly the same frequency, with the L1 site being used slightly less than the other four (NEVINS and DARNELL 1978). These same studies demonstrated that 3' end formation could occur soon after the poly(A) site was transcribed, before the RNA polymerase had reached the end of the MLTU. Thus, the promoter proximal sites should have a temporal advantage and be used more frequently than the distal sites. The reason they do not have such an advantage is not clear, but it has been proposed that if the promoter-distal poly(A) sites were stronger sites, this could offset the effect of distance (NEVINS and DARNELL 1978). Support for this model comes from the recent finding that the L3 site competes more effectively for processing factors than the L1 site (PRESCOTT and FALCK-PEDERSEN 1992; MANN et al. 1993). Early studies also indicated that use of the L1 poly(A) site might be mechanistically different than use of the other sites (MANLEY et al. 1982). In these experiments, nuclei isolated from late-infected cells were pulsed with radiolabeled uridine triphosphate (UTP), transcription was blocked, and the RNA was chased into products. What was found was that the L1 site differed from the other sites in that its use required ongoing transcription, thus appearing to be more of a cotranscriptional than post-transcriptional event.

3.3 Expression from the Major Late Transcription Unit Early in Infection

Another layer of regulation of poly(A) site choice became apparent when it was discovered that the major late promoter is also active during the early phase of infection, giving rise to a different pattern of poly(A) site use (SHAW and ZIFF 1980; AKUSJÄRVI and PERSSON 1981; NEVINS and WILSON 1981). These studies indicated that the L1 poly(A) site is used almost exclusively prior to the onset of DNA replication. Moreover, the L4 and L5 regions are not transcribed during the early infection by RNA polymerases initiating at the major late promoter; instead, these polymerases terminate just upstream of the E3 promoter (IWAMOTO et al. 1986).

The significance of this change in poly(A) site choice is not completely understood. One can speculate, however, that early in the infection the virus uses the major late promoter to drive synthesis of the L1 52/55K protein, which is not a structural protein yet is required for viral assembly (HASSON et al. 1989), in order to set the cell up to produce progeny virions. Late in the infection, then, the virus switches to production of the virion components and assembles them. In addition, it is possible that transcription termination early in the infection is required to prevent occlusion of the E2 and E3 promoters by polymerases initiating at the major late promoter.

Recently, a change in poly(A) site use at intermediate times after infection has been described (LARSSON et al. 1992). In these experiments, protein synthesis inhibitors were used to block translation late in the infection. Under these conditions, only L1 and L4 mRNAs accumulate, implying that the use of the L2, L3 and L5 poly(A) sites may require one or more late viral factors.

3.4 Competition Between Polyadenylation Sites

The simple explanation for how the L1 site is used predominantly early in the infection is the "first come, first served" mechanism described above. The finding that a miniature MLTU (mini-MLTU), containing only the L1 and L3 poly(A) sites under the control of the major late promoter, exhibits a similar switch in poly(A) site choice implies that regulation is accomplished by simple competition between the poly(A) sites, with no contribution from splice sites (FALCK-PEDERSEN and LOGAN 1989). This fits with the observation that 3' end processing precedes splicing in vivo (NEVINS and DARNELL 1978). These results have also led to the analysis of the *cis*-acting sequences that are responsible for the predominant use of the L1 site. A mutational study of the L1 region indicated that there are sequences both upstream of the AAUAAA signal and downstream of the G+U-rich element that are required for its preferential use early in the infection (Fig. 4; DeZazzo and Imperiale 1989). These same sequences are necessary for a change in relative use late in the infection (DeZazzo et al. 1991). Deletion of either of these regulatory elements abolishes both effects. Although the mechanism by which these sequences act is not fully understood, progress is being made. For example, these sequences only function in *cis*: if the L1 poly(A) site is the only poly(A) site on the assay construct, deletion of these elements has no effect on 3' end processing (Hales et al. 1988; DeZazzo and Imperiale 1989). This result is in agreement with the finding that if a mini-MLTU containing only the L1 site is cloned onto one adenovirus chromosome, a second mini-MLTU containing only the L3 site is cloned onto another, and coinfection with the two viruses is carried out, regulation is not seen (Falck-Pedersen and Logan 1989). Moreover, these same studies demonstrated that if cells are infected, the infection allowed to enter the late phase, and a second strain then superinfected, the superinfecting virus still gives an early pattern of poly(A) site use. Thus, the switch from early to late requires expression from a replicated template and, if it is mediated by *trans*-acting factors, these factors are not freely diffusible in the nucleus. In this regard, the compartmentalization of viral DNA during replication may be involved in regulation of processing in that the local concentration of 3' end processing factors may be altered (Moyne et al. 1978; Bodnar et al. 1989; Walton et al. 1989; Moen et al. 1990). A redistribution of splicing factors does appear to occur during infection (Bridge et al. 1993).

3.5 Possible Mechanisms of the Early-to-Late Switch

These *cis*-acting elements are also functional in vitro. When an L1 poly(A) site containing the regulatory sequences is competed in *cis* in a combined in vitro transcription-processing reaction against either another L1 site without these sequences or against the L3 site, the L1 site with the regulatory sequences is used predominantly (Wilson-Gunn et al. 1992). Interestingly, however, when the L1 and L3 sites are competed against each other in *trans*, or in *cis* on a presynthesized pre-mRNA, the L3 site is preferred (Prescott and Falck-Pedersen 1992; . Kipatrick and M.J. Imperiale, unpublished work). Therefore, it appears that when both sites are simultaneously presented to the processing machinery, L3 is the stronger site, but if the L1 site is presented first, as it would be in the combined transcription-processing reaction, the temporal advantage overcomes the lower inherent strength of the L1 site. This model is supported by the recent findings that the L3 site has a higher affinity for 3' end processing factors than does the L1 site (Prescott and Falck-Pedersen 1992; Mann et al. 1993). The studies described earlier indicating that use of the L1 site might be more tightly linked to transcription than use of the other sites might be of relevance in this regard (Manley et al. 1982). For example, perhaps differences in the template early and late in the infection result in differences in polymerase elongation rates and, subsequently, L1 site use.

The question, then, is the following: what are the *trans*-acting processing factors that either recognize these sequences directly or whose binding to other sequences is influenced by these regulatory elements? The L3 and L1 sites clearly compete for common factors (PRESCOTT and FALCK-PEDERSEN 1992; MANN et al. 1993), but there is no evidence to date for specific factors that interact exclusively with the L1 elements. A recent report looked at the concentration of processing factors early and late in an adenovirus infection (MANN et al. 1993). It was found that the level of CStF changes during the infection. These investigators also showed a slight decrease in the stability of processing complexes bound to a mutant L1 site, but it is unclear whether this can account for the more dramatic switch in poly(A) site choice. The continued use of in vitro systems will likely gain us more insight into these issues.

4 mRNA Transport and Stability

4.1 mRNA Transport: General Considerations

The term "RNA transport" embraces a number of events in eukaryotic gene expression that intervene between the completion of mRNA maturation in the nucleus and RNA entry into the actively translating cytoplasmic mRNA pool. These transport events include changes in the profile of proteins bound to the RNA (KUMAR and PEDERSON 1975; reviewed by DREYFUSS et al. 1993) and changes in the interaction of the RNA (as ribonucleoprotein) with a structure known as the nuclear matrix (BEREZNEY and COFFEY 1977; CIEJEK et al. 1982; reviewed by SCHRODER et al. 1987), as well as the translocation of the RNA from its site of synthesis to and then through a nuclear pore to enter the cytoplasmic compartment. Clearly, changes affecting any of these events could result in an altered rate of accumulation of RNA in the cytoplasmic pool of translatable mRNA. Furthermore, if there were sequence preference or specificity in any aspect of RNA handling through the transport pathway or if individual transcription units differed in their linkage to the transport apparatus, then this phase of gene expression would be a possible target for differential control. As the following discussion demonstrates, such control is evident during adenovirus infection, serving to facilitate the expression of certain viral genes.

4.2 E1B Mutant Viruses Deficient in Late Gene Expression

The adenovirus E1B gene, which encodes two major proteins of 55K and 19K (Fig. 7; PERRICAUDET et al. 1980; VAN ORMONDT et al. 1980; BOS et al. 1981), was an early target of directed mutagenesis experiments, largely because of its role in adenovirus-mediated cell transformation. The resulting mutants have also been studied to determine the role of E1B proteins in lytic infection. Deletion mutations

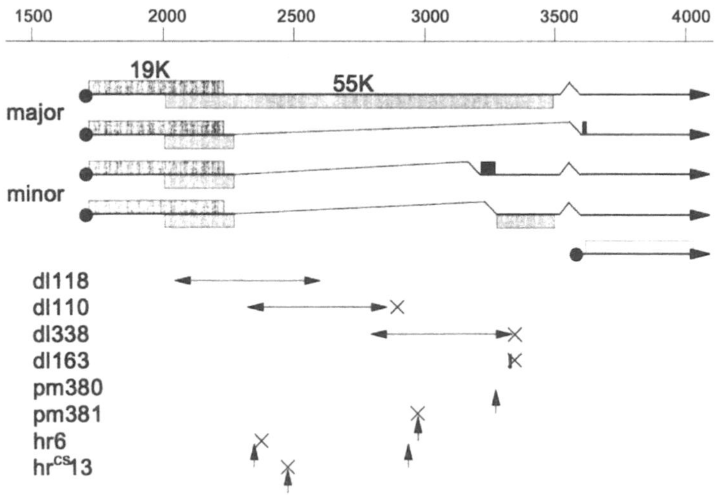

Fig. 7. Physical map of the E1B transcription unit. The linear genome is shown with numbering in base pairs from the left end of the genome. *Filled circles* indicate transcription start sites. A single primary transcript gives rise to multiple mRNA species that are indicated as *bold lines*. A second promoter (IX), nested within the E1B region, is also indicated. *Boxes* indicate the coding regions in each mRNA: *hatched*, E1B 55K protein or related sequences; *shaded*, E1B 19K protein; *black*, other potential E1B coding segments; *open*, polypeptide IX (pIX). Below the transcripts the genome lesions are indicated in various E1B mutants used in the characterization of the E1B 55K protein (see text for references). The point of translation termination resulting from nonsense or frame-shift mutations is indicated by an *X*. The absence of this symbol indicates a missense or in-frame mutation

specifically disrupting expression of the 55K protein were shown to cause an impaired growth phenotype in HeLa cells (Babiss and Ginsberg 1984; Logan et al. 1984). This defect was attributed directly to reduced late viral protein synthesis, since neither early gene expression nor viral DNA replication was affected by these mutations. Analysis of cytoplasmic mRNAs levels further showed that this defect in late protein synthesis was a reflection of reduced levels of the relevant cytoplasmic mRNAs; the reduced rate of hexon protein synthesis matched the reduction in hexon mRNA levels (Babiss et al. 1985). Classical Ad5 point mutants (Harrison et al. 1977; Ho et al. 1982), shown subsequently to carry specific E1B 55K mutations, gave similar results (Williams et al. 1986). The locations of mutations within the E1B 55k protein used in the above studies are summarized in Fig. 7.

Contemporaneous with these basic mutational analyses, E1B gene expression was shown to be somewhat more complex than had been thought. A previously described 18K protein was shown to comprise the N and C termini of E1B 55K with the central portion removed (Anderson et al. 1984), and mRNAs able to encode this and a further predicted 55K-related protein were detected (Virtanen and Pettersson 1985). Some of the specific E1B 55K mutations, including *dl*338, on which much work has been done subsequently, affect synthesis of one or both of these proteins as well as 55K. However, a point nonsense mutant *pm*381, which affects only 55K and is otherwise isogenic with *dl*338 (Fig. 7), showed an identical phenotype to *dl*338 (LEPPARD et al. 1987), while splice site mutants unable to make some or all of the 55K-related E1B polypeptides showed no growth defect (MONTELL et al. 1984; PILDER et al. 1986a). Thus, the E1B 55K-related proteins apparently have no essential unique role in lytic infection, and the lack of these proteins does not contribute to the phenotype of those E1B 55K deletion mutants in which their synthesis is affected.

4.3 A Defect in RNA Transport Underlies the E1B 55K Mutant Phenotype

Three studies, each employing different E1B 55K mutants, have concluded that the requirement for 55K in viral late gene expression occurs at the level of mRNA transport or RNA stabilization postmaturation (BABISS et al. 1985; PILDER et al. 1986b; WILLIAMS et al. 1986). In each case, the rate of entry of specific newly synthesized, late viral mRNA into the cytoplasm was found to be reduced by the 55K mutation and this effect was seen from the earliest stages of viral late gene expression; while these findings could have resulted from effects on preceding steps in the pathway of mRNA production rather than from a direct effect on mRNA transport, such alternative explanations were excluded by other experiments. In particular, the transcription rate across the MLTU was only minimally reduced by the absence of the 55K protein, at a stage of infection at which a considerable reduction in cytoplasmic late viral mRNA levels was already apparent (BABISS et al. 1985; PILDER et al. 1986b); those transcription rate differences that do emerge during the late phase of infection are likely, therefore, to be secondary consequences of the mutant phenotype rather than its primary cause. Also, steady state levels of selected mature poly(A)+ viral late mRNA in the nucleus, as determined by Northern analysis, were unaltered in 55K-mutant infections, suggesting that RNA processing is not the primary site of E1B 55K action (BABISS et al. 1985; PILDER et al. 1986b; WILLIAMS et al. 1986).

The rate of accumulation of an mRNA in the cytoplasmic pool depends on its stability as well as its intrinsic rate of transport out of the nucleus. PILDER et al. (1986b) showed that L5 cytoplasmic mRNA stability was unaltered in the absence of 55K at the beginning of the late phase, but was reduced somewhat later in the late phase. WILLIAMS et al. (1986), using different methodology and conditions, found some reduction in hexon mRNA half-life during the late phase. In neither case could these effects account, either in magnitude or in timing, for the observed reduction in cytoplasmic mRNA levels or rates of accumulation, thus sustaining the conclusion that the principal target of E1B 55K action was mRNA transport.

One study has sought to pinpoint directly the site of action of E1B 55K within the transport pathway (LEPPARD and SHENK 1989). Here, a reduced temperature incubation was used, both to slow down the events of the transport pathway and to exacerbate the 55K-negative phenotype (certain E1B 55K mutants show cold sensitivity for lytic growth, Ho et al. 1982; *dl*338 was found to display an extended eclipse phase and reduced final yield at 32°C in comparison with wild-type virus, LEPPARD and SHENK 1989). The movement of specific viral mRNA through various cellular compartments in either wild type- or mutant-infected HeLa cells was monitored by linking a standard in vivo pulse labeling and chase to a fractionation protocol designed to separate cytoplasm, nuclear membrane, soluble nucleoplasm, and nuclear matrix. In the absence of E1B 55K, late viral RNAs were released more slowly from the nuclear matrix (the site of transcription and processing) and arrived later in the nuclear membrane fraction. There was also a striking failure of specific late RNA accumulation in the soluble nuclear fractions, taken to indicate dramatically increased turnover of these RNAs in either this or preceding compartments in the kinetic pathway. From these observations it was concluded that E1B 55K facilitated an early step in the transport pathway for late viral RNA, possibly the release of mature mRNA from the nuclear matrix (i.e., entry into the transport pathway), and that in the absence of such release, RNA was degraded within the nucleus. A role for E1B 55K in the early stages of the transport pathway is in agreement with data localizing the protein to discrete intranuclear sites associated with viral replication and transcription rather than with nuclear pore complexes (ORNELLES and SHENK 1991).

4.4 Selectivity of the E1B 55K RNA Transport Function

A central feature of the E1B 55K mutant virus phenotype is the absence of any effect on early gene expression other than that directly consequent upon the E1B mutation. Steady state levels of cytoplasmic mRNA from selected early genes at early times were not affected by the absence of 55K function (BABISS and GINSBERG 1984; PILDER et al. 1986b; WILLIAMS et al. 1986) and a systematic analysis, by RNase protection, of the cytoplasmic levels of most of the individual mRNA species produced at early times has confirmed the generality of this result (LEPPARD 1993 and unpublished data). Interestingly, this finding also extends to L1 52/55K mRNA, the only major late promoter-derived mRNA produced at early times, although at later times postinfection, accumulation of this mRNA becomes strongly E1B 55K-dependent. All mRNAs expressed from the adenovirus genome during the early phase of infection seem, therefore, to be transported to the cytoplasm in an E1B 55K-independent manner.

The original demonstrations of defective late mRNA accumulation in E1B 55K mutants employed Northern analysis with probes specific for selected segments of the MLTU (BABISS and GINSBERG 1984; PILDER et al. 1986b; WILLIAMS et al. 1986). In addition to the MLP, three other promoters, IX, IVa2, and E2-L, are activated in the late phase of infection (Fig. 1). The cytoplasmic levels of mRNAs derived from each of these promoters also show some dependence on E1B 55K function (LEPPARD 1993), although the reduction in level when 55K is absent is less than for the MLTU-derived mRNAs. Therefore, dependence on the E1B 55K function is not restricted to MLTU-derived RNAs.

Many adenovirus recombinants which express heterologous gene constructs during late infection have been isolated in various contexts. In general, these heterologous genes have not been characterized for expression in an E1B 55K-deficient background. However, PILDER et al. (1986b) showed that viruses expressing a second copy of the adenovirus L5 region under heterologous major late promoter control did so in an E1B 55K-dependent manner. In contrast, MOORE and SHENK (1988) noted that herpes simplex virus *tk* gene expression from a recombinant virus under major late promoter control did not show any E1B 55K-dependence. Further examples will need to be analyzed before any general conclusion can be drawn concerning the expression of such heterologous constructs in the absence of E1B 55K.

The initial studies of E1B 55K mutants suggested that the levels of defect in expression of RNA from specific MLTU segments might not all be equivalent (PILDER et al. 1986b; WILLIAMS et al. 1986). Further analysis of the levels of specific mRNA species from L1, L2, and L3 revealed that within each segment it was the longest mRNA of the 3' coterminal set that was most strongly dependent on 55K function, with the steady state levels of the shorter members being correspondingly less dependent on, or even independent of, this activity (LEPPARD 1993). Further evidence for the selective dependence of particular mRNA species on normal E1B 55K function came from a study of the regulated expression of the E4 region. This gene transcript is differentially spliced to give at least 14 mRNAs (Fig. 3) encoding possibly seven distinct proteins (FREYER et al. 1984; TIGGES and RASKAS 1984; VIRTANEN et al. 1984). Certain of these mRNAs, such as C, D, G and J, are produced during the early phase, while others (A, E, H, K, L) appear later, dependent on DNA replication (TIGGES and RASKAS 1984; ROSS and ZIFF 1992; DIX and LEPPARD 1993); the latest of these to appear is mRNA A. The cytoplasmic level of this mRNA, alone among the multiple E4 mRNA species examined, was strongly dependent on normal E1B 55K function (DIX and LEPPARD 1993). In summary, these data indicate that, while most mRNAs derived from the adenovirus genome at late times show some degree of dependence on E1B 55K for cytoplasmic accumulation, there is considerable variation in the extent of that dependence, even between mRNAs closely related in sequence (for example E4 mRNAs A and L).

What determines whether a late viral mRNA will be strongly or weakly dependent on E1B 55K function? One hypothesis is that it is the presence in a mature mRNA of either unused splice donor and/or acceptor sites or potential intron sequences that renders an mRNA strongly dependent on E1B function (LEPPARD 1993; DIX and LEPPARD 1993). mRNAs retaining such features may be recognized as incompletely processed and thus retained in the nucleus by host cell systems; such systems certainly operate in the uninfected cell (CIEJEK et al. 1982), and viruses that need such mRNAs to be translated might have to specify a function specifically to overcome this retention mechanism. This model fits nicely the observed pattern of selectivity in both E4 and the MLTU, since all the mRNAs from these genes that are strongly dependent on E1B 55K retain one or more unused splice sites and hence also retain potential intron sequences. In contrast, the less dependent mRNAs, which are expressed at late times from these genes, retain no or fewer splice signals, with those that are present being generally recognized inefficiently during infection (i.e., the resulting mRNAs are low in abundance). The observation that, in human immunodeficiency virus gene expression, the cytoplasmic accumulation of mRNAs containing unused splice sites critically depends on a viral *trans*-acting function (reviewed in CHANG and SHARP 1990) supports the idea that similar mechanisms may be required in other systems.

If E1B 55K is needed to counteract the effects of a nuclear retention system on certain viral mRNAs which contain splicing/intron sequences, why do viral mRNAs without these features (e.g., IVa2; Fig. 1) show a low level of dependence on this function? One hypothesis, which unites all the above observations, is that late viral transcription, in contrast to early transcription, occurs in an environment which is poorly coupled to the RNA transport pathway. This circumstance might result from the saturation of the capacity of the infected cell for gene expression through the normal, efficient pathways in the late phase of infection. It is known that replication and late gene expression occur within viral inclusions within the nucleus in which viral DNA templates reach high concentrations (MOYNE et al. 1978; BODNAR et al. 1989; WALTON et al. 1989; MOEN et al. 1990). Under such citcumstances, most or all mRNAs produced from late viral transcription complexes would be released from the nuclear matrix into the transport pathway inefficiently and so their rate of cytoplasmic accumulation would be enhanced by a viral function which facilitated that release. Those mRNAs most strongly retained (above) would depend most on that function, suggested to be provided by E1B 55K. Clearly, much further investigation is needed to test the validity of this model.

4.5 Reciprocal Effects on Host and Viral Late mRNA Transport

In the normal course of adenovirus infection, viral protein synthesis dominates total synthetic activity during the late phase of infection and host cell protein synthesis ceases (BELLO and GINSBERG 1967). In the same time frame, host mRNA in general (BELTZ and FLINT 1979), from randomly selected unidentified genes (BABICH et al. 1983), or from specific genes (FLINT et al. 1983; PILDER et al. 1986b) ceases to reach the cytoplasm, although transcription continues unchecked (however, the block to mRNA export from the nucleus is not believed to be the cause of the shutoff in protein synthesis; see Sect. 5.2). By contrast, cells infected by E1B 55K mutant viruses show continued host protein synthesis through the late phase of the infectious cycle (BABISS and GINSBERG 1984; LOGAN et al. 1984; WILLIAMS et al. 1986) and continued entry of host mRNA into the cytoplasm (BABISS et al. 1985; PILDER et al. 1986b; WILLIAMS et al. 1986). Thus, the E1B 55K protein, in addition to having a positive regulatory role in late viral mRNA accumulation, is also required for negative regulation of cellular mRNA accumulation during the late phase of infection.

It remains to be determined whether the apparently reciprocal effects of E1B 55K on viral and cellular mRNA accumulation are both direct, independent consequences of 55K activity or whether a direct effect of 55K on one aspect of gene expression leads indirectly to the other effect. Considered in the context of the model for E1B 55K dependence discussed earlier (Sect. 4.4), the block to cell mRNA accumulation could reflect 55K actively disconnecting cell mRNAs from a saturated transport apparatus to give viral mRNAs greater access. Alternatively, cell mRNAs might be passively excluded from transport by the overwhelming mass of viral mRNA that is competent for transport in the presence of E1B 55K activity, although it is hard to envisage such a mechanism giving the 100% shutoff of transport of specific mRNAs that is seen. This latter mechanism also fails to account for the observation that certain cell mRNAs, products of genes subject to transcriptional induction after virus infection, are exempt from the block to export (.Moore et al. 1987). This result can be accommodated more readily in an active displacement model, making the assumption that transcription induction increases the strength of linkage of a transcription complex to the transport machinery. Again, further work is needed to understand fully the relationship between the effects of E1B 55K protein on viral and cell gene expression.

4.6 Involvement of E4 Gene Products in Regulated mRNA Transport

So far, this discussion has been restricted to consideration of the role of the adenovirus E1B 55K protein in RNA transport regulation. However, other viral proteins also play a role in regulating this or closely related processes in the infected cell. In particular, studies of frame-shifting insertion or deletion mutations in each of the candidate E4 coding regions (Fig. 3; see also Sect. 2.4) showed that loss of E4-ORF6 function created a late viral and host cell protein synthesis phenotype very similar to that of E1B 55K mutants (Halbert et al. 1985). Since a molecular interaction between E1B 55K and E4-ORF6 proteins had been previously described (Sarnow et al. 1984), these observations were taken to indicate that this molecular complex was the functional form of the two proteins. This idea gained subsequent support from the isolation of an E1B 55K/E4-ORF6 double mutant which showed a pattern of late protein synthesis identical to that of each of the two single mutants from which it derived (Cutt et al. 1987). Studies of other E4 mutants deficient in ORF6 have given similar results (Bridge and Ketner 1989; Huang and Hearing, 1989).

Although the phenotype of E4-ORF6 mutants is similar to that of E1B 55K mutants, it may not be identical. Some E4-ORF6 mutations show a small reduction in the rate of viral DNA replication in addition to larger effects on late gene expression (Halbert et al. 1985; Huang and Hearing 1989), while other ORF6-

specific mutants have been reported to replicate DNA at normal or near-normal levels (BRIDGE and KETNER 1989); in contrast, E1B 55K mutants show no detectable DNA replication defect. It is still unclear whether this apparent difference in phenotype reflects subtly distinct roles for the E4-ORF6 and E1B 55K proteins within the complex or whether the E4-ORF6 protein has separate additional functions outside the complex.

In the original genetic analysis of region E4, no phenotype was associated with inactivating E4-ORF3 (HALBERT et al. 1985), unlike ORF6, above. However, double mutants of ORF3 and ORF6 have a very severe late gene expression and replication defect, suggesting that the functions of ORF3 and ORF6 are largely redundant and can compensate for one another (BRIDGE and KETNER 1989; HUANG and HEARING, 1989). A similarly severe phenotype is seen for an E1B 55K and E4-ORF3 double mutant (BRIDGE and KETNER 1990). These findings implicated E4-ORF3, as well as ORF6, in the control of mRNA transport and suggested a model in which two parallel pathways for adenovirus late mRNA biogenesis exist in the infected cell, one facilitated by E4-ORF3 and the other by E4-ORF6 in complex with E1B 55K, with either one being sufficient for late gene expression to occur at significant levels. This model was refined by BRIDGE et al. (1991), who showed that E4-ORF6, but not E4-ORF3, stimulated cytoplasmic pIX mRNA accumulation without having any dramatic effect on nuclear pIX accumulation. This result may be informative, since the pIX gene encodes an unspliced mRNA (ALESTRÖM et al. 1980) whose expression would not be expected to be stimulated by the E4-ORF3 protein if it regulated nuclear RNA accumulation at the level of RNA splicing.

Kinetic analyses of cytoplasmic mRNA accumulation, analogous to those performed for E1B 55K mutants, have not been reported for either E4-ORF3 or ORF6 mutants. However, analysis of the steady state levels of a late viral mRNA or its unspliced precursors in the nucleus and cytoplasm of either E4-ORF3 or ORF3/E1B 55K double mutant-infected cells showed no evidence for a defect in cytoplasmic mRNA accumulation due to the absence of ORF3 that could not be fully accounted for by the observed reduction in levels of the relevant RNA in the nucleus (BRIDGE and KETNER 1990). Thus, the E4-ORF3 protein may function solely as a splicing regulator, with consequent effects on RNA stability in the nucleus and/or mRNA recruitment into transport pathways. By similar analyses, mutants lacking E4-ORF6 showed a defect in cytoplasmic mRNA levels that could not be accounted for by reduced nuclear RNA levels (BRIDGE and KETNER 1990). Although the precise quantitation of RNA in these analyses depends crucially on the allowance made for changes in DNA copy number (also consequent upon the mutation and having a direct effect on the level of late gene expression), these findings probably indicate that E4-ORF6 does function directly in nuclear RNA metabolism postmaturation. The collectively available data indicate that E4-ORF6 serves two functions in RNA metabolism: an effect on nuclear levels of MLTU-derived RNAs (e.g., stability, probably through a pathway involving RNA splicing) that can also be provided by E4-ORF3 and an effect on RNA transport, presumably provided in conjunction with E1B 55K.

4.7 Relationship of RNA Transport Regulation to Transformation Functions

The adenovirus E1B 55K protein is known to be required for the full transformation of rodent cells by infectious virus (BABISS et al. 1984). A review of the basis of this activity is outside the scope of this chapter; however, it is relevant to consider whether this activity of the protein is related to its known role in lytic infection. This point is best addressed by attempts to separate the two functions of E1B 55K by mutation. YEW et al. (1990) reported the isolation and characterization of a series of in-frame linker insertion mutants of Ad2 E1B 55K. Although their data did not allow the precise delineation of 55K domains to which specific functions could be assigned, certain mutants showed strongly discordant phenotypes in lytic growth versus transformation analyses. These data suggest that the biochemical functions of E1B 55K in these two situations are not identical.

5 Translational Controls

5.1 Virus-Associated RNAs Preserve Translational Activity During Late Infection

This aspect of adenovirus infection has been reviewed in detail elsewhere (MATHEWS and SHENK 1991; ZHANG and SCHNEIDER 1993) and will be summarized only briefly here. Wild-type adenovirus expresses a short RNA pol III transcript, virus-associated RNA VAI, which has been shown by mutational analysis to be required for late viral protein synthesis (THIMMAPPAYA et al. 1982). This RNA blocks the activation of the interferon-inducible protein kinase, DAI, by double-stranded (ds) RNA that forms due to symmetrical transcription of the viral genome and hence prevents the inactivation, by phosphorylation, of the essential translation factor eIF2-α (SCHNEIDER et al. 1984, 1985; REICHEL et al. 1985). Thus in the absence of VAI, little or no translation of viral or cellular mRNA occurs in the late-infected cell.

5.2 Virus-Mediated Shutoff of Host Protein Synthesis

Although the effects of virus infection on cellular mRNA transport described earlier (Sect. 4.5) apparently correlate with changes in protein synthesis, a direct cause and effect relationship between the two effects is unlikely, because the block to mRNA transport makes only minimal impact on the cytoplasmic level of specific host mRNAs at times postinfection when the shutoff of host protein synthesis is virtually complete (BABICH et al. 1983). Also, those host mRNAs which escape the transport block nevertheless fail to be translated efficiently (.MOORE et al. 1987). Studies of the inactivation of eIF2-α by the protein kinase DAI in adenovirus-infected cells led to the suggestion that this inactivation was the basis of reduced host cell translation (O'MALLEY et al. 1989). More recently, the inactivation of cap-binding protein (CBP) has been shown to correlate with the shutoff of host cell protein synthesis during wild-type adenovirus infection (HUANG and SCHNEIDER 1991; reviewed in ZHANG and SCHNEIDER 1993). Although the shutoff of host protein synthesis clearly fails in the absence of E1B 55K, it remains to be demonstrated that either CBP or eIF2-α inactivation is altered in an E1B 55K mutant infection. However, it is likely that the continued cell protein synthesis in such infections reflects a failure to produce a viral component needed to cause one or both of these modifications to the translation apparatus. This component might be 55K itself or another viral protein whose synthesis is impaired in a 55K mutant infection.

5.3 Selective Translation of Viral Proteins

As noted above, CBP, an essential translation factor, is inactivated in a late adenovirus infection. Despite this inactivation, synthesis of viral proteins continues efficiently in the infected cell. The ability of late viral mRNA to be translated in the absence of active CBP is conferred by the tripartite leader sequence (DOLPH et al. 1988). The tripartite leader is a 201-nucleotide RNA segment present at the 5' end of virtually all late adenovirus mRNAs, and mRNAs carrying it had been shown earlier to display enhanced translation in a late adenovirus-infected cell (LOGAN and SHENK 1984). Its lack of secondary structure is believed to promote ribosome binding (DOLPH et al. 1990).

A second viral function, the L4 100K protein, is also required for efficient viral protein synthesis during late infection (HAYES et al. 1990). In its absence, a reduced proportion of specific cytoplasmic viral mRNA is found associated with polysomes. L4 100K is known to bind mRNA in vivo (ADAM and DREYFUSS 1987) and in vitro (RILEY and FLINT 1993); the altered 100K protein from a *ts* mutant in L4 that displays reduced late protein synthesis at elevated temperature showed a corresponding reduction in RNA binding at that restrictive temperature in vitro (RILEY and FLINT 1993). The L4 100K protein therefore appears to activate late translation via a direct interaction with RNA. Although there is no evidence for sequence-specific binding of 100K to the tripartite leader, it is possible that this protein in some way mediates the translational enhancement attributed to the presence of the tripartite leader; alternatively, since the translation of viral RNA not containing tripartite leader is also enhanced by L4 100K (HAYES et al. 1990), it may operate through an independent mechanism. Further experiments will be needed to determine this point.

6 Conclusions

In order to be able to maintain a compact genome size yet remain an effective pathogen, adenovirus has taken full advantage of the host cell's capacity to regulate gene expression post-transcriptionally. Differential splicing, 3' end processing, transport of mRNA, and stability of mRNA are all employed by the virus to maximize coding capacity and to tightly control the expression of its genes as well as those of the host.

The common thread that seems to join these different processes together is the observation that some change occurs during the switch from the early to the late phase of the infection which results in the alteration of mRNA expression patterns. Exactly how the viral genome is assembled into chromatin, where it resides in the nucleus, and how efficiently each template is transcribed may all bear on this issue. The task in the future, then, is to continue to elucidate the molecular mechanisms of how the virus interacts with the cell's macromolecular synthetic machinery, from both a biochemical and structural standpoint, in order to accomplish its goals. Undoubtedly, as has been the case in the past, these studies will lend important insights not only into how the virus replicates and causes disease, but also into the functioning of the cell itself and the possible use of the virus as a tool in treating disease.

References

Adam SA, Dreyfuss G (1987) Adenovirus proteins associated with mRNA and hnRNA in infected HeLa cells. J Virol 61: 3276–3283

Adami G, Babiss LE (1991) DNA template effect on RNA splicing: two copies of the same gene in the same nucleus are processed differently. EMBO J 11: 3457–3465

Akusjärvi G, Persson H (1981) Controls of RNA splicing and termination in the major late adenovirus transcription unit. Nature 292: 420–426

Akusjärvi G, Pettersson U, Roberts RJ (1986) Structure and function of the adenovirus -2 genome. In: Doerfler W (ed) Adenovirus DNA: the viral genome and its expression. Martin Nijhoff, Boston, pp 53–95

Aleström P, Akusjärvi G, Perricaudet M, Mathews MB, Klessig DF, Pettersson U (1980) The gene for polypeptide IX of adenovirus type 2 and its unspliced messenger RNA. Cell 19: 671–681

Anderson CW, Schmitt RC, Smart JE, Lewis JB (1984) Early region 1B of adenovirus 2 encodes two coterminal proteins of 495 and 155 amino acid residues. J Virol 50: 387–396

Babich A, Feldman LT, Nevins JR, Darnell JE, Weinberger C (1993) Effects of adenovirus on metabolism of specific host mRNAs: transport control and specific translational discrimination. Mol Cell Biol 3: 1212–1221

Babiss LE, Ginsberg HS (1984) Adenovirus type 5 early region 1b gene product is required for efficient shutoff of host protein synthesis. J Virol 50: 202–212

Babiss LE, Fisher PB, Ginsberg HS (1984) Effect on transformation of mutations in the early region 1b-encoded 21- and 55-kilodalton proteins of adenovirus 5. J Virol 52: 389–395

Babiss LE, Ginsberg HS, Darnell JE Jr (1985) Adenovirus E1B proteins are required for accumulation of late viral mRNA and for effects on cellular mRNA translation and transport. Mol Cell Biol 5: 2552–2558

Bello LJ, Ginsberg HS (1967) Inhibition of host protein synthesis in type 5 adenovirus-infected cells. J Virol 1: 843–850

Beltz GA, Flint SJ (1979) Inhibition of HeLa cell protein synthesis during adenovirus infection: restriction of cellular messenger RNA sequences to the nucleus. J Mol Biol 131: 353–373

Berezney R, Coffey DS (1977) Nuclear matrix: isolation and characterization of a framework structure from rat liver nuclei. J Cell Biol 73: 616–637

Berget SM, Moore C, Sharp PA (1977) Spliced segments at the 5' terminus of adenovirus 2 late mRNA. Proc Nat Acad Sci USA 74: 3171–3175

Berk AJ, Sharp PA (1978) Structure of the adenovirus 2 early mRNAs. Cell 14: 695–711
Bhat BM, Wold WS (1986) Genetic analysis of mRNA synthesis in adenovirus region E3 at different stages of productive infection by RNA-processing mutants. J Virol 60: 54–63
Bodnar JW, Hanson PI, Polvino-Bodnar M, Zempsky W, Ward DC (1989) The terminal regions of adenovirus and minute virus of mice DNAs are preferentially associated with the nuclear matrix in infected cells. J Virol 63: 4344–4353
Bos JL, Polder LJ, Bernards R, Scrier PI, van den Elsen PJ, van der Eb AJ, van Ormondt H (1981) The 2.2 kb E1b mRNA of human Ad12 and ad5 codes for two tumor antigens starting at different AUG triplets. Cell 27: 121–131
Brady HA, Wold WSM (1988) Competition between splicing and polyadenylation reactions determines which adenovirus region E3 mRNAs are synthesized. Mol Cell Biol 8: 3291–3297
Brady HA, Scaria A, Wold WSM (1992) Map of cis-acting sequences that determine alternative premessenger-RNA processing in the E3 complex transcription unit of adenovirus. J Virol 66: 5914–5923
Bridge E, Ketner G (1989) Redundant control of adenovirus late gene expression by early region 4. J Virol 63: 631–638
Bridge E, Ketner G (1990) Interaction of adenoviral E4 and E1b products in late gene expression. Virology 174: 345–353
Bridge E, Hemstrom C, Pettersson U (1991) Differential regulation of adenovirus late transcription units by the products of early region 4. Virology 183: 260–266
Bridge E, Carmo-Fonseca M, Lamond A, Pettersson U (1993) Nuclear organization of splicing small nuclear ribonucleoproteins in adenovirus-infected cells. J Virol 67: 5792–5802
Chang DD, Sharp PA (1990) Messenger RNA transport and HIV rev regulation. Science 249: 614–615
Chebli K, Gattoni R, Schmitt P, Hildwein G, Stevenin J (1989) The 216-nucleotide intron of the E1A pre-mRNA contains a hairpin structure that permits utilization of unusually distant branch acceptors. Mol Cell Biol 9: 4852–4861
Chow LT, Broker TR (1978) The spliced structures of adenovirus 2 fiber message and the other late mRNAs. Cell 15: 497–510
Chow LT, Gelinas RE, Broker TR, Roberts RJ (1977) An amazing sequence arrangement at the 5' ends of adenovirus 2 messenger RNA. Cell 12: 1–8
Chow LT, Broker TR, Lewis JB (1979) Complex splicing patterns of RNAs from the early regions of adenovirus-2. J Mol Biol 134: 265–303
Chroboczek J, Jacrot B (1987) The sequence of adenovirus fiber: similarities and differences between serotypes 2 and 5. Virology 161: 549–554
Chroboczek J, Bieber F, Jacrot B (1992) The sequence of the genome of adenovirus type 5 and its comparison with the genome of adenovirus type 2. Virology 186: 280–285
Ciejek EM, Nordstrom JL, Tsai M-J, O'Malley BW (1982) Ribonucleic acid precursors are associated with the chick oviduct nuclear matrix. Biochemistry 21: 4945–4953
Cladaras C, Wold WSM (1985) DNA sequence of the early E3 transcription unit of adenovirus 5. Virology 140: 28–43
Cutt JR, Shenk T, Hearing P (1987) Analysis of adenovirus early region 4-encoded polypeptides synthesized in productively infected cells. J Virol 61: 543–552
Delsert C, Morin N, Klessig DF (1989) Cis-acting elements and a trans-acting factor affecting alternative splicing of adenovirus L1 transcripts. Mol Cell Biol 9: 4364–4371
DeZazzo JD (1990) Poly(A) site selection in the adenovirus type 2 major late transcription unit: processing and regulatory signals. PhD thesis, University of Michigan
DeZazzo JD, Imperiale MJ (1989) Sequences upstream of AAUAAA influence poly(A) site selection in a complex transcription unit. Mol Cell Biol 9: 4951–4961
DeZazzo JD, Falck-Pedersen E, Imperiale MJ (1991) Sequences regulating temporal poly(A) site switching in the adenovirus major late transcription unit. Mol Cell Biol 11: 5977–5984
Dix I, Leppard KN (1993) Regulated splicing of adenovirus type 5 E4 transcripts and regulated cytoplasmic accumulation of E4 mRNA. J Virol 67: 3226–3231
Dolph PJ, Racaniello V, Villamarin A, Palladino F, Schneider RJ (1988) The adenovirus tripartite leader eliminates the requirement for cap binding protein during translation initiation. J Virol 62: 2059–2066
Dolph PJ, Huang J, Schneider RJ (1990) Translation by the adenovirus tripartite leader: elements which determine independence from cap-binding protein complex. J Virol 64: 2669–2677
Domenjoud L, Gallinaro H, Kister L, Meyer S, Jacob M (1991) Identification of a specific exon sequence that is a major determinant in the selection between a natural and a cryptic 5' splice site. Mol Cell Biol 11: 4581–4590

Dressler GR, Fraser NW (1987) DNA sequences downstream of the adenovirus type 2 fiber polyadenylation site contain transcription termination signals. J virol 61: 2770–2776
Dreyfuss G, Matunis MJ, Pinol-Roma S, Burd CG (1993) hnRNP proteins and the biogenesis of mRNA. Annu Rev Biochem 62: 289–321
Dunn AR, Hassel JA (1977) A novel method to map transcripts: evidence for homology between an adenovirus mRNA and discrete multiple regions of the viral genome. Cell 12: 23–36
Evans R, Weber J, Ziff E, Darnell JE (1979) Premature termination during adenovirus transcription. Nature 278: 367–370
Falck-Pedersen E, Logan J (1989) Regulation of poly(A) site selection in adenovirus. J Virol 63: 532–541
Flint SJ, Beltz G, Linzer DIH (1983) Synthesis and processing of simian virus 40-specific RNA in adenovirus-infected, simian virus 40-transformed human cells. J Mol Biol 167: 335–359
Fraser NW, Nevins JR, Ziff E, Darnell JE Jr (1979) The major late adenovirus type-2 transcription unit: termination is downstream from the last poly(A) site. J Mol Biol 129: 643–656
Freyer GA, Katoh Y, Roberts RJ, (1984) Characterization of the major mRNAs from adenovirus 2 early region 4 by cDNA cloning and sequencing. Nucleic Acids Res 12: 3503–3519
Gattoni R, Schmitt P, Stevenin J (1988) In vitro splicing of adenovirus E1A transcripts: characterization of novel reactions and of multiple branch points far from the 3' splice site. Nucleic Acids Res 16: 2389–2409
Gattoni R, Chebli K, Himmelspach M, Stevenin J (1991) Modulation of alternative splicing of adenoviral E1A transcripts: factors involved in the early-to-late transition. Genes Dev 5: 1847–1858
Halbert DN, Cutt JR, Shenk T (1985) Adenovirus early region 4 encodes functions required for efficient DNA replication, late gene expression and host cell shutoff. J Virol 56: 250–257
Hales KH, Birk JM, Imperiale MJ (1988) Analysis of adenovirus type 2 L1 RNA 3'-end formation in vivo and in vitro. J Virol 62: 1464–1468
Harper JE, Manley JL (1991) A novel protein factor is required for use of distal alternative 5' splice sites in vitro. Mol Cell Biol 11: 5945–5953
Harrison T, Graham F, Williams J (1977) Host-range mutants of adenovirus type 5 defective for growth in HeLa cells. Virology 77: 319–329
Hasson TB, Soloway PD, Ornelles DA, Doerfler W, Shenk T (1989) Adenovirus L1 52- and 55-kilodalton proteins are required for assembly of virions. J Virol 63: 3612–3621
Hayes BW, Telling GC, Myat MM, Williams JF, Flint SJ (1990) The adenovirus L4 100-kilodalton protein is necessary for efficient translation of viral late mRNA species. J Virol 64: 2732–2742
Hemström C, Nordqvist K, Pettersson U, Virtanen A (1988) Gene product of region E4 of adenovirus type 5 modulates accumulation of certain viral polypeptides. J Virol 62: 3258–3264
Ho YS, Galos R, Williams J (1982) Isolation of type 5 adenovirus mutants with a cold sensitive host range phenotype: genetic evidence of an adenovirus transformation maintenance function. Virology 122: 109–124
Huang J, Schneider RJ (1991) Adenovirus inhibition of cellular protein synthesis involves inactivation of cap binding protein. Cell 65: 271–280
Huang M-M, Hearing P (1989) Adenovirus early region 4 encodes two gene products with redundant effects in lytic infection. J Virol 63: 2605–2615
Iwamoto S, Eggerding F, Falck-Pedersen E, Darnell JE Jr (1986) Transcription unit mapping in adenovirus: regions of termination. J Virol 59: 112–119
Johnston JM, Anderson KP, Klessig DF (1985) Partial block to transcription of human adenovirus type 2 late genes in abortively infected monkey cells. J Virol 56: 378–385
Ketner G, Bridge E, Virtanen A, Hemström C, Pettersson U (1989) Complementation of adenovirus E4 mutants by transient expression of E4 cDNA and deletion plasmids. Nucleic Acids Res 17: 3037–3048
Kinloch R, Mackay N, Mautner V (1984) Adenovirus hexon: sequence comparision of subgroup C serotypes 2 and 5. J Biol Chem 259: 6431–6436
Klessig D (1977) Two adenovirus messenger RNAs have a common 5' terminal leader sequence encoded at least 10kb upstream form their main coding regions. Cell 12: 9–12
Kreivi J-P, Akusjärvi G (1994) Regulation of adenovirus alternative RNA splicing at the level of commitment complex formation. Nucleic Acids Res 22: 332–337
Kreivi J-P, Zerivitz K, Akusjärvi G (1991) Sequences involved in the control of adenovirus L1 alternative RNA splicing. Nucleic Acids Res 19: 2379–2386
Kumar A, Pederson T (1975) Comparison of proteins bound to heterogeneous nuclear RNA and messenger RNA in HeLa cells. J Mol Biol 96: 353–365

Larsson S, Kreivi J-P, Akusjärvi G (1991) Control of adenovirus alternative RNA splicing: effect of viral DNA replication on splice site choice. Gene 107: 219–227
Larsson S, Svensson C, Akusjärvi G (1992) Control of adenovirus major late gene expression at multiple levels. J Mol Biol 225: 287–298
Le Moullec JM, Akusjärvi G, Stalhandske P, Pettersson U, Chambraud B, Gilardi P, Nasri M, Perricaudet M (1983) Polyadenylic acid addition sites in the adenovirus type 2 major late transcription unit. J Virol 48: 127–134
Leppard KN (1993) Selective effects on adenovirus late gene expression of deleting the E1b 55K protein. J Gen Virol 74: 575–582
Leppard KN, Shenk T (1989) The adenovirus E1B 55kd protein influences mRNA transport via an intranuclear effect on RNA metabolism. EMBO J 8: 2329–2336
Leppard KN, Pilder S, Moore M, Logan J, Shenk T (1987) An adenovirus oncogene post-transcriptionally modulates mRNA accumulation. In: Kjeldgaard N, Forchhammer J (eds) Viral carcinogenesis. Alfred Benzon Symposium 24. Munksgaard, Copenhagen, pp 196–208
Lewis JB, Anderson CW, Atkins JF (1977) Further mapping of late adenovirus genes by cell-free translation of RNA selected by hybridization to specific DNA fragments. Cell 12: 37–44
Logan J, Shenk T (1984) Adenovirus tripartite leader sequence enhances translation of mRNAs late after infection. Proc Natl Acad Sci USA 81: 3655–3659
Logan J, Pilder S, Shenk T (1984) Functional analysis of the adenovirus type-5 early region 1B. Cancer Cells 2: 527–532
Mandel JL (1989) Dystrophin: the gene and its product. Nature 339: 584–586
Manley JL (1988) Polyadenylation of mRNA precursors. Biochim Biophys Acta 950: 1–12
Manley JL, Sharp PA, Gefter ML (1982) RNA synthesis in isolated nuclei. Processing of adenovirus serotype 2 late messenger RNA precursors. J Mol Biol 159: 581–599
Mann KP, Weiss EA, Nevins JR (1993) Alternative poly(A) site utilization during adenovirus infection coincides with a decrease in the activity of a poly(A) site processing factor. Mol Cell Biol 13: 2411–2419
Mathews MB, Shenk T (1991) Adenovirus virus-associated RNA and translational control. J Virol 65: 5657–5662
Mayeda A, Krainer AR (1992) Regulation of alternative pre-mRNA splicing by hnRNP A1 and splicing factor SF2. Cell 68: 365–375
Mayeda A, Helfman DM, Krainer AR (1993) Modulation of exon skipping and inclusion by heterogeneous nuclear ribonucleoprotein A1 and pre-mRNA splicing factor SF2/ASF. Mol Cell Biol 13: 2993–3001
McNally MT, Beemon K (1992) Intronic sequences and 3' splice sites control Rous sarcoma virus RNA splicing. J Virol 66: 6–11
Moen PT Jr, Fox E, Bodnar JW (1990) Adenovirus and minute virus of mice DNAs are localized at the nuclear periphery. Nucleic Acids Res 18: 513–520
Montell C, Fisher EF, Caruthers MH, Berk AJ (1984) Control of adenovirus E1B mRNA synthesis by a shift in the activities of RNA splice sites. Mol Cell Biol 4: 966–972
Moore MA, Shenk T (1988) The adenovirus tripartite leader sequence can alter nuclear and cytoplasmic metabolism of a non-adenovirus mRNA within infected cells. Nucleic Acids Res 16: 2247–2262
Moore M, Schaack J, Baim SB, Morimoto RI, Shenk T (1987) Induced heat shock mRNAs escape the nucleocytoplasmic transport block in adenovirus-infected HeLa cells. Mol Cell Biol 7: 4505–4512
Moore MJ, Query CC, Sharp PA (1993) Splicing of precursors to mRNAs by the spliceosome. In: Gesteland RF, Atkins JF (eds) The RNA world. Cold Spring Harbor Laboratory, Cold Spring Harbor, NY, pp 303–357
Moyne G, Pichard E, Bernhard W (1978) Localization of simian adenovirus 7 (SA 7) transcription and replication in lytic infection. An ultracytochemical and autoradiographical study. J Gen Virol 40: 77–92
Nevins JR, Darnell JE Jr (1978) Steps in the processing of Ad2 mRNA: poly(A)+ nuclear sequences are conserved and poly(A) addition precedes splicing. Cell 15: 1477–1493
Nevins JR, Wilson MC (1981) Regulation of adenovirus-2 gene expression at the level of transcriptional termination and RNA processing. Nature 290: 113–118
Nevins JR, Ginsberg HS, Blanchard JM, Wilson MC, Darnell JE (1979) Regulation of the primary expression of the early adenovirus transcription units. J Virol 32: 727–733
Nordqvist K, Akusjärvi G (1990) Adenovirus early region 4 stimulates mRNA accumulation via 5' introns. Proc Natl Acad Sci USA 87: 9543–9547

Nordqvist K, Ohman K, Akusjärvi G (1994) Human adenovirus encodes two proteins which have opposite effects on accumulation of alternatively spliced mRNAs. Mol Cell Biol 14: 437–445
Ohman K, Nordqvist K, Akusjärvi G (1993) Two adenovirus proteins with redundant activities in virus growth facilitates tripartite leader mRNA accumulation. Virology 194: 50–58
O'Malley RP, Duncan RF, Hershey JWB, Mathews MB (1989) Modification of protein synthesis initiation factors and the shut-off of host protein synthesis in adenovirus infected cells. Virology 168: 112–118
Ornelles DA, Shenk T (1991) Localization of the adenovirus early region 1B 55-kilodalton protein during lytic infection: association with nuclear viral inclusions requires the early region 4 34-kilodalton protein. J Virol 65: 424–439
Perricaudet M, Akusjärvi G, Virtanen A, Pettersson U (1979) Structure of two spliced mRNAs from the transforming region of human subgroup C adenoviruses. Nature 281: 694–696
Perricaudet M, Le Moullec JM, Pettersson U (1980) Predicted structure of two adenovirus tumor antigens. Proc Natl Acad Sci USA 77: 3778–3782
Pilder S, Leppard K, Logan J, Shenk T (1986a) Functional analysis of the adenovirus E1B 55K polypeptide. Cancer Cells 4: 285–290
Pilder S, Moore M, Logan J, Shenk T (1986b) The adenovirus E1B-55K transforming polypeptide modulates transport or cytoplasmic stabilization of viral and host cell mRNAs. Mol Cell Biol 6: 470–476
Popielarz M, Gattoni R, Stevenin J (1993) Contrasted cis-acting effects of downstream 5' splice sites on the splicing of a retained intron: the adenoviral E1A pre-mRNA model. Nucleic Acids Res 21: 5144–5151
Prescott JC, Falck-Pedersen E (1992) Varied poly(A) site efficiency in the adenovirus major late transcription unit. J Biol Chem 267: 8175–8181
Reichel PA, Merrick WC, Skiekierka J, Mathews MB (1985) Regulation of a protein synthesis initiation factor by adenovirus virus-associated RNA. Nature 313: 196–200
Riley D, Flint SJ (1993) RNA binding properties of a translational activator, the adenovirus L4 100-kilodalton protein. J Virol 67: 3586–3595
Roberts RJ, O'Neill K, Yen CT (1984) DNA sequences from the adenovirus 2 genome. J Biol Chem 259: 13968–13975
Ross D, Ziff E (1992) Defective synthesis of early region 4 mRNAs during abortive adenovirus infections in monkey cells. J Virol 66: 3110–3117
Sandler AB, Ketner G (1989) Adenovirus early region 4 is essential for normal stability of late nuclear RNAs. J Virol 63: 624–630
Sarnow P, Hearing P, Anderson CW, Halbert DN, Shenk T, Levine AJ (1984) Adenovirus early region 1B 58,000-dalton tumor antigen is physically associated with an early region 4 25 000-dalton protein in productively infected cells. J Virol 49: 692–700
Scaria A, Wold WSM (1994) Fine-mapping of sequences that suppress splicing in the E3 complex transcription unit of adenovirus. Virology 205: 406–416
Schmitt P, Gattoni R, Keohavong P, Stevenin J (1987) Alternative splicing of E1A transcripts of adenovirus requires appropriate ionic conditions in vitro. Cell 50: 31–39
Schneider RJ, Weinberger C, Shenk T (1984) Adenovirus VA1 RNA facilitates the initiation of translation in virus infected cells. Cell 37: 291–298
Schneider RJ, Safer B, Munemitsu S, Samuel CE, Shenk T (1985) Adenovirus VA1 RNA prevents phosphorylation of the eukaryotic initiation factor 2 alpha subunit subsequent to infection. Proc Natl Acad Sci USA 82: 4321–4324
Schroder HC, Bachmann M, Diehl-Seifert B, Muller WEG (1987) Transport of mRNA from nucleus to cytoplasm. Prog Nucleic Acid Res Mol Biol 45: 98–142
Shaw AR, Ziff EB (1980) Transcripts from the adenovirus-2 major late promoter yield a single early family of 3' coterminal mRNAs and five late families. Cell 22: 905–916
Smith CWJ, Patton JG, Nadal-Ginard B (1989) Alternative splicing in the control of gene expression. Annu Rev Genet 23: 527–577
Soloway PD, Shenk T (1990) The adenovirus type 5 i-leader open reading frame functions in cis to reduce the half-life of L1 mRNAs. J Virol 64: 551–558
Stephens C, Harlow E (1987) Differential splicing yields novel adenovirus 5 E1A mRNAs that encode 30kd and 35kd proteins. EMBO J 6: 2027–2035
Svensson C, Akusjärvi G (1986) Defective RNA splicing in the absence of adenovirus-associated RNA I. Proc Natl Acad Sci USA 83: 4690–4694
Symington JS, Lucher LA, Brackmann KH, Virtanen A, Pettersson U, Green M (1986) Biosynthesis of adenovirus type 2 i-leader protein. J Virol 57: 848–856

Thimmappaya B, Weinberger C, Schneider RJ, Shenk T (1982) Adenovirus VAI RNA is required for efficient translation of viral mRNAs at late times after infection. Cell 31: 543–351
Thomas GP, Mathews MB (1980) DNA replication and the early to late transition in adenovirus infection. Cell 22: 523–532
Tigges MA, Raskas HJ (1984) Splice junctions in adenovirus 2 early region 4 mRNAs: multiple splice sites produce 18 to 24 RNAs. J Virol 50: 106–117
Tollefson AE, Scaria A, Saha SK, Wold WSM (1992) The 11 600-MW protein encoded by region E3 of adenovirus is expressed early but is greatly amplified at late stages of infection. J Virol 66: 3633–3642
Ulfendahl PJ, Linder S, Kreivi JP, Nordqvist K, Svensson C, Hultberg H, Akusjärvi G (1987) A novel adenovirus-2 E1A mRNA encoding a protein with transcription activation properties. EMBO J 6: 2037–2044
Van Ormondt H, Maat H, Van Beveren CP (1980) The nucleotide sequence of the transforming region E1 of adenovirus type 5 DNA. Gene 11: 299–309
Virtanen A, Pettersson U (1985) Organization of early region 1B of human adenovirus type 2: identification of four differentially spliced mRNAs. J Virol 54: 383–391
Virtanen A, Gilardi P, Naslund A, LeMoullec JM, Pettersson U, Perricaudet M (1984) mRNAs from human adenovirus 2 early region 4. J Virol 51: 822–831
Wahle E, Keller W (1992) The biochemistry of 3'-end cleavage and polyadenylation of messenger RNA precursors. Annu Rev Biochem 61: 419–440
Walton TH, Moen PT Jr, Fox E, Bodnar JW (1989) Interactions of minute virus of mice and adenovirus with host nucleoli. J Virol 63: 3651–3660
Weinberg DH, Ketner G (1986) Adenoviral early region 4 is required for efficient viral DNA replication and for late gene expression. J Virol 57: 833–838
Wickens M (1990) How the messenger got its tail: addition of poly(A) in the nucleus. Trends Biochem Sci 15: 277–281
Williams J, Karger BD, Ho YS, Castiglia CL, Mann T, Flint SJ (1986) The adenovirus E1B-495R protein plays a role in regulating the transport and stability of the viral late messages. Cancer Cells 4: 275–284
Wilson-Gunn SI, Kilpatrick JE, Imperiale MJ (1992) Regulated adenovirus mRNA 3' end formation in a coupled in vitro transcription/processing system. J Virol 6: 5418–5424
Wold WS, Gooding LR (1991) Region E3 of adenovirus: a cassette of genes involved in host immunosurveillance and virus-cell interactions. Virology 184: 1–8
Yew Y, Kao CC, Berk AJ (1990) Dissection of functional domains in the adenovirus 2 early 1B 55K polypeptide by suppressor-linker insertional mutagenesis. Virology 179: 795–805
Zamore PD, Patton JG, Green MR (1992) Cloning and domain structure of the mammalian splicing factor U2AF. Nature 355: 609–614
Zerivitz K, Kreivi J-P, Akusjärvi G (1992) Evidence for a HeLa cell splicing activity that is necessary for activation of a regulated adenovirus 3' splice site. Nucleic Acids Res 20: 3955–3961
Zhang Y, Schneider RJ (1993) Adenovirus inhibition of cellular protein synthesis and the specific translation of late viral mRNAs. Semin Virol 4: 229–236

Structure, Function, and Evolution of Adenovirus Virus-Associated RNAs

M.B. MATHEWS

1 Background

One of the charms of adenovirus is its unique blend of orthodoxy and idiosyncrasy. Although the virus operates largely via host cell mechanisms, at least for transcriptional and translational purposes, it coopts these mechanisms by using a variety of viral products to modify cellular activities for its own benefit. Nowhere is this better illustrated than in the virus-associated (VA) RNA genes, which are transcribed by RNA polymerase III (pol III) while the rest of the viral genome is grist for pol II. More abundant in the infected cell than any messenger RNA, the VA RNAs encode no protein but are essential for efficient translation, a feat that is accomplished through a protein kinase that is regulated by highly structured RNA molecules. The protein kinase is a key element in the cellular antiviral defenses that are induced by interferon, and it is probable that one of the roles of the VA RNAs in natural infections is to spike the defensive guns.

VA RNA is a short RNA molecule with a long history. It was first reported in 1966 by Weissman's laboratory (REICH et al. 1966) as a remarkably abundant new transcript that accumulated in cells infected with adenovirus type 2 (Ad2) and was termed VA RNA before hybridization studies demonstrated its viral origin. A few years later, work by MATHEWS (1975) and PETTERSSON and PHILIPSON (1975) revealed the existence of two Ad2 VA RNA species – a major species, VA RNA_I, equivalent

---

Cold Spring Harbor Laboratory, P.O. Box 100, Cold Spring Harbor, New York, NY 11724, USA

to the original VA RNA, and a minor species, VA RNA_{II} – although not all adenoviruses are endowed with two such RNAs. The VA RNAs were among the first viral RNAs to be sequenced and have attracted a great deal of attention ever since as prototypical members of an expanding class of small virus-encoded RNA species, including the EBERs RNAs of Epstein-Barr virus, TAR RNA of human immuno–deficiency virus (HIV), the U-RNAs of herpesvirus saimiri, the leader RNA of vesicular stomatitis virus (VSV), and the polads of vaccinia. This expanding group is heterogeneous in structure, and undoubtedly in function, although in many cases the functions are not well understood. For VA RNA_I, the discovery of its function came through a series of experiments conducted by Shenk's group. Having defined the gene's promoter as an internal control region lying within the transcribed sequences, a mutant virus (*dl* 331) was produced which carried a defective version of the gene from which the promoter had been excised (THIMMAPPAYA et al. 1982). This virus displayed a lesion in protein synthesis at late times of infection as well as an increased sensitivity to interferon. Both of these phenomena were traced to the activation of the protein kinase DAI through a series of studies in the Shenk and Mathews laboratories (reviewed by MATHEWS and SHENK, 1991). Intriguingly, the equivalent VA RNA_{II} mutant displayed no obvious phenotype, and the role of this minor species remains a subject of conjecture.

Like many other viruses, adenovirus activates DAI, the double-stranded RNA (dsRNA)-activated inhibitor of protein synthesis. Also known by other names (PKR, P1, p68 etc.), the kinase phosphorylates the protein synthesis initiation factor eIF-2 on its α-subunit, leading to inhibition of a recycling factor (GEF or eIF-2B) and consequently bringing translational initiation in the cell to a halt (HERSHEY 1991). This pathway is depicted in Fig. 1. The activator of DAI appears to be dsRNA that is produced by symmetrical transcription of the two viral DNA strands: presumably, introns (or even exons) generated in large quantity by the high rate of viral transcription that occurs late in the infectious cycle anneal and overwhelm the cellular mechanisms that normally dispose of unwanted dsRNA (MARAN and MATHEWS 1988). With the increased nuclear permeability that accompanies the transition to the late phase, the dsRNA could then emerge into the cytoplasm where it encounters DAI. Present in many cells and tissues in a "latent" or inactive form, DAI becomes an active eIF-2 kinase through the act of autophosphorylation, which is mediated by dsRNA (HOVANESSIAN 1991; SAMUEL 1991). It is at this step that VA RNA intervenes, by preventing the autophosphorylation and activation of DAI, hence averting the shutdown of protein synthesis (MATHEWS 1993).

In principle, at least, the dsRNA could also leak out of infected cells and trigger the induction of interferon leading to the induction of DAI synthesis at the transcriptional level (Fig.1). Thus, dsRNA appears to sensitize the cell and organism to virus infection and switch on a mechanism that limits virus multiplication. The role of VA RNA – at least its best-known role – is to intercede in the interest of viral protein synthesis. Evidence implicating these small RNAs in additional functions also exists. This chapter will summarize the current state of

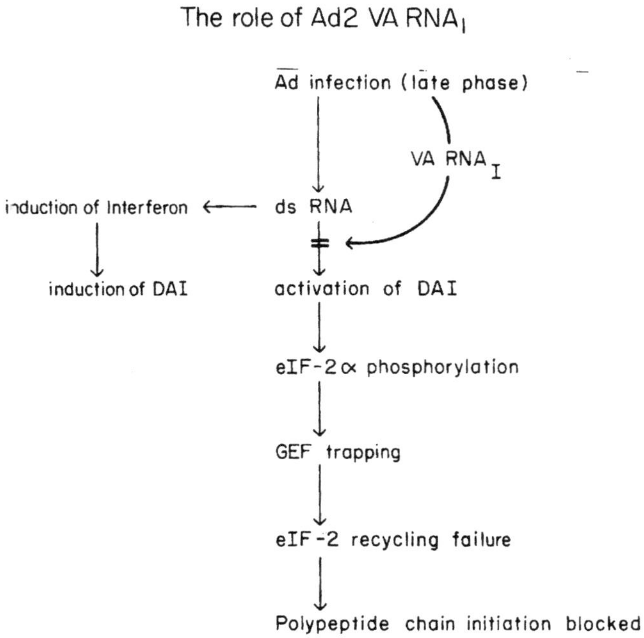

Fig. 1. The role of adenovirus type 2 (Ad2) virus-associated (VA) RNA I in infection (Summarizes data reviewed by MATHEWS and SHENK (1991)). *ds*, double-stranded; *GEF*, guanosine nucleotide exchange factor (eIF -2B); *eIF*, eucaryotic initiation factor

knowledge in the field. For a comprehensive account of earlier work, the reader can consult the review by MATHEWS and SHENK (1991).

2 Genes for Virus-Associated RNA

All members of the *Adenoviridae* examined to date, including those infecting humans, monkeys, and birds, have genes encoding VA RNA, so it seems likely that VA RNA arose early in adenovirus evolution and will be found in viruses infecting other species too. Simian adenovirus 7 (SA7) and some of the human adenovirus serotypes (those in groups A and F, as well as some group B viruses) possess a single VA RNA gene, usually 150–170 nucleotides in length, located at approximately map unit 30 on the genetic map (for references, see MA and MATHEWS 1993). Transcribed in the rightward direction as the genome is conventionally oriented, it is sandwiched between exons of protein-coding genes. To its left is the second coding exon of the E2B gene,which is transcribed to the left and encodes the precursor to the terminal protein (pTP); to its right is the first coding exon of the L1 family, transcribed to the right and encoding the 52–55-kDa polypeptide (L1-52,55K). It is unclear whether the location of the VA RNA genes at this particular genomic crossroads is of any significance, and whether the VA RNA sense or antisense sequences present in mRNA precursors play any role in the viral life cycle is also unknown.

Most of the human adenovirus serotypes have two VA RNA genes, VA RNA_I and VA RNA_{II}, within the same interval (see Fig. 2). They are oriented in the same direction and are separated by about 100 nucleotides of spacer sequence. It would appear that they originated by gene duplication sometime during the evolution of the human adenoviruses, followed by sequence divergence and functional specialization as discussed below. Half of the group B viruses, Ad11 Ad14, Ad34, and Ad35, seem to have discarded their VA RNA_{II} genes at a late stage of their evolution. They possess a single VA RNA which is very similar to the VA RNA_I of the other group B viruses and different from their VA RNA_{II} species. Thus, in this subgroup of the B viruses, the single VA RNA species is of the VA RNA_I type. Although it has been studied well only in the group C viruses Ad2 and Ad5, and in Ad7 (a representative of group B), the VA RNA_I species is more efficiently transcribed and accumulates to higher levels (SÖDERLUND et al. 1976). All of the genes retain sequences corresponding to the A and B boxes that comprise the promoter for pol III transcription, as well as the run of T residues that signals transcriptional termination.

Exceptionally, the avian virus CELO (chick embryo lethal orphan) contains a VA RNA gene that maps near position 90 on the viral chromosome (LARSSON et al. 1986). It is transcribed in the leftward direction, giving two RNAs that are substantially shorter than their mammalian counterparts: the two forms, of 90 and 110 nucleotides, respectively, share a common 5' end but differ at their 3' ends as a result of the presence of two T-rich sequences which serve as alternative terminal signals. Thus, the CELO VA RNA gene differs sharply from the genes of simian and human adenoviruses in its location, length, and transcription; it does not display great homology with the mammalian virus except in its B box, so it may also differ in its evolutionary origin and in its function.

3 Synthesis and Accumulation

Most of the work on VA RNA synthesis, as well as its function, has been done on Ad2 or Ad5, but it seems reasonable to extend the conclusions since comparative studies of other serotypes are in general agreement insofar as they have been conducted. The VA RNA genes are transcribed copiously, by virtue of their strong pol III promoters, and the RNAs are stable and accumulate to high levels, possibly because of their compact secondary structure. Transcription begins in the early phase and accelerates dramatically late in infection. When there are two VA RNA genes, the VA RNA_{II} species is made more slowly, in part because its promoter is intrinsically weaker than that of VA RNA_I but also because of promoter competition (BHAT and THIMMAPPAYA 1984); consequently,

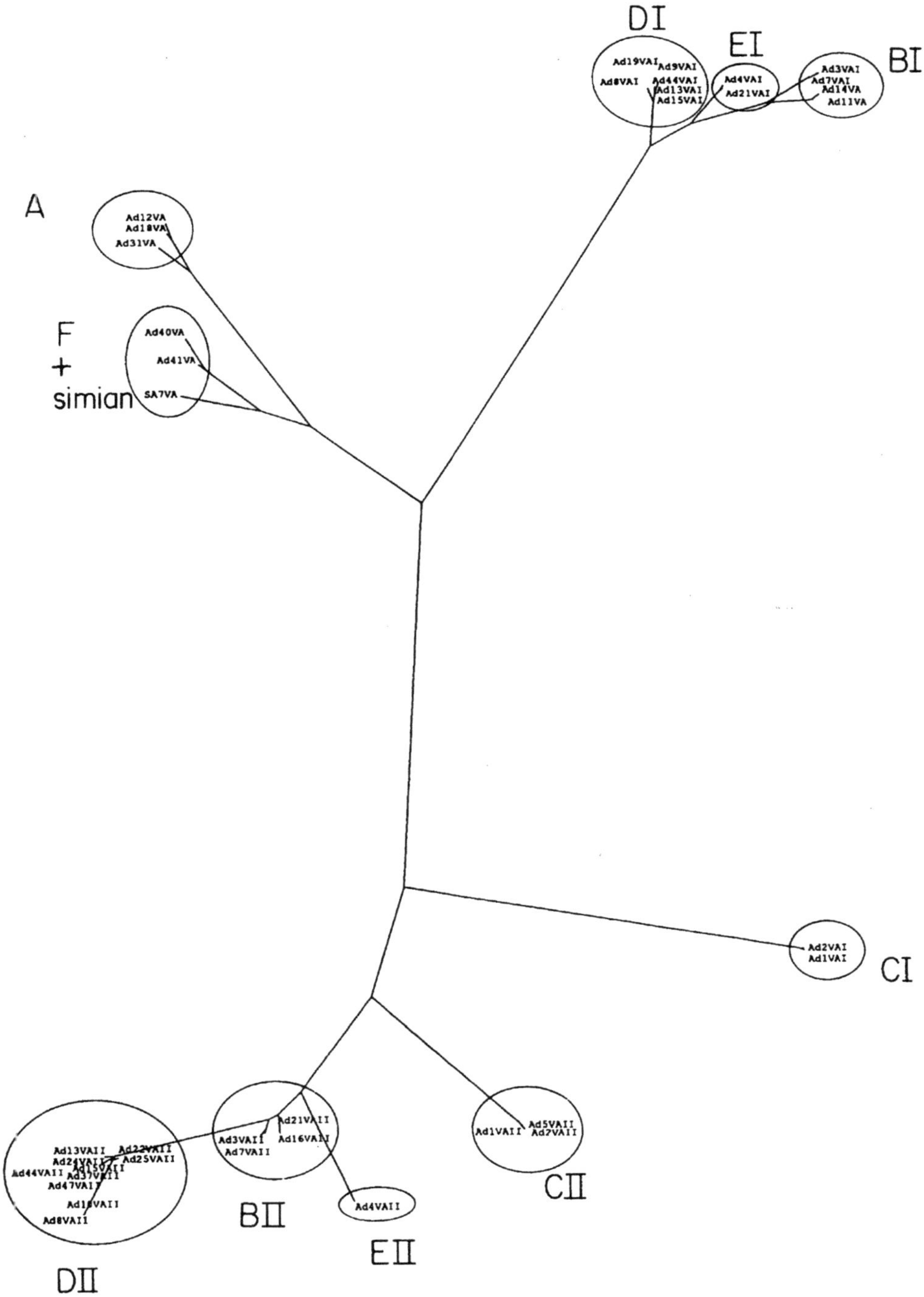

Fig. 2. Unrooted tree showing the sequence relationships among the virus-associated (VA) RNAs of human and simian adenoviruses. The ovals surround clusters of closely similar sequences and are labeled according to the virus group (A–F) and VA RNA type (I or II) in the cluster. When a sequence is found in more than one serotype, only a single name is shown to simplify the diagram (e.g., Ad5 VA_I = Ad1 VA_I). (From unpublished data of Y. MA and M.B. MATHEWS)

ten- to 20-fold less VA RNA_{II} accumulates. Nevertheless, bearing in mind that the VA RNA_{I} concentration exceeds 10^8 copies per cell (roughly equivalent to the number of ribosomes), VA RNA_{II} is still one of the most abundant RNA in the cell at late times of infection.

The VA RNAs display a limited heterogeneity at their 5' and 3' ends. Start site selection is influenced by sequences at the 5' end of the gene as well as a short distance upstream (THIMMAPPAYA et al. 1979). The precise position of the 3' end within the run of T residues, as well as the choice of T-rich runs when there is more than one available, is determined by factors not fully understood. The functional significance, if any, of the heterogeneity is also a mystery. Consistent with the known function of VA RNA_{I}, at least a fraction of it is found in association with ribosomes upon biochemical fractionation of infected cell extracts (SCHNEIDER et al. 1984). DAI is also in part ribosome associated (LANGLAND and JACOBS 1992). At the ultrastructural level, VA RNA is distributed rather uniformly throughout the cytoplasm, as expected. It also appears in the nucleoplasm as fine dots which grow larger during the late phase of infection (JIMÉNEZ-GARCIA et al. 1993). DAI, on the other hand, is seen in the nucleolus and diffusely distributed in the nucleoplasm, so the two moieties do not colocalize in the nucleus. Conceivably, this merely reflects their sites of synthesis (in the case of VA RNA) or assembly (in the case of DAI), but alternative explanations cannot be ruled out: for example, one or both of them may have an additional function, independent of the other, in the nucleus. Binding to a nuclear component, such as the La antigen, an RNA-binding protein that interacts with the run of 3' uridylate residues on VA RNA (MATHEWS and FRANCOEUR 1984) and is at least partly nuclear, could be responsible for the appearance of the RNA in this compartment of the cell.

4 Secondary Structure

Ever since the sequence of the Ad2 VA RNA became available, it was evident that the molecule is highly folded, and attempts were made to predict its secondary structure from thermodynamic parameters. However, none of the most stable computer-generated structures was entirely consistent with the experimental results obtained using nucleases to probe for regions of single- or double-strandedness. Accordingly, second-generation models were created to take account of the nuclease sensitivity data as well as the base-pairing potential inherent in the sequence (MELLITS and MATHEWS 1988; FURTADO et al.1989). An example of such a model, one of several that fits the data to a large degree, is shown in Fig. 3 (left). This particular structure proved quite robust, serving as a working model for several years. It consists of three regions: a terminal stem, in which the 5' and 3' ends of the molecule are paired; an apical stem-loop, also containing an extended region of duplexed RNA; and a connecting portion, dubbed the central domain, composed of a series of stems and loops.

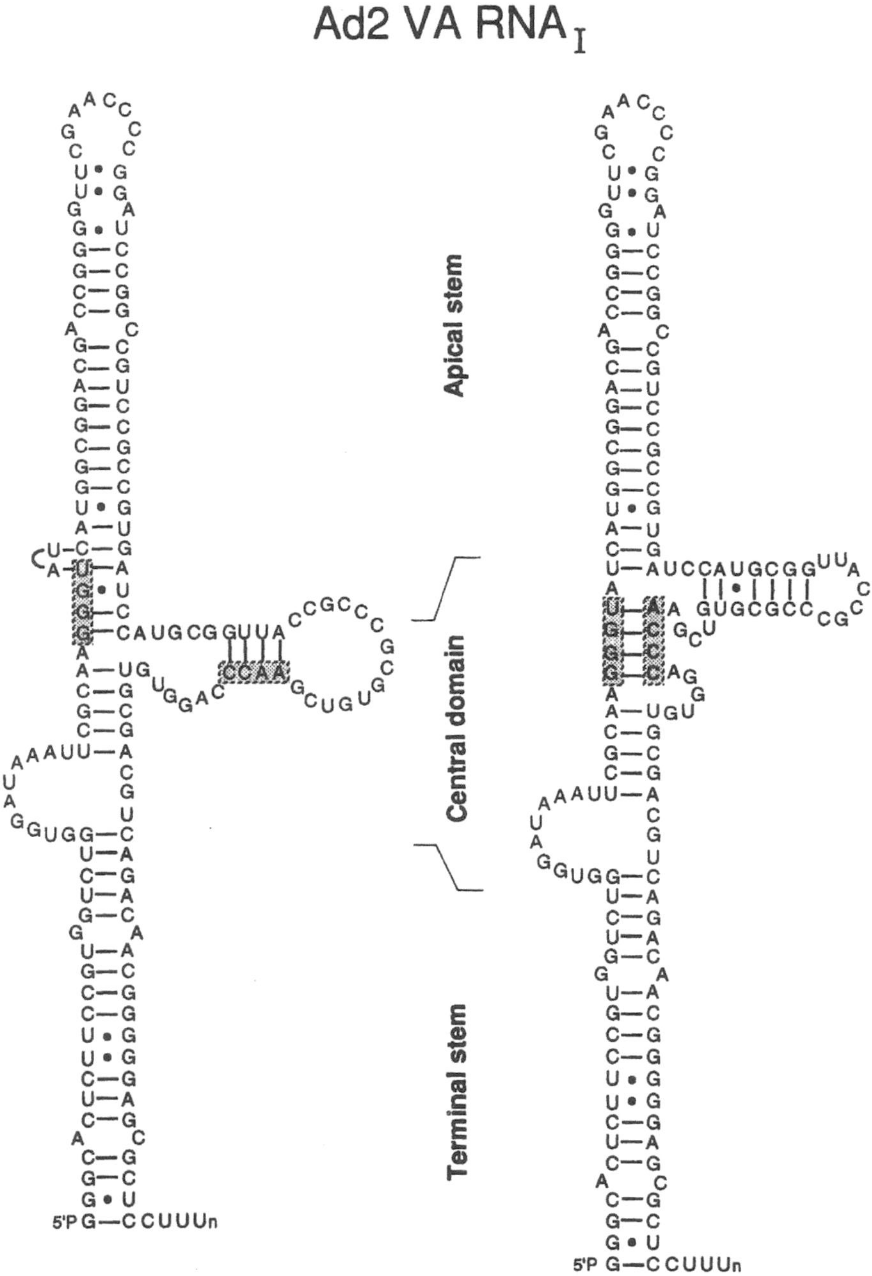

Fig. 3. Structure of adenovirus type 2 (Ad2) virus-associated (VA) RNA$_I$. Left, earlier structure, based largely on nuclease sensitivity analysis (MATHEWS and MELLITS 1988). Right, recent model, incorporating pairing of the conserved tetranucleotides GGGU and ACCC (MA and MATHEWS 1993 and unpublished data; CLARKE and MATHEWS 1994). The three principal regions of the molecule are marked

While the structure withstood the test of mutagenesis in the apical stem (Mellits et al.1992) and (to the limited degree that is was examined) in the terminal stem, it succumbed to a detailed mutagenic analysis of the central domain that called into question a short but critical stem in this part of the molecule (Pe'ery et al. 1993). The groundwork for a revised central domain structure was laid by a comparison of the sequences and nuclease sensitivity patterns of the VA RNA from Ad2 and Ad7 (two species each) and Ad12 and SA7 (each with a single VA RNA). This study disclosed the existence of two tetranucleotide sequences that are conserved in all six RNAs (Ma and Mathews 1993). Suggestive of an important structural role, the two tetranucleotides are mutually complementary (GGGU: ACCC), are present in equivalent positions in the several VA RNA species, and are insensitive to single-strand-specific nucleases as if they are indeed paired. The revised structural model (Fig. 3, right) incorporates this conserved tetranucleotide pair as a short stem within the central domain and is supported by data from other sources. First, extension of the phylogenetic comparison to include all of the human adenoviruses revealed only two exceptions to the conservation of the GGGU:ACCC pair, and in these cases (Ad40 and Ad41) the change (to GGGU:ACCU) was minor and still compatible with pairing (Y. Ma and M.B. Mathews, manuscript in preparation). Second, mutagenesis of the tetranucleotides bolsters the idea that they pair together (Y. Ma and M.B. Mathews, manuscript in preparation). Third, more refined examination of Ad2 VA RNA_I with both nucleases and chemical probes (Clarke and Mathews 1995) indicates that the central domain probably adopts the conformation shown in Fig. 3 (right).

The revised structure is also compatible overall with the sequences determined for the VA RNA of other serotypes; although their structures have not yet been tested experimentally, it is apparent from sequence comparisons that the blueprint for the VA RNAs – with apical and terminal stems and a central domain – is evolutionarily durable, despite considerable sequence variation. The next challenge is to determine the molecule's tertiary structure. We presently have few clues as to the higher-order structure, but nuclease sensitivity analysis of central domain mutants suggests that folding in this region brings together loops that, in the two-dimensional representation, are depicted as lying far apart (Pe'ery et al. 1993). In view of the functional importance of the central domain, a deeper understanding of the conformation of this part of the molecule is critical for a complete understanding of its interactions with DAI.

5 Interactions Between Virus Associated RNA and DAI

DAI binds selectively to dsRNA, rather than single-stranded RNA, with a marked preference for perfectly duplexed molecules. It also displays a pronounced size dependence: duplexes of less than 30 bp bind poorly; binding efficiency increases steadily between 30 and 85 bp; and longer duplexes bind with high affinity (MANCHE et al. 1992). The RNA-binding domain of DAI lies in its N-terminal third, while the conserved kinase domain is in the C-terminal half of the protein (reviewed by MATHEWS 1993). The RNA-binding domain is composed of two copies of a motif that characterizes a number of cellular and viral RNA-binding proteins (ST JOHNSTON et al. 1992; GREEN and MATHEWS 1992); each copy of the motif is about 67 residues long, rich in basic residues and containing a predicted α-helical region at its C terminus. Both copies of the motif are required for effective RNA binding, and all mutations studied to date seem to affect the binding of dsRNA and VA RNA equally (GREEN and MATHEWS 1992). Because dsRNA and VA RNA compete with one another for binding, they appear to bind at the same site or at sites that are very closely related. This being the case, it would not be surprising if duplex structure in VA RNA played a role in its binding to DAI.

The relationship between VA RNA_I structure and function has been addressed in several ways, ranging from direct binding assays conducted in vitro to functional assays performed in tissue culture cells. Contrary to expectation, these assays unanimously point to the central domain as a key element in VA RNA function, while the contribution of the stems has been disputed. Thus, mutations in the central domain have been shown to reduce the binding of VA RNA_I to DAI in vivo and in vitro, to reduce its ability to block DAI activation by dsRNA in vitro, to reduce the efficiency of virus infection, and to reduce its ability to stimulate expression of a reporter gene in vivo and in vitro (see, for example, GHADGE et al. 1994; CLARKE et al. 1994; reviewed in MATHEWS and SHENK 1991; MATHEWS 1993). Indeed, to the author's knowledge, no central domain mutant has been isolated which displays a wild-type level of function in all assays. Presumably this is because most central domain mutations have been relatively large, while the structure is subtle and intricate; however, even some relatively small alterations, changing as few as two nucleotides, ablate some functions (PE'ERY et al. 1993).

The influence of mutations in stem sequences is less clear-cut, probably because the nature of the requirement is less stringent. Some studies have gone as far as to conclude that there is no stem requirement at all (GHADGE et al. 1991), but a detailed analysis—including the generation of mutations and compensating mutations which destroy and restore pairing in the apical stem—led to the conclusion that the existence of a stem is important, whereas its sequence is less critical (MELLITS et al.1992). This view is supported by recent data (CLARKE and MATHEWS 1995) showing that DAI, as well as the isolated RNA-binding domain of the protein, protects a region of VA RNA encompassing the central domain and adjacent apical stem from attack by nucleases and chemical probes (Fig. 4). It could be argued that the apical stem requirement is more structural than functional, perhaps serving as a scaffold to stabilize the central domain (GHADGE et al. 1994). The absence of a defined sequence requirement in the region of the apical stem proximal to the central domain would be consistent with this view. However, the existence of close contacts in the minor groove of the helix (CLARKE and

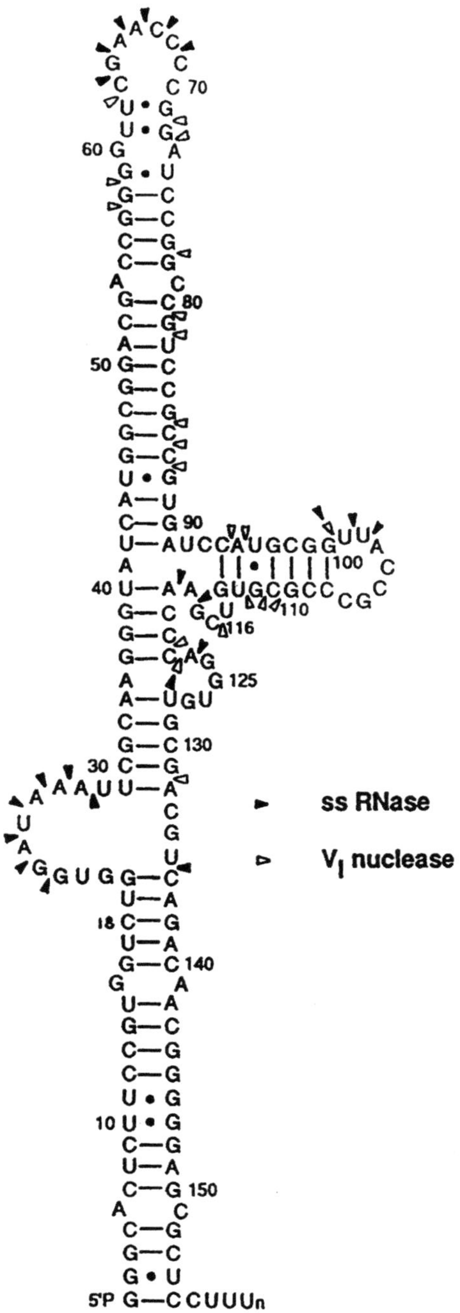

Fig. 4. Binding of DAI to virus-associated (VA) RNA I. Sites of nuclease or chemical attack on adenovirus type 2 (Ad2) VA RNA I that are protected by DAI are indicated by arrowheads: *white*, chemical probes; *black*, RNase digestion (from CLARKE and MATHEWS 1995)

MATHEWS, 1995), together with the absence of discernible changes in the central domain structure contingent on some apical stem alterations which nevertheless hamper function, makes it more likely that the stem plays a direct role in DAI binding.

How does VA RNA_I inhibit DAI activation? The simplest notion – that it serves as a competitive inhibitor of binding, much like short or imperfect RNA duplexes which fail to activate the enzyme – is rendered unlikely by the finding that the central domain, rather than the stems, plays a dominant role in inhibiting DAI. For the same reason, a second otherwise attractive hypothesis seems less plausible. Autophosphorylation and activation of DAI appear to occur by an intramolecular mechanism in which two DAI molecules dimerize on a single dsRNA strand. This could explain the dsRNA length dependence of DAI activation and, by the same token, could explain why the relatively short stems of VA RNA block DAI activation. Unless the central domain contrives to simulate a short stem, this idea cannot account for the role of the central domain, however.

A third possibility, which places the central domain firmly at the focus of the action, is that this part of the molecule interacts with DAI in such a way as to preclude dsRNA binding. This could be accomplished by a conformational change which leads to a distortion of the RNA-binding domain. The chief drawback to this idea is that it fails to take into account the growing body of data suggesting that the binding sites on DAI for VA RNA and dsRNA are the same. Since not all possible mutations in DAI have been made, it remains possible that the sites are similar but not identical, but this point of view is becoming less tenable as more mutants are tested. As a modification of this hypothesis, one might imagine that the central domain binds to the same site as dsRNA; once bound, because of its different structure, this ligand would failto activate the kinase and instead distorts the enzyme in such a way as to preclude activation. Such a view fits most of the known facts, but it is open to the criticism that it excludes the apical stem from playing any significant role in the process.

If all the pieces of information are to be accommodated in a single model, the most comprehensive account would hold that parts of both the apical stem and central domain interact with the RNA-binding domain of DAI and that this inhibits either by causing a distortion (as in the previous model) or, alternatively, by positioning some portion of the VA RNA molecule – presumably part of the central domain – in a sensitive region of the enzyme such that its functioning is inhibited. Not only does this model explain the data described up to this point, but it also could explain the slight differences between DAI and its isolated RNA-binding domain in protection experiments (CLARKE and MATHEWS 1995). However, it will probably require sophisticated biophysical analysis of the VA RNA–DAI complex and its separate components, providing high-resolution structures, to describe the mechanism precisely.

6 Additional Virus-Associated RNA Functions

It seems clear that the predominant, if not the sole, role of Ad2 VA RNA_I, is to prevent the activation of DAI. Other functions which have been postulated – in splicing (SVENSSON and AKUSJÄRVI, 1985), in mRNA stabilization (STRIJKER et al. 1989; SVENSSON and AKUSJÄRVI 1990); in transient expression assays (SVENSSON and AKUSJÄRVI 1985; KAUFMAN and MURTHA 1987), and possibly in the preferential translation of viral versus host mRNA as part of the host cell shutoff phenomenon (reviewed by ZHANG and SCHNEIDER 1993) – are all likely to reflect the inhibition of kinase activity as an underlying mechanism. On the other hand, the significance of VA RNA_{II} is an enigma. It substitutes only poorly for VA RNA_I in functional assays both in vivo and in vitro, so one might be led to believe that VA RNA_{II} is an inert copy, analogous to a pseudogene. The absence of a second VA RNA species from several human adenoviruses, as well as from SA7, could be taken as evidence in support of this view.

Circumstantial evidence speaks in favor of VA RNA_{II} as a useful viral gene, however. First there is the contention that a virus with as compact a genome as adenovirus – sporting many overlapping genes and differentially spliced gene products – would not be expected to waste genetic space on a functionless gene. This admittedly weak argument is bolstered by a much stronger one, drawn from phylogenetic data. The evolutionary tree depicted in Fig. 2 indicates that the VA RNA_{II} sequences are no more diverged from one another than are the VA RNA_I sequences; in fact, they are less divergent because the group C VA RNA_I sequences are relatively distant from those of groups B, E, and D. If there were no function associated with VA RNA_{II}, there would be little pressure to conserve its sequence, so the sequences would be expected to drift apart. Because this has not occurred, there must be some pressure constraining the drift, most likely a VA RNA_{II} function. Third, in a comparison of the single VA RNA of Ad2 and SA7 with the two VA RNAs of Ad2 and Ad7, the VA RNA_I species of Ad2 and Ad7 worked best, their VA RNA_{II} species registered essentially no activity, and the single VA RNA species displayed intermediate efficacy (MA and MATHEWS 1993). This hints that the VA RNAs might serve two distinct functions, DAI inhibition and some function presently unknown. Thus, the VA RNA I and II species might be specialized effectors for DAI inhibition and for the unknown function, respectively, while the single VA RNA species of the group A and F viruses and SA7 could serve both functions. Since it does double duty, the structure of the single VA RNA would represent a compromise between the demands of each function, and in consequence it might not do as well as VA RNA_I in inhibiting the kinase.

What might this unknown function be? A cardinal principle of adenoviral organization is that related functions are clustered on the genome: for example, E1 is concerned with transcriptional regulation, E2 with replication, E3 with neutralizing host defense mechanisms, and so on. Therefore, in imagining a role for VA RNA_{II}, it would be reasonable to consider pathways that are related to the DAI pathway in some way. A front-running candidate involves the enzyme 2'–5' oligoadenylate synthetase. Like DAI, 2–5A synthetase is induced by interferon and is activated by dsRNA (SAMUEL 1991; HOVANESSIAN 1991). It produces a short oligonucleotide of unusual structure, containing 2'–5' linked adenylate residues, which in turn activates a ribonuclease, RNase L, thereby limiting infection by certain viruses. Just as with VA RNA_I and DAI, a mechanism for interfering with this antiviral defense would be beneficial to a virus, and we speculate that VA RNA_{II} may interfere with the dsRNA-mediated activation of 2–5A synthetase, as diagrammed in Fig. 5. Although there is no direct experimental evidence for this hypothesis at present, it could explain why some adenoviruses in group B (e.g., Ad11 and Ad14) seem to have lost their VA RNA_{II} genes secondarily. Perhaps in certain cell types or environments the 2–5A pathway is not a threat to the virus, so a mechanism to neutralize it would be superfluous: in such a case, there would be no advantage to expressing VA RNA_{II}, and the pressures to compact the viral genome might lead to loss of the gene. Further investigations will be required to establish or refute this hypothesis.

7 Concluding Remarks

Investigation of VA RNA structure and function has shed light on the cellular processes that involve this viral gene product as well as on the viral genes

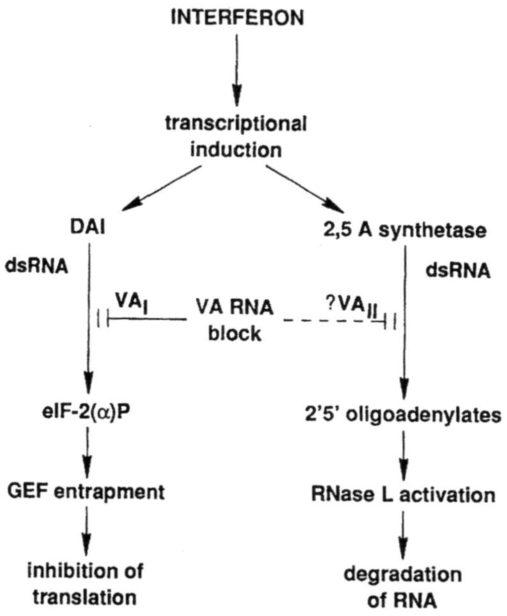

Fig. 5. Hypothetical role of virus-associated (VA) RNA_{II}. By analogy with the role of VA RNA_I in blocking DAI activation (*left*), VA RNA_{II} may block activation of the interferon-induced enzyme 2'–5' oligoadenylate synthetase (see text). *ds*, double-stranded; *GEF*, guanosine nucleotide exchange factor (eIF-2B)

themselves. Because the adenovirus genome is limited in size, the virus confines itself to the essentials and targets important cellular processes for its attention. The study of VA RNA_I has illuminated the functioning of the protein kinase DAI, translational control, and aspects of the interferon response – and continues to do so. Whether the discovery of the function of VA RNA_{II} will be equally instructive remains to be seen.

Acknowledgments. The author is grateful for the contributions of friends and collaborators to this work and to Debbie Taylor for discussions. Work in his own laboratory was supported by grants CA 13106 and AI 34552 from the National Institutes of Health.

References

Bhat RA, Thimmappaya B (1984) Adenovirus mutants with DNA sequence perturbations in the intragenic promoter of VAI RNA gene allow the enhanced transcription of VAII RNA gene in HeLa cells. Nucleic Acids Res 12: 7377–7388

Clarke PA, Pe'ery T, Ma Y, Mathews MB (1994) Structural features of adenovirus 2 virus-associated RNA required for binding to the protein kinase DAI. Nucleic Acids Res 22/21: 4364–4374

Clarke PA, Mathews MB (1995) The binding site for the interferon-induced protein kinase DAI on adenovirus VA RNA_I RNA 1: 7–20

Furtado MR, Subramanian S, Bhat RA, Fowlkes DM, Safer B, Thimmappaya B (1989) Functional dissection of adenovirus VAI RNA. J Virol. 63: 3423–3434

Ghadge GD, Swaminathan S, Katze MG, Thimmappaya B (1991) Binding of the adenovirus VAI RNA to the interferon-induced 68 kDA protein kinase correlates with function. Proc Natl Acad Sci USA 88: 7140–7144.

Ghadge GD, Malhotra P, Furtado MR, Dhar R, Thimmappaya B (1994) In vitro analysis of virus-associated RHAI (VAI RNA): inhibition of the double-stranded RNA-activated protein kinase PKR by the VAI RNA mutants correlates with the In vivo phenotype and the structural integrity of the central domain. J Virol 68: 4137–4151

Green SR, Mathews MB (1992) Two RNA binding motifs in the double-stranded RNA activated protein kinase, DAI. Genes Dev. 6: 2478–2490

Hershey JWB (1991) Translational control in mammalian cells. Annu Rev Biochem 60: 717–755

Hovanessian AG (1991) Interferon-induced and double-stranded RNA-activated enzymes: a specific protein kinase and 2',5'-oligoadenylate synthetases. J Interferon Res 11: 199–205

Jimnez-Garcia LF, Green SR, Mathews MB, Spector DL (1993) Organization of the double-stranded RNA-activated protein kinase DAI and virus-associated VA RNA_I in adenovirus-2 infected HeLa cells. J Cell Sci 106: 11–22

Kaufman RJ, Murtha P (1987) Translational control mediated by eukaryotic initiation factor-2 is restricted to specific mRNAs in transfected cells. Mol Cell Biol 7: 1568–1571

Langland JO, Jacobs BL (1992) Cytosolic double-stranded RNA-dependent protein kinase is likely a dimer of partially phosphorylated M_r=66,000 subunits. J Biol Chem 267: 10729–10736

Larsson S, Bellett A, Akusjärvi G (1986) VA RNAs from avian and human adenoviruses: dramatic differences in length, sequence, and gene location. J Virol 58: 600–609

Ma Y, Mathews MB (1993) Comparative analysis of the structure and function of adenovirus virus-associated RNAs. J Virol 67: 6605–6617

Manche L, Green SR, Schmedt C, Mathews MB (1992) Interactions between double-stranded RNA regulators and the protein kinase DAI. Mol Cell Biol 12: 5238–5248

Maran A, Mathews MB, (1988) Characterization of the double-stranded RNA implicated in the inhibition of protein synthesis in cells infected with a mutant adenovirus defective for VA RNA_I. Virology 164: 106–113

Mathews MB (1975) Genes for VA-RNA in adenovirus 2. Cell 6: 223–229

Mathews MB (1993) Viral evasion of cellular defense mechanisms: regulation of the protein kinase DAI by RNA effectors. Semin Virol 4: 247–257

Mathews MB, Francoeur AM (1984) La antigen recognizes and binds to the 3'-oligouridylate tail of a small RNA. Mol Cell Biol 4: 1134–1140
Mathews MB, Shenk T (1991) Adenovirus virus-associated RNA and translational control. J Virol 65: 5657–5662
Mellits KH, Mathews MB, (1988) Effects of mutations in stem and loop regions on the structure and function of adenovirus VA RNA_I. EMBO J 7: 2849–2859
Mellits KH, Pe'ery T, Mathews MB (1992) Role of the apical stem in maintaining the structure and function of adenovirus virus-associated RNA stem. J Virol 66: 2369–2377
Pe'ery T, Mellits KH, Mathews MB (1993) Mutational analysis of the central domain of adenovirus VA RNA mandates a revision of the proposed secondary structure. J Virol 67: 3534–3543
Pettersson U, Philipson L (1975) Location of sequences on the adenovirus genome coding for the 5.5S RNA. Cell 6: 1–4
Reich PR, Rose J, Forget B, Weissman SM (1966) RNA of low molecular weight in KB cells infected with Ad2. J Mol Biol, 17: 428–439
Samuel CE (1991) Antiviral actions of interferon: interferon-regulated cellular proteins and their surprisingly selective antiviral activities. Virology 183: 1–11
Schneider RJ, Weinberger C, Shenk T (1984) Adenovirus VAI RNA facilitates the initiation of translation in virus-infected cells. Cell 37: 291–298
Söderlund H, Pettersson U, Vennström B, Philipson L, Mathews MB (1976) A new species of virus-coded low molecular weight RNA from cells infected with Ad2. Cell 7: 585–593
St Johnston D, Brown NH, Gall JG, Jantsch M (1992) A conserved double-stranded RNA binding domain. Proc Natl Acad Sci USA 89: 10979–10983
Strijker R, Fritz DT, Levinson AD (1989) Adenovirus VAI-RNA regulates gene expression by controlling stability of ribosome-bound RNAs. EMBO J 8: 2669–2675
Svensson C, Akusjärvi G (1985) Adenovirus VA RNA_I mediates a translational stimulation which is not restricted to the viral mRNAs. EMBO J 4: 957–964
Svensson C, Akusjärvi G (1990) A novel effect of adenovirus VA RNA_I on cytoplasmic mRNA abundance. Virology, 174: 613–617
Thimmappaya B, Jones N, Shenk T (1979) A mutation which alters initiation of transcription by RNA polymerase III on the Ad5 chromosome. Cell 18: 947–954
Thimmappaya B, Weinberger C, Schneider, RJ, Shenk T (1982) Adenovirus VAI RNA is required for efficient translation of viral mRNAs at late times after infection. Cell 31: 543–551
Zhang Y, Schneider RJ (1993) Adenovirus inhibition of cellular protein synthesis and the specific translation of late viral mRNAs. Semin Virol 4: 229–236

The Complete Nucleotide Sequence of the DNA of Human Adenovirus Type 12

J. SPRENGEL[1], B. SCHMITZ[1], D. HEUSS-NEITZEL[1,2], and W. DOERFLER[1]

1 Introduction

In earlier work from this laboratory, we used human adenovirus type 12 (Ad12) extensively for basic research on the mechanism and the consequences of foreign DNA integration into mammalian genomes (DOERFLER 1968, 1970; for recent reviews DOERFLER 1991, 1992, 1993, 1995; FECHTELER et al. 1995). The investigations on Ad12 DNA integration led to a long-standing interest in the biological significance of DNA methylation, its role in the long-term silencing of eukaryotic promoters, and the mechanism of de novo methylation of foreign DNA in mammalian cells (SUTTER et al. 1978; SUTTER and DOERFLER 1980; DOERFLER 1981, 1983, 1991, 1993, 1995). In the course of these studies it became necessary to perform nucleotide sequence determinations in various parts of the Ad12 genome. Recently, the entire nucleotide sequence was completed (SPRENGEL et al. 1994) and made generally available under EMBO Accession Number X73487. Complete nucleotide sequences were thus available for the DNAs of adenovirus type 2 (Ad2) (ROBERTS et al. 1986), Ad5 (CHROBOCZEK et al. 1992), Ad12 (SPRENGEL et al. 1994), and Ad40 (DAVISON et al. 1993; MAUTNER et al. 1995).

During the productive infection of human cells in culture with Ad2 or Ad12, it was shown that the viral genome (Ad2) could recombine with human cellular DNA

[1] Institut für Genetik, Universität zu Köln, Weyertal 121, 50931 Cologne, Germany
[2] Present address: Qiagen, Max-Volmer-Str. 4, 40724 Hilden, Germany

(Burger and Doerfler 1974; Schick et al. 1976) and that parts of the Ad12 genome were able to form symmetric recombinant (SYREC) molecules with human cell DNA (Deuring et al. 1981; Deuring and Doerfler 1983). These SYREC molecules (Ad12) represented huge palindromes, in one instance with the left terminal 2081 nucleotide pairs (nt.p.) from the left end of Ad12 DNA at either terminus and two symmetric stretches of cellular DNA between them. Upon denaturation and renaturation, half-length molecules could be demonstrated by electron microscopy (Deuring et al. 1981).

In constructing the *Pst*I clones of Ad12 DNA in the pBluescript KS vector, it was therefore mandatory to observe special precautions in order not to include Ad12-cellular DNA recombinants. Ad12 DNA was therefore prepared from the second passage of Ad12 on human embryonic kidney (HEK) cells of a newly purchased Ad12 inoculum from the American Type Culture Collection (ATCC VR-863). The *Pst* I fragments from this Ad12 DNA preparation were immediately cloned. Experimental details were published previously (Sprengel et al. 1994). The entire nucleotide sequence in the double-stranded form was presented at the end of this paper (Fig. 10). The 17 *Pst*I cleavage sites used for cloing were indicated in the nucleotide sequence.

It was reported previously that DNA sequences in the nucleic acid data bases GenBank or EMBL were not infrequently contaminated with DNA sequences from the vector (Kristensen et al. 1992; Lopez 1992). We paid particular attention to this possibility before reporting the Ad12 nucleotide sequence and were confident that the Ad12 DNA sequence was free of vector contaminations.

The total nucleotide sequence of 34 125 nucleotide pairs was determined independently for both DNA strands by using appropriate synthetic oligodeoxyribonucleotide primers and a 373A DNA Synthesizer of Applied Biosystems. At randomly selected sites, a total of 3513 nt.p. (10.3%) of this nucleotide sequence was also determined "by hand" in a simple gel electrophoresis apparatus. Discrepancies between these results and those of the automated procedure were not found.

When we compared our nucleotide sequence with partial Ad12 DNA sequences reported earlier from several other laboratories, occasional deviations were apparent, as reported (Sprengel et al. 1994). Considering the different origins and/or passage histories of various Ad12 isolates used in nucleotide sequence work, such variations were to be expected.

2 Regions of Repetitive Sequences

The interest in repetitive DNA sequences, especially in their functional significance and in their facility to expand triplet repeats, led us to search for such structures in the Ad12 genome. For this search, a synthetic query DNA sequence was used with the programs COMPARE/DOTPLOT from the GCG program package (DEVEREUX et al. 1984). An artificial query sequence was constructed by creating multiple repeats for all 64 triplet sequences, regardless of the redundancy which was introduced by the complementarity of DNA and the conversion of one triplet into another due to frame shifts.

The relatively high copy number of 200 was chosen to simplify annotations. The advantage of visualizing repeats in this way was based on the ability to modify the stringency of the repeat regions. Application of this method revealed repeats and repeat-like segments as indicated by vertical lines in Fig. 1. The projections of vertical lines to the abscissa marked the positions in the Ad12 genome, the projections to the ordinate individual triplets.

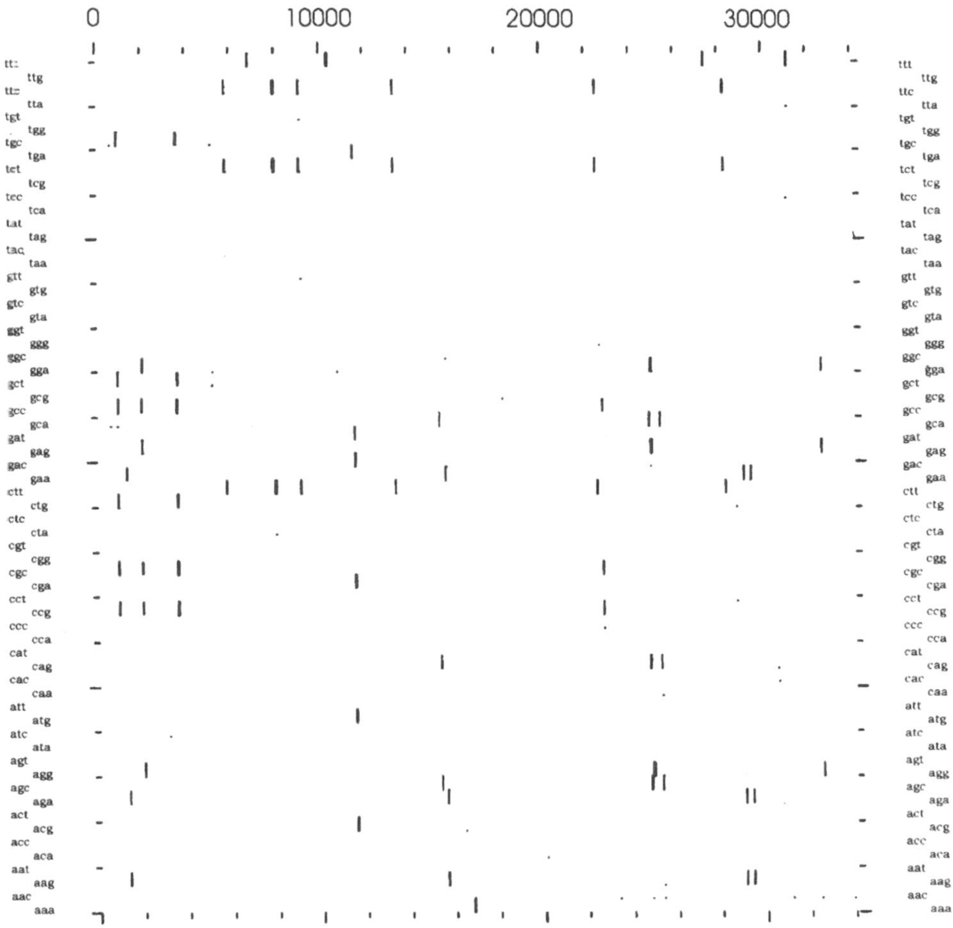

Fig. 1. Repeat-like regions in the human adenovirus type 12 genome. This dot plot was generated by the programs COMPARE/DOTPLOT. Window size and stringency were set at 21 and 16.0, respectively. The query sequence used to probe for repeat-like structures contained all 64 possible triplets, each with a copy number of 200. Note that part of the information is redundant. *Vertical bars* indicate the positions of repeat-like regions in the Ad12 genome

One of these repeat-like regions was located in the *Pst*I-E fragment close to the region of the E2-promoter of the Ad12 genome. The identified GGA repeat also presented a purine-rich DNA sequence motif which was thought to contribute to enhancer activities in SV40 and lymphotropic papovavirus DNAs (PETTERSON and SCHAFFNER 1987).

The alignment in Fig. 2 was produced with the program lfasta (PEARSON and LIPMAN 1988) and revealed a multiple repeat of the triplet GGA in its core region. If the alignment was transposed/into the usual notation in which the sequence patterns were presented, the repetitive structure at position 23 885 in Ad12 DNA became distinct:

$$(\mathit{GGA})(\mathrm{nGA})\mathrm{n}_3(\mathit{GGA})_4(\mathit{GGn})(\mathit{GGA})\mathrm{n}_6(\mathrm{nGA})_2(\mathit{GGA})\mathrm{n}_3(\mathrm{nGA})(\mathit{GGA})_4$$

Although non-GGA triplets were also part of the segment, the interspersed bases were in-frame and the whole segment could be presented as an $(\mathrm{nnn})_{20}$ repeat. The $(\mathrm{nnn})_{20}$ sequence stretch carried 11 GGA-triplets together with five less stringent nGA and GGn triplets and was interrupted by only four triplets with different compositions. This striking repeat alignment could also be extended upstream to nucleotide 24 828.

Constructing a "more perfect" $(\mathrm{GGA})_{17}$ triplet segment by mutating the corresponding nucleotides, we easily could demonstrate the effect of an adeno-viral-human interaction correlated to the fidelity of the (GGA) repeat in the virus. Exchanging the internal GGC to a (GGA) would result in a $(\mathrm{GGA})_6$ repeat without affecting the amino acid translation in this frame. All triplets code for the amino acid glycine.

```
Tripl        GGAGGAGGAGGAGGAGGAGGAGGAGGAGGAGGAGGAGGAGGAGGAGGAGGAGG
             |||      |||||| ||      ||||    |||  ||||| ||||  ||||
Ad12         GGAAATCTGGGAGGAAGAATCTTGGGAGAGTCAGGCAGAGGACGAGGTCGAGG
           24830     24840     24850     24860     24870     24880

Tripl  AGGAGGAGGAGGAGGAGGAGGAGGAGGAGGAGGAGGAGGAGGAGGAGGAGGAGGAGGAGG
       |   ||| ||   |||||||||||||| |||       || |||||    ||||||||||
Ad12   ACTTGGAAGAATGGGAGGAGGAGGAGGCGGACAGCCTAGACGAGGATCCAGAGGAGGAGG
            24890     24900     24910     24920     24930     24940

Tripl  AGGAGGAGGAGGAGGA-GGAG
       |||| |      |||| ||||
Ad12   AGGAAGGTTCCAAGGACGGAG
```

Fig. 2. Core region of the alignment of the query GGA-repeat sequence to Ad12 DNA

3 Nucleotide Sequence Comparison to Other Adenovirus Genomes

Up to data base release 41 of December 1994 in the EMBL data base, the complete nucleic acid sequences of adenovirus type 2 (ROBERTS et al. 1986), type 5 (CHROBOCZEK et al. 1992), type 12 (SPRENGEL et al. 1994), and type 40 (DAVISON et al. 1993) were published. Table 1 summarizes general information about these sequences.

To elucidate similarities or, perhaps more interestingly, differences between these genomes, we constructed a multiple alignment of all four sequences using the program CLUSTAL V (HIGGINS et al. 1992). A similarity profile based on these multiple alignments is shown in Fig. 3.

Although the profile provides interesting information about locations and sizes of divergent or similar regions, one should be careful not to overinterpret these results for the following reasons:

1. Alignment of four different DNA sequences of slightly different lengths of about 3.5×10^4 nt.p. requires an "overall best solution", which could be compromised by local translocations, inversions, and duplications in certain sequences.
2. Due to the excessive search space and the resulting exorbitant memory requirements and computing power, the alignment had to be constructed using the default parameters inherent in the CLUSTAL V algorithm.
3. The close similarity between Ad2 and Ad5 DNA of nearly 90% biased the profile towards this genome pair.

Nevertheless, Fig. 3 presents some general properties of the sequence similarity profile. This profile was affected by sequence motifs, similarities and divergent regions with a length of at least one third of the window size. Depending on the window in which similarities were determined, short-, mid-, and long-range phenomena had to be distinguished. For a general overview, we screened the profile with a 4000-nucleotide window (insert in Fig. 3). Nucleotide numbers reflect positions of the multiple alignment but not positions in one of the analyzed adenovirus genomes.

Table 1. Completely sequenced adenovirus genomes; lengths of genomes are given in nucleotides

Type	Group	Cons[a]	Length	ID[b]	AC[b]	
2	E	Yes	35 937	AD2	J01917	ROBERTS et al. (1986)
5	E	Yes	35 935	ADRCOMPGE	M73260	CHROBOCZEK et al.(1992)
12	A	No	34 125	AT12CGA	X73487	SPRENGEL et al. (1994)
40	F	No	34 214	ADRGENOME	L19443	DAVISON et al. (1993)

[a]Indicates whether a sequence was determined as a consensus sequence.
[b]ID and AC, data base identifiers and accession numbers, respectively.

The average similarity between the DNA sequences of the adenoviral genomes was estimated to be about 63%. The type-specific inverted terminal repeats (ITR) at the left and right ends of all adenovirus genomes exhibited a similarity well below the average of about 63%. The remarkable increase in the profile of up to 76% in the region between nucleotides 2 000 and 10 000 reflected the highest similarity. The genes for DNA polymerase, terminal protein, and hexon-associated protein were located in this area. From nucleotide positions

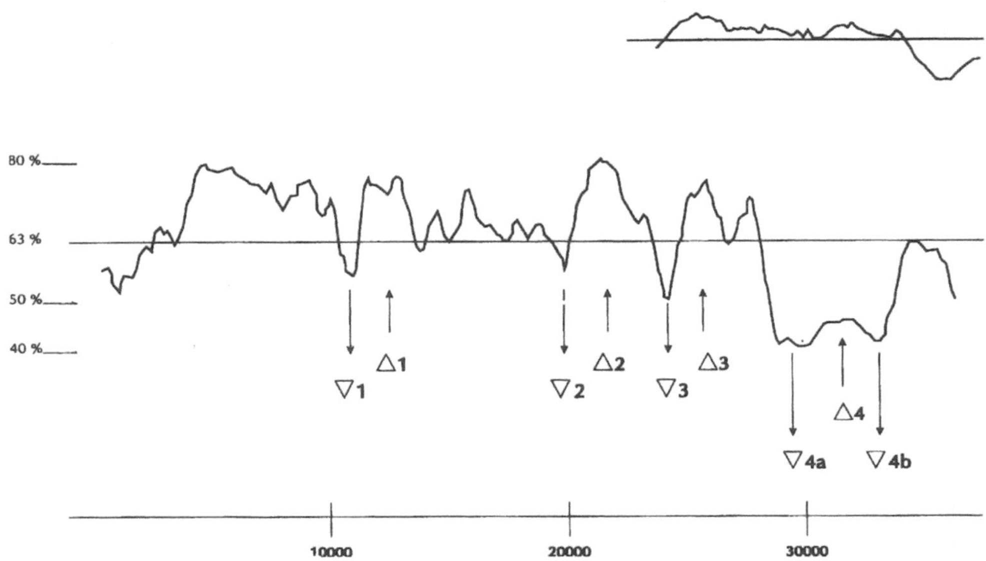

Fig. 3. Similarities among the completely determined adenovirus DNA sequences. Similarity profile is derived from a multiple alignment of all the published, complete sequences of adenoviral genomes. Window size was 1000 nucleotides. *Insert* shows profile determined using a window size of 4000 nucleotides. Numbering reflects positions in the alignment not in the sequence itself. Sequence segments flanking regions of similarity (Δ1, Δ2, Δ3, and Δ4) and of remarkable divergence (∇1, ∇2, ∇3, ∇4a, and ∇4b) were used to probe and to identify the corresponding areas in the adenovirus genomes.

Δ1 This region (positions 11 800–13 200 in the alignment) belonged to the coding regions for the agno DNA binding protein, the LI 52K protein, the DNA polymerase, and the terminal protein and was found in all four genomes with a similarity close to 80%.

Δ2 This peak region (positions 20 400–22 400 in the alignment) contains the hexon gene similar in all adenoviruses.

Δ3 The region around nucleotide 26 000 belongs to the L100 gene and to E2A transcripts.

Δ4 In the remarkable depression of the profile at the *right end* of the genomes, a moderately increased similarity was identified as the region which coded for the fiber protein.

∇1 This depression resulted from a lack of sequence in Ad12 DNA. It was, however, represented in Ad2 and Ad5 DNAs and was identified as the region coding for VAII RNA, which did not exist in Ad12.

∇2, ∇3 These regions of divergence were not directly correlated to a certain function.

∇4a, ∇4b The whole segment (positions 28 000–34 000 in the alignment) corresponded to the intensely spliced regions E3 and E4, the most divergent regions in the adenovirus genomes

10 000 to 21 000 the similarity decreased but was still close to the average value. After an increase of the profile to about 70% at approximately position 21000, a marked depression to 40% similarity correlated with the intensively spliced regions E3 and E4.

For a more elaborate profile analysis we utilized a window size of 1000 nucleotides (Fig. 3). Further narrowing resulted in too much noise to provide useful information. For short-range comparisons, the "overall alignment" was not suitable.

4 Restriction Maps of Ad12 DNA

Table 2 summarizes cleavage sites and the resulting fragment lengths of a number of selected restriction enzymes for Ad12 DNA. In this sequencing project, the cloned *Pst*I fragments of Ad12 DNA were used to determine the entire Ad12 DNA sequence.

Adenovirus research in this laboratory has focused on the status and dynamics in the development of DNA methylation patterns of integrated Ad12 (foreign) DNA in mammalian host cells. Figure 4 presents a graph with all the cleavage sites of the commonly used enzymes *Pst*I, *Eco*RI, *Bam*HI, and *Hind*III, and, in addition, the cleavage sites of the enzymes *Hha*I and *Hpa*II (*Msp*I). *Hha*I and *Hpa*II are methylation-sensitive restriction endonucleases frequently used in research on DNA methylation patterns. To illustrate one major drawback in investigating

Table 2. Restriction maps of human adenovirus type 12

*Pst*I	from - to	size
A	15544 -20884	5341
B	3584 - 8558	4875
C	1 - 3583	3583
D	20885 -24053	3169
E	24054 -26725	2672
F	11108 -12862	1755
G	26726 -28354	1629
H	28355 -29967	1613
I	30629 -32121	1493
J	8559 - 9856	1298
K	14254 -15543	1290
L	33001 -34125	1125
M	32122 -33000	879
N	9857 -10698	842
O	13417 -14253	837
P	29968 -30628	661
Q	12863 -13416	554
R	10699 -11107	409

*Hind*III	from - to	size
A	22719 -27974	5256
B	6233 -10874	4642
C	10955 -14590	3636
D	14591 -18157	3567
E	30828 -34125	3298
F	3708 - 6232	2525
G	1 - 2319	2319
H	27975 -29760	1786
I	2320 - 3707	1388
J1	19051 -20159	1109
J2	20160 -21199	1040
K	29836 -30827	992
L	22043 -22718	676
M	18158 -18695	538
N	21200 -21682	483
O	21683 -22042	360
P	18696 -19050	355
Q	10875 -10954	80
R	29761 -29835	75

*Eco*RI	from - to	size
A	22021 -34125	12105
B	9587 -18804	9218
C	1 - 5574	5574
D	5575 - 9586	4012
E	18805 -21299	2495
F	21300 -22020	721

*Bam*HI	from - to	size
A	1 - 7999	7999
B	24925 -30121	5197
C	20362 -24924	4563
D	11612 -16090	4479
E	30122 -34125	4004
F	17461 -20361	2901
G	8000 -10331	2332
H	10332 -11611	1280
I	16091 -17224	1134
J	17225 -17460	236

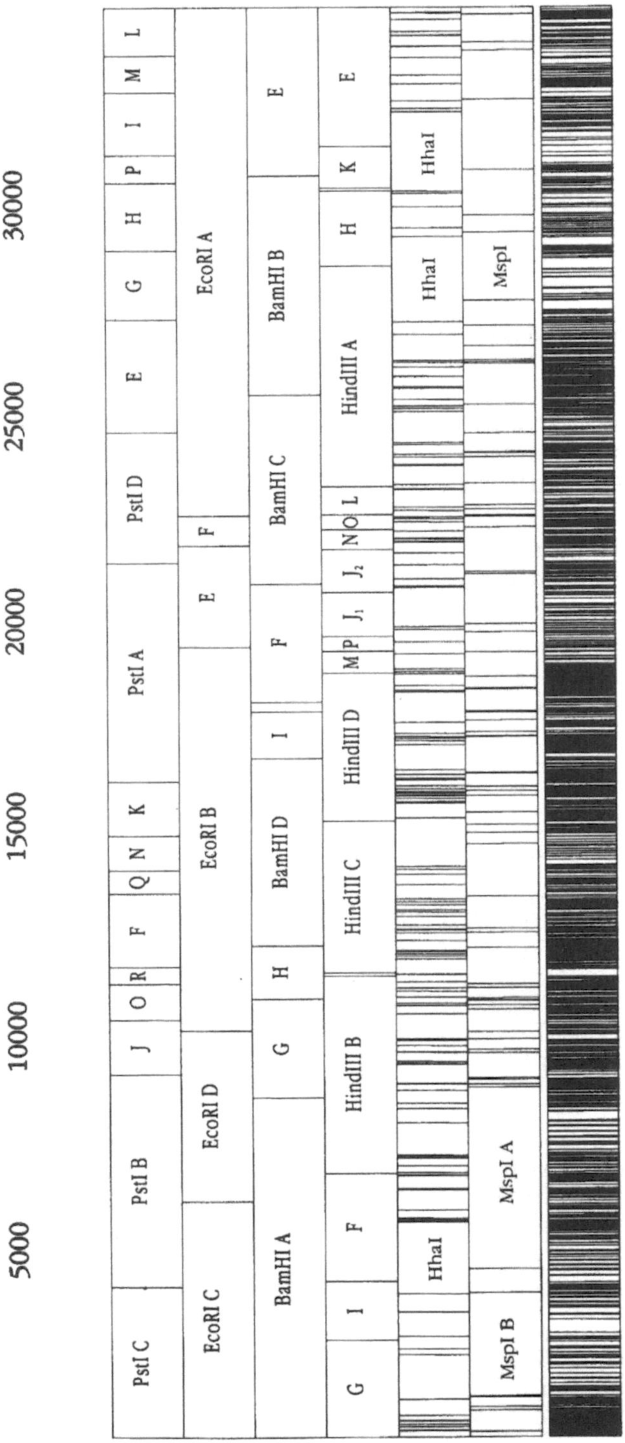

Fig. 4. Restriction maps of Ad12 DNA. *Top* to *bottom*: the *Pst*I, *Eco*RI, *Bam*HI, and *Hind*III cleavage sites are shown. For details on nucleotide numbers and fragment sizes, see Table 2. The three bottom-most maps represent all recognition sites of *Hha*I and *Msp*I-*Hpa*II sites, as well as the 1500 5'-CG-3' dinucleotides found in the Ad12 genome

the target sites of the DNA methyltransferase system (5'-CG-3') with a few methylation-sensitive enzymes, Fig. 4 contrasts all 1500 5'-CG-3' dinucleotides in the Ad12 genome with the 75 *HpaII-MspI* (5'-CCGG-3') and the 181 *HhaI* (5'-GCGC-3') sites. Thus, of the 1500 sequences that could become methylated only 256, i.e., about 17%, could be assessed for their methylation status by the two methylation-sensitive restrictases.

5 Open Reading Frames in the Ad12 Genome

Table 3 summarizes start signals, reading frames, and lengths in amino acids of the putative proteins deduced from the open reading frames in the Ad12 DNA sequence. Moreover, the identifiers of all known homologous adenoviral proteins published in the SwissProt protein data base are included in the table. The proteins in the E3 and E4 regions, which are subject to intensive splicing, are not specified in detail. We did not try to predict native proteins in these regions. Figure 5 presents the organization of and landmarks in the Ad12 genome, i.e., the locations of prominent reading frames, the matching protein identifiers, and the location of the adenoviral promoters, which are indicated by triangles pointing in the direction of transcription.

The open reading frame for the Ad12-encoded fiber protein (FIBP) (Table 4) was located in the intensely spliced region E3. The classification of the adenovirus subgroups has been based on the specificity of serological reactions. Heterogeneity in the amino acid sequences (Fig. 6) reflected serological divergence. This observation is consistent with the structurally and biologically exposed location of the fiber protein on the virus surface. The fiber is predominantly recognized as antigen by the host's immunological defense.

6 The Viral Endoprotease

The important function of the endoprotease VPRT in virus maturation possibly explains the much more homogeneous similarity profile as compared with that of the fiber protein (Figs. 6, 7).

The open reading frame of the Ad12 endoprotease consists of 257 amino acids, whereas all other published adenoviral endoprotease sequences have a total length of around 210 amino acids. We repeatedly redetermined the nucleotide sequence in this region of the Ad12 genome and ascertained this striking length difference of > 50 amino acids. So far, the Ad12 endoprotease has not yet been isolated and analyzed from Ad12-infected cells. Sequence analyses for a cryptic splice site or a presumptive processing signal for the N-terminal 51 amino acid

Table 3. Adenovirus proteins

Human adenovirus type 12 reading frame Start		Length	Protein	SwissProt entry[a] Prefix	Serotype
nt		aa			
503	B	198	E1A Protein	E1A	12, S7, 7, 4, 2, 5, 40, NT, M1
1542	C	163	E1B small antigen	E1BS	12, S7, 4, 40, 41, 2, 5, 7, NT, M1
1847	B	482	EIB large antigen	E1BL	12, 7, 2, 5, 40, 41, NT, C2, M1
3374	B	144	Hexon-associated protein IX	HEX 9	12, 2, 5, 7, NT, C2
6142	A	52		Y115	7, 2
7602	C	373	DNA-binding protein	DNBI	2, 7
8438	B	43		Y145	7
10428	C	205	Late L1 52-kD protein	L52	2, 5, 7
11570	B	582	Hexon-associated protein	HEX3	2, 5
13394	B	497	Penton, protein III	PEN3	5, 2
14902	A	188	Major core, protein VII	VCO7	2, 5
15500	B	347	Minor core, protein V	VCOM	2, 5
16568	B	72	11kD protein	11KD	2
16843	A	265	Protein VI, precursor	PIV6	2, 5, 41
17740	A	919	Hexon, late L2 protein	HEX	41, 40, 5, 2, B3, 12
20372	B	51	(see text)		
20525	B	206	Endoprotease, late L3	VPRT	12, 41, 40, 2, 5, 3, B3, 4, B7
22695	C	782	Late 100-kD	L100	5, 2, 41, 40
22927	A	100		YL11	41
25202	B	118	33-kD phosphoprotein	V33P	2, 5, 41, 40
25612	A	233	Hexon-associated protein VIII	HEX8	41, 2, 5, 3, M1, C1, 40
26313	C	105	Protein of the E3 region	E312	3, 2, 5
28207	A	91	Protein of the E3 region	E310	3, 7, 2, 5
28479	C	110	Protein of the E3 region	E315	2, 5, 3, 7
28804	A	128	Protein of the E3 region	E314	5, 2, 7, 3
29368	A	587	Fiber protein	FIBP	2, 5, 41, 40, 40, 41, 3, M1, 7, C1
5202	F	452	Maturation protein IVa2	PIV2	5, 2, 7, M1, 12
8138	E	1061	DNA polymerase	DPOL	12, 2, 5, 7
10132	D	606	Terminal protein	TERM	12, 2, 5, 7
22669	D	484	E2A DNA-binding protein	DNB2	12, 40, 41, 5, 2, 4, 7
23753	E	43		YL12	41
24067	D	53		YL13	41
31407	F	74	Protein of the E4 region	E417	2, 3
32311	D	291	Protein of the E4 region	E434	2, M1, 2
32606	E	120	Protein of the E4 region	E413	2
32963	E	116	Protein of the E4 region	E411	5, 2
33355	D	131	Protein of the E4 region	EXXK	2
33770	E	127	Protein of the E4 region	EXXK	2

nt, Nucleotides; *aa*, amino acids.
[a]Known adenoviral proteins published in the SwissProt database. In addition to the human adenoviruses, viruses with the following host ranges were taken into account: *NT*, Tupaja; *M1*, mouse type1; *C1*, *C2*, canine types 1 and 2; *B3*, *B7* bovine types 3 and 7; *S7*, simian type 7.

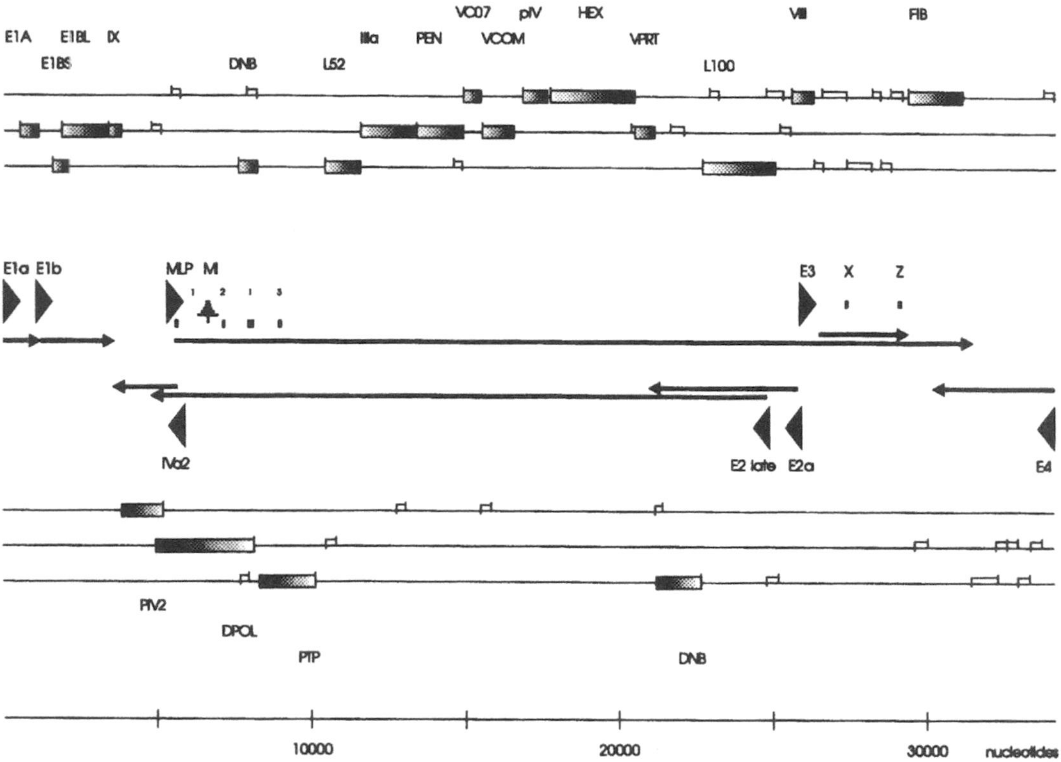

Fig. 5. Organization of the Ad12 genome. *Shaded boxes* represent identified proteins; *open boxes* indicate reading frames of 80 amino acids minimum. Promoters are designated by *filled triangles*

Table 4. Adenoviral fiber proteins

Type	Length (amino acids)	Weight (KD)	Reference
2	582	61.9	HÉRISSÉ et al. (1981)
3	319	34.8	SIGNÄS et al. (1985)
5	581	61.6	CHROBOCZEK and JACROT (1987)
7	343	37.3	HONG et al. (1988)
12	587	61.7	SPRENGEL et al. (1994)
40	547	59.1	KIDD and ERASMUS (1989)
41	562	60.6	PIENIAZEK et al. (1989)
C1	543	57.0	DRAGULEV et al. (1991)
M1	613	66.8	RAVIPRAKASH et al. (1989)

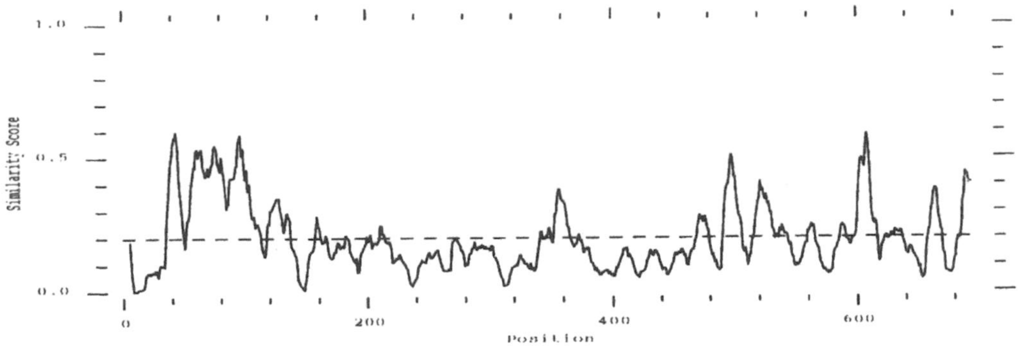

Fig. 6. Similarity profile of adenoviral fiber proteins, deduced from the multiple alignment shown in Fig. 7. Numbering reflects positions in the alignment not in the nucleotide sequence

peptide did not reveal any clues. This N-terminal peptide did not exhibit significant similarities to known proteins or protein motifs taken from the Swiss-Prot data base, suggesting that this peptide might not belong to the native protein.

Table 5 compares published amino acid sequences, their lengths, molecular weights, and similarities of adenoviral endoproteases (Fig. 8) with the same parameters of the Ad12 endoprotease. We observed one deviation in the amino acid sequence deduced for the endoprotease from our Ad12 DNA sequence to the previously published Ad12 endoprotease sequence (HOUDE and WEBER 1988b) (see highlighted amino acid in Fig. 9, at position 191 in the alignment sequence): Serine^{180} in VPRT-Ad12 (HOUDE and WEBER 1988b) was found to be an asparagine^{180}.

The adenoviral endoprotease has been shown to cleave specifically alanine-glycine-alanine residues in the virus precursor proteins TERM, pVI, pVII, IIa and 11kD (AKUSJÄRVI et al. 1981) and is thus essential for virus maturation. The catalytically active residues histidine^{60}, aspartic^{108}, and serine^{180}, as well as the

Table 5. Adenoviral endoproteases

Type	Length amino acids	Weight (KD)	Similarity (%)	Reference
2	204	23.1	75.9	AKUSJÄRVI et al. (1981)
3	209	23.8	80.1	HOUDE and WEBER (1988a)
4	201	22.8	70.6	HOUDE and WEBER (1987)
5	204	23.0	75.9	KRUIJER et al. (1980)
12[a]	206	23.5	99.5	HOUDE and WEBER (1988b)
*12	206	23.5	100	SPRENGEL et al. (1994)
40	205	23.3	77.1	VOS et al. (1988)
41	214	24.5	79.9	(VOS et al. (1988)
B3	204	23.3	67.2	(CAI et al. (1990a)
B7	202	23.3	39.1	CAI et al. 1990b)

[a]The previously published protein sequence of the Ad12 endoprotease (HOUDE and WEBER, 1988b) and our deduced protein sequence (*12) differed in one position. The similarity was related to the *12 VPRT protein sequence.

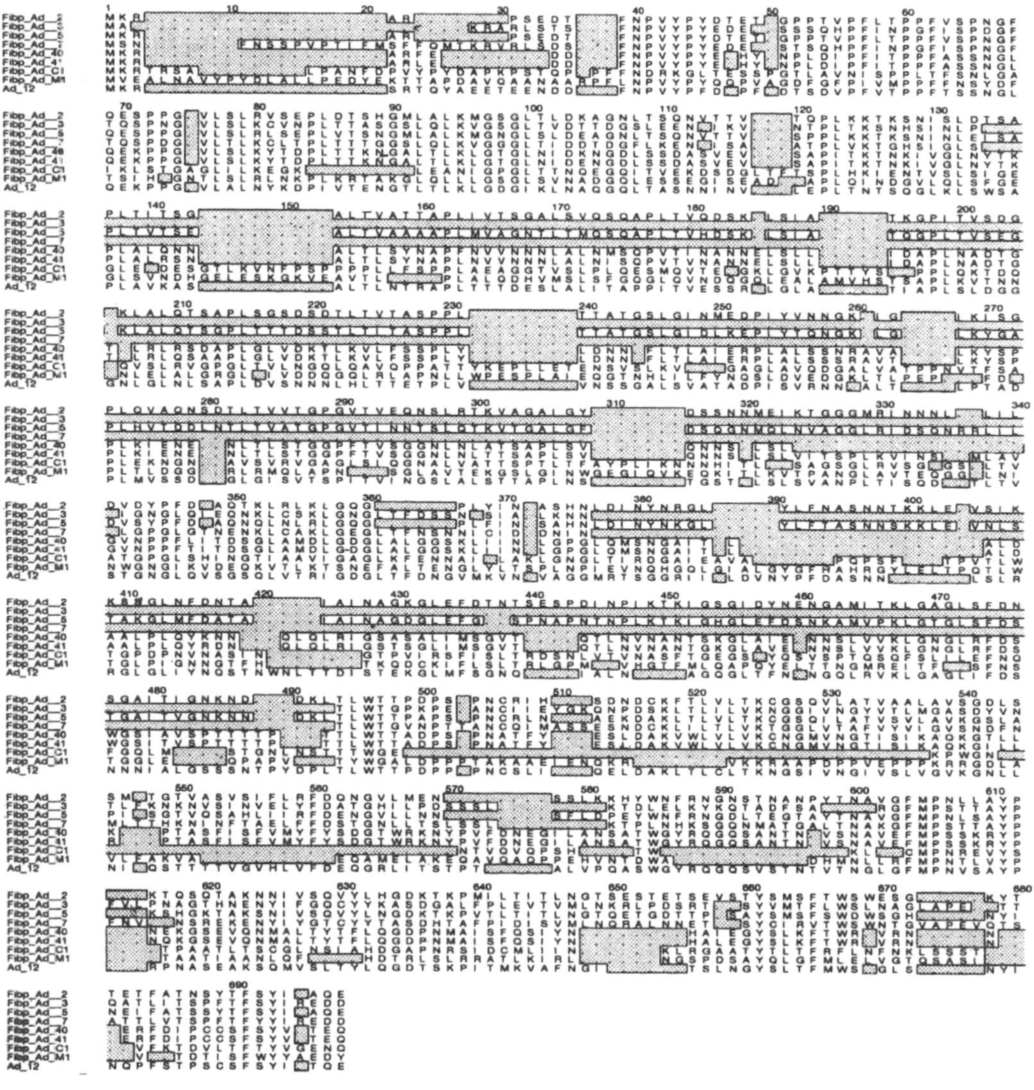

Fig. 7. Multiple alignment of the amino acid sequences of known adenovirus fiber proteins (*FIBP*). Gaps introduced by the alignment program are *shaded*. The following adenovirus serotypes are included: 2, 3, 5, 7, 40, 41, C1, M1 and 12. Bottom-most entry shows the protein sequence derived from the completely sequenced human adenovirus type 12 genome

internal cleavage site alanine-glycine[52–53], are parts of the highly conserved regions in these adenovirus proteins (see Figs. 8, 9).

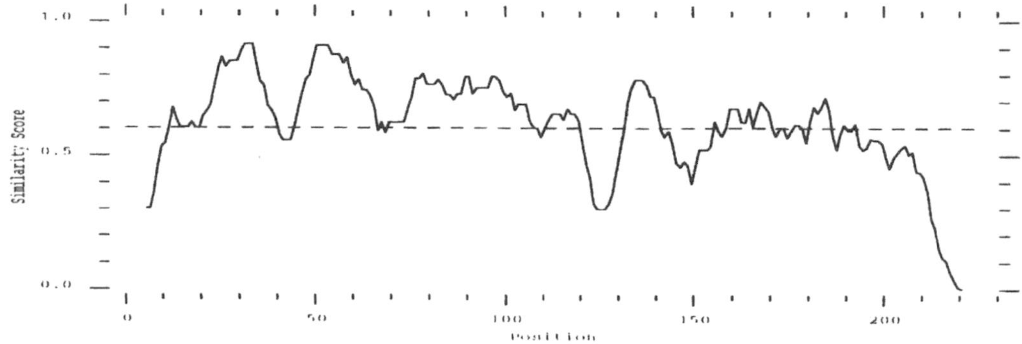

Fig. 8. Similarity profile of adenoviral endoproteases, deduced from the multiple alignment shown in Fig. 9, considers identical amino acid residues. Charge similarities of the amino acids are not taken into consideration. Numbering reflects positions in the alignment, not in the nucleotide sequence

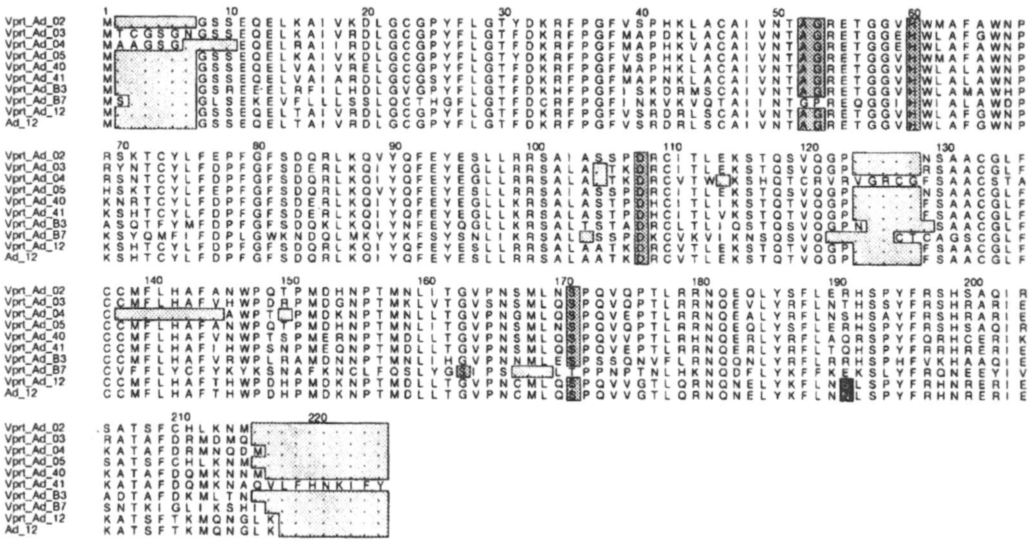

Fig. 9. Multiple alignments of the amino acid sequences of the known endoproteases (*VRT*) from adenovirus types 2, 3, 4, 5, 40, 41, B3, B7, and 12. Gaps introduced by the alignment program are *light gray*. Bottom-most entry shows the protein sequence derived from our Ad12 DNA sequence data

7 Presumptive Locations of Ad12 Promoters

We identified positions of the viral promoter regions in the Ad12 genome by utilizing the Eucaryotic Promoter Database (EPD) (Bucher and Trifonov 1986). As mentioned before, only biochemical data identified the exact position of a promoter as known for the major late promoter (MLP).

8 The Human Adenovirus Type 12 DNA Sequence

In closing, it should be mentioned that we failed to detect motifs in the Ad12 DNA nucleotide sequence that resembled previously published integrase-like or DNA methyltransferase-like sequences at the protein level. Such motifs would have been of interest for two of the main lines of research in the authors´ laboratory.

The complete double-stranded DNA sequence of Ad12 DNA is reproduced in Fig. 10. The following structural pecularities were emphasized, using the program ALSCRIPT (Barton 1993):

- The recognition sites of the restriction endonuclease *Pst*I were marked by lines at the corresponding cleavage sites (see Table 2 for the fragment identifiers and their sizes).
- Every TATA-like sequence motif in the leftward or rightward transcribed DNA strand was marked (boxes). Due to the variations in the consensus motifs published in the Transcription Factor Database (TFD, Gosh 1992), overlapping boxes were possible.
- Presumptive promoter sequences were underlined (leftward transcribed DNA strand) or marked by a line drawn on top of the corresponding sequence (rightward transcribed DNA strand).
- The triplet repeat region $(GGA)_n$ was designated by a shaded background (p. 253).

1
c c t a t c t a a t a a t a t a c c t t a t a c t g g a c t a g t g c c a a t a t t a a a a t g a a
g g a t a g a t t a t t a t a t g g a a t a t g a c c t g a t c a c g g t t a t a a t t t t a c t t

100
g t g g g c g t a g t g t g t a a t t t g a t t g g g t g g a g g t g t g g c t t t g g c g t g c t
c a c c c g c a t c a c a c a t t a a a c t a a c c c a c c t c c a c a c c g a a a c c g c a c g a

t g t a a g t t t g g g c g g a t g a g g a a g t g g g g c g c g g c g t g g g a g c c g g g c g c
a c a t t c a a a c c c g c c t a c t c c t t c a c c c c g c g c c g c a c c c t c g g c c c g c g

200
g c c g g a t g t g a c g t t t t a g a c g c c a t t t t a c a c g g a a a t g a t g t t t t t t g
c g g c c t a c a c t g c a a a a t c t g c g g t a a a a t g t g c c t t t a c t a c a a a a a a c

g g c g t t g t t t g t g c a a a t t t t g t g t t t t a g g c g c g a a a a c t g a a a t g c g g
c c g c a a c a a a c a c g t t t a a a a c a c a a a a t c c g c g c t t t t g a c t t t a c g c c

300
a a g t g a a a a t t g a t g a c g g c a a t t t t a t t a t a g g c g c g g a a t a t t t a c c g
t t c a c t t t t a a c t a c t g c c g t t a a a a t a a t a t c c g c g c c t t a t a a a t g g c

a g g g c a g a g t g a a c t c t g a g c c t c t a c g t g t g g g t t t c g a t a c g t g a g c g
t c c c g t c t c a c t t g a g a c t c g g a g a t g c a c a c c c a a a g c t a t g c a c t c g c

400
a c g g g g a a a c t c c a c g t t g g c g c t c a a a g g g c g c g t t t a t t g t t c t g t c a
t g c c c c t t t g a g g t g c a a c c g c g a g t t t c c c g c g c a a a t a a c a a g a c a g t

g c t g a t c g t t t g g g t a t t t a a t g c c g c c g t g t t c g t c a a g a g g c c a c t c t
c g a c t a g c a a a c c c a t a a a t t a c g g c g g c a c a a g c a g t t c t c c g g t g a g a

500
t g a g t g c c a g c g a g a a g a g t t t t c t c t g c c a g c t c a t t t t c a c g g c g c c a
a c t c a c g g t c g c t c t t c t c a a a a g a g a c g g t c g a g t a a a a g t g c c g c g g t

```
t t a t g a g a a c t g a a a t g a c t c c c t t g g t c c t g t c g t a t c a g g a a g c t g a c
a a t a c t c t t g a c t t t a c t g a g g g a a c c a g g a c a g c a t a g t c c t t c g a c t g

                                                                                                  600
g a c a t a t t g g a g c a t t t g g t g g a c a a c t t t t t t a a c g a g g t a c c c a g t g a
c t g t a t a a c c t c g t a a a c c a c c t g t t g a a a a a a t t g c t c c a t g g g t c a c t

t g a t g a t c t t t a t g t t c c g t c t c t t t a c g a a c t g t a t g a t c t t g a t g t g g
a c t a c t a g a a a t a c a a g g c a g a g a a a t g c t t g a c a t a c t a g a a c t a c a c c

                                                                                                  700
a g t c t g c c g g t g a a g a t a a t a a t g a a c a g g c g g t g a a t g a g t t t t t t c c c
t c a g a c g g c c a c t t c t a t t a t t a c t t g t c c g c c a c t t a c t c a a a a a a g g g

g a a t c g c t t a t t t t a g c t g c c a g t g a g g g g g t t g t t t t t a c c g g a g c c t c c
c t t a g c g a a t a a a a t c g a c g g t c a c t c c c c a a c a a a a a t g g c c t c g g a g g

                                                                                                  800
t g t a c t t t c t c c t g t c t g t g a g c c t a t t g g g g g c g a a t g t a t g c c a c a a c
a c a t g a a a g a g g a c a g a c a c t c g g a t a a c c c c c g c t t a c a t a c g g t g t t g

t g c a c c c t g a a g a t a t g g a t t t a t t g t g c t a c g a g a t g g g c t t t c c c t g t
a c g t g g g a c t t c t a t a c c t a a a t a a c a c g a t g c t c t a c c c g a a a g g g a c a

                                                                                                  900
a g c g a t t c g g a a g a c g a g c a a g a c g a g a a c g g a a t g g c g c a t g t t t c t g c
t c g c t a a g c c t t c t g c t c g t t c t g c t c t t g c c t t a c c g c g t a c a a a g a c g

a t c c g c a g c t g c t g c t g c c g c t g a t a g g g a a c g t g a g g a g t t t c a g t t a g
t a g g c g t c g a c g a c g a c g g c g a c t a t c c c t t g c a c t c c t c a a a g t c a a t c

                                                                                                  1000
a c c a t c c a g a g t t g c c c g g a c a c a a t t g t a a g t c c t g t g a g c a c c a c c g g
t g g t a g g t c t c a a c g g g c c t g t g t t a a c a t t c a g g a c a c t c g t g g t g g c c
```

```
a a t a g t a c t g g a a a t a c t g a c t t a a t g t g c t c t t t g t g c t a t c t g c g a g c
t t a t c a t g a c c t t t a t g a c t g a a t t a c a c g a g a a a c a c g a t a g a c g c t c g

c t a c a a c a t g t t c a t t t a c a g t a a g t g t g c t a t g g g a g g t g g g a g g t g a t 1100
g a t g t t g t a c a a g t a a a t g t c a t t c a c a c g a t a c c c c t c c c c c t c c a c t a

t t t t t t t t c t t a a g c a g t g a a a a a t a a t a t t t t g t t g t t t t t a g g t c c t g
a a a a a a a a g a a t t c g t c a c t t t t t a t t a t a a a a c a a c a a a a a t c c a g g a c

t t t c c g a t a a t g a g c c t g a a c c t a a t a g c a c t t t g g a t g g c g a t g a g c g a 1200
a a a g g c t a t t a c t c g g a c t t g g a t t a t c g t g a a a c c t a c c g c t a c t c g c t

c c c t c a c c c c c g a a a c t a g g a a g t g c g g t t c c a g a a g g a g t a a t a a a a c c
g g g a g t g g g g g c t t t g a t c c t t c a c g c c a a g g t c t t c c t c a t t a t t t t g g

t g t g c c t c a g c g g g t g a c t g g g a g g c g t a g a t g t g c t g t g g a a a g c a t t t 1300
a c a c g g a g t c g c c c a c t g a c c c t c c g c a t c t a c a c g a c a c c t t t c g t a a a

t g g a t t t g a t t c a a g a g g a a g a a a g a g a a c a a a c a g t g c c t g t t g a t c t g
a c c t a a a c t a a g t t c t c c t t c t t t c t c t t g t t t g t c a c g g a c a a c t a g a c

t c a g t g a a a c g c c c t a g a t g t a a t t a a t g g a c t t t g a g c a c c t g g g c a a t 1400
a g t c a c t t t g c g g g a t c t a c a t t a a t t a c c t g a a a c t c g t g g a c c c g t t a

a a a a t a g g g g t a a t g t g g t t t t t g t g a g t c a t g t a t a a t a a a a c t g g t t t
t t t t a t c c c c a t t a c a c c a a a a a c a c t c a g t a c a t a t t a t t t t g a c c a a a

c g g t t g a a g t g t c t t g t t a a t g t t t g t t t g g g c g t g g t t a a a c a g g g a t a 1500
g c c a a c t t c a c a g a a c a a t t a c a a a c a a a c c c g c a c c a a t t t g t c c c t a t
```

```
t a a a g c t g g g t t g g t g t t g c t t t g a a t a g t t c a t c t t a g t a a t g g a g t t g
a t t t c g a c c c a a c c a c a a c g a a a c t t a t c a a g t a g a a t c a t t a c c t c a a c

                                                                                              1600
g a a a c t g t g c t g c a a a g t t t t c a g a g c g t t c g c c a g c t c t t g c a g t a t a c
c t t t g a c a c g a c g t t t c a a a a g t c t c g c a a g c g g t c g a g a a c g t c a t a t g

c t c t a a a a a c a c t t c a g g t t t t t g g a g g t a t c t g t t t g g c t c t a c c t t a a
g a g a t t t t t g t g a a g t c c a a a a a c c t c c a t a g a c a a a c c g a g a t g g a a t t

                                                                                              1700
g c a a g g t g g t a a a t a g g g t g a a a g a a g a c t a t a g a g a g g a a t t t g a a a a c
c g t t c c a c c a t t t a t c c c a c t t t c t t c t g a t a t c t c t c c t t a a a c t t t t g

a t a t t g g c c g a c t g t c c a g g g c t t t t g g c t t c a c t a g a c c t t t g t t a c c a
t a t a a c c g g c t g a c a g g t c c c g a a a a c c g a a g t g a t c t g g a a a c a a t g g t

                                                                                              1800
c t t g g t g t t t c a g g a a a a a g t g g t c a g a t c c t t a g a t t t t t c a t c t g t g g
g a a c c a c a a a g t c c t t t t t c a c c a g t c t a g g a a t c t a a a a a g t a g a c a c c

g a c g a a c g g t t g c t t c t a t t g c t t t t t t g g c a a c c a t a t t g g a t a a a t g g
c t g c t t g c c a a c g a a g a t a a c g a a a a a a c c g t t g g t a t a a c c t a t t t a c c

                                                                                              1900
a g c g a g a a a t c c c a c c t g a g t t g g g a t t a c a t g c t g g a t t a c a t g t c a a t
t c g c t c t t t a g g g t g g a c t c a a c c c t a a t g t a c g a c c t a a t g t a c a g t t a

g c a g c t g t g g a g g g c a t g g c t g a a g a g g a g g g t t t g c a t t t a c t c g c t g g
c g t c g a c a c c t c c c g t a c c g a c t t c t c c t c c c a a a c g t a a a t g a g c g a c c

                                                                                              2000
c g c g g c c t t t g a c c a t g c c g c c g c t g c c g a c g t t g c a a g a g g a g a a g g a g
g c g c c g g a a a c t g g t a c g g c g g c g a c g g c t g c a a c g t t c t c c t c t t c c t c
```

g a g g a g c g g a a c c c t g c g g t g g t g g a g a a g t a a a c a t g g a a c a a c a g g t g
c t c c t c g c c t t g g g a c g c c a c c a c c t c t t c a t t t g t a c c t t g t t g t c c a c

2100

c a a g a a g g c c a t g t a c t t g a c t c t g g c g a a g g g c c t a g t t g c g c a g a t g a
g t t c t t c c g g t a c a t g a a c t g a g a c c g c t t c c c g g a t c a a c g c g t c t a c t

t a g a g a t a a g c a g g a a a a a a a a a g a a a g t t t a a a g g a a g c t g c t g t t c t t a
a t c t c t a t t c g t c c t t t t t t t t t c t t t c a a a t t t c c t t c g a c g a c a a g a a t

2200

g t a g g c t a a c t g t t a a t c t g a t g t c c c g c c c g c g t t t g g a a a c t g t a t a t
c a t c c g a t t g a c a a t t a g a c t a c a g g g c g g g c g c a a a c c t t t g a c a t a t a

t g g c a g g a g t t g c a g g a t g a a t t t c a g c g g g g t g a t a t g c a t t t a c a g t a
a c c g t c c t c a a c g t c c t a c t t a a a g t c g c c c c a c t a t a c g t a a a t g t c a t

2300

c a a a t a c a g t t t t g a a c a a t t a a a a a c c c a c t g g t t a g a g c c a t g g g a g g
g t t t a t g t c a a a a c t t g t t a a t t t t t g g g t g a c c a a t c t c g g t a c c c t c c

a t a t g g a g t g t g c t a t t a a a g c t t t t g c t a a a t t g g c c t t a c g t c c t g a t
t a t a c c t c a c a c g a t a a t t t c g a a a a c g a t t t a a c c g g a a t g c a g g a c t a

2400

t g t a g c t a c a g a a t t a c t a a a a c a g t a a c c a t t a c t t c a t g c g c c t a t a t
a c a t c g a t g t c t t a a t g a t t t t g t c a t t g g t a a t g a a g t a c g c g g a t a t a

t a t a g g t a a c g g g g c a a t a g t t g a g g t a g a t a c a a g c g a c a g a g t t g c t t
a t a t c c a t t g c c c c g t t a t c a a c t c c a t c t a t g t t c g c t g t c t c a a c g a a

2500

t t a g a t g t c g a a t g c a g g g t a t g g g c c c a g g g g t g g t g g g t t t g g a t g g a
a a t c t a c a g c t t a c g t c c c a t a c c c g g g t c c c c a c c a c c c a a a c c t a c c t

a t t a c a t t t a t a a a t g t t a g g t t t g c t g g a g a t a a g t t t a a a g g c a t t a t
t a a t g t a a a t a t t t a c a a t c c a a a c g a c c t c t a t t c a a a t t t c c g t a a t a

2600
g t t c g a a g c t a a t a c c t g t c t t g t c t t g c a t g g t g t t t a c t t t c t t a a c t
c a a g c t t c g a t t a t g g a c a g a a c a g a a c g t a c c a c a a a t g a a a g a a t t g a

t t a g t a a c a t t t g t g t a g a g t c t t g g a a t a a g g t t t c t g c t a g g g g c t g t
a a t c a t t g t a a a c a c a t c t c a g a a c c t t a t t c c a a a g a c g a t c c c c g a c a

2700
a c t t t t t a t g g a t g t t g g a a g g g t t t g g t g g g t a g a c c a a a a a g t a a a c t
t g a a a a a t a c c t a c a a c c t t c c c a a a c c a c c c a t c t g g t t t t t c a t t t g a

g t c t g t a a a a a a g t g t t t g t t t g a a a a a t g t g t a c t t g c t t t a a t t g t a g
c a g a c a t t t t t t c a c a a a c a a a c t t t t t a c a c a t g a a c g a a a t t a a c a t c

2800
a g g g g g a t g c a c a t a t t a g g c a t a a t g c a g c t t c a g a a a a t g c c t g t t t t
t c c c c c t a c g t g t a t a a t c c g t a t t a c g t c g a a g t c t t t t a c g g a c a a a a

g t a t t a t t g a a g g g a a t g g c t a t t t t a a a g c a t a a t a t g g t t t g t g g g g t
c a t a a t a a c t t c c c t t a c c g a t a a a a t t t c g t a t t a t a c c a a a c a c c c c a

2900
g t c t g a t c a a a c t a t g c g a c g t t t t g t t a c c t g t g c t g a t g g a a a t t g t c
c a g a c t a g t t t g a t a c g c t g c a a a a c a a t g g a c a c g a c t a c c t t t a a c a g

a t a c c t t a a a a a c t g t t c a t a t t g t g a g c c a c a g t a g a c a t t g t t g g c c t
t a t g g a a t t t t t g a c a a g t a t a a c a c t c g g t g t c a t c t g t a a c a a c c g g a

3000
g t a t g t g a t c a t a a c a t g t t t a t g c g c t g t a c c a t a c a t t t a g g c t t a a g
c a t a c a c t a g t a t t g t a c a a a t a c g c g a c a t g g t a t g t a a a t c c g a a t t c

g c g g g g t a t g t t t a g a c c t t c c c a a t g t a a c t t c a g c c a c t c a a a c a t t a
c g c c c c a t a c a a a t c t g g a a g g g t t a c a t t g a a g t c g g t g a g t t t g t a a t

3100

t g c t g g a a c c t g a a g t g t t t t c t a g a g t g t g t t t a a a t g g g g t a t t t g a t
a c g a c c t t g g a c t t c a c a a a a g a t c t c a c a c a a a t t t a c c c c a t a a a c t a

t t a t c t g t g g a a t t a t g t a a g g t t a t a a g a t a t a a t g a t g a t a c t c g a c a
a a t a g a c a c c t t a a t a c a t t c c a a t a t t c t a t a t t a c t a c t a t g a g c t g t

3200

t c g t t g c c g a c a g t g t g a g t g t g g t a g c a g t c a t c t a g a a c t t c g t c c c a
a g c a a c g g c t g t c a c a c t c a c a c c a t c g t c a g t a g a t c t t g a a g c a g g g t

t t g t g c t a a a t g t a a c t g a g g a g c t g a g a a g t g a c c a c c t t a c c c t g t c t
a a c a c g a t t t a c a t t g a c t c c t c g a c t c t t c a c t g g t g g a a t g g g a c a g a

3300

t g c c t g c g g a c t g a c t a t g a g t c a a g t g a t g a a g a c g a c a a c t g a g g t a a
a c g g a c g c c t g a c t g a t a c t c a g t t c a c t a c t t c t g c t g t t g a c t c c a t t

g t g g g t g g a g c t a g g t g g g a t t a t a a a a g g c t g g a a g t c a a c t a a a a a t t
c a c c c a c c t c g a t c c a c c c t a a t a t t t t c c g a c c t t c a g t t g a t t t t t a a

3400

g t t t t t g t t c t t t t a a c a g c a c g a t g a a c g g a a c t a c t c a g a a c a a c g c t
c a a a a a c a a g a a a a t t g t c g t g c t a c t t g c c t t g a t g a g t c t t g t t g c g a

g c g c t t t t t g a t g g a g g g g t t t t t a g c c c t t a t t t g a c t t c c a g g t t a c c
c g c g a a a a a c t a c c t c c c c a a a a a t c g g g a a t a a a c t g a a g g t c c a a t g g

3500

a t a t t g g g c c g g a g t a c g t c a g a a t g t g g t a g g a t c t a c a g t g g a c g g t c
t a t a a c c c g g c c t c a t g c a g t c t t a c a c c a t c c t a g a t g t c a c c t g c c a g

g a c c t g t g g c a c c t g c a a a t t c a t c a a c a t t a a c c t a t g c a a c t a t t g g a
c t g g a c a c c g t g g a c g t t t a a g t a g t t g t a a t t g g a t a c g t t g a t a a c c t

3600

c c c t c g c c t t t g g a t a c c g c c g c c g c c g c t g c a g c t t c c g c g g c c g c t t c
g g g a g c g g a a a c c t a t g g c g g c g g c g g c g a c g t c g a a g g c g c c g g c g a a g

t a c g g c t c g c a g t a t g g c a g c t g a t t t c a g c t t c t a c a a t c a c t t g g c t t
a t g c c g a g c g t c a t a c c g t c g a c t a a a g t c g a a g a t g t t a g t g a a c c g a a

3700

c g a a t g c t g t g a c a c g c a c c g c a g t t c g a g a g g a c a t t c t g a c t g t t a t g
g c t t a c g a c a c t g t g c g t g g c g t c a a g c t c t c c t g t a a g a c t g a c a a t a c

c t t g c c a a g c t t g a a a c t c t a a c t g c t c a g c t g g a a g a g c t a t c g c a a a a
g a a c g g t t c g a a c t t t g a g a t t g a c g a g t c g a c c t t c t c g a t a g c g t t t t

3800

g g t t g a g g a a t t a g c t g a t g c t a c t a c c c a t a c c c c a g c c c a a c c t g t a a
c c a a c t c c t t a a t c g a c t a c g a t g a t g g g t a t g g g g t c g g g t t g g a c a t t

c c c a a t a a a g a a a a a a c t t a a a t t g a g a t g g t g t t a t g a a t c t t t a t t g a
g g g t t a t t t c t t t t t t g a a t t t a a c t c t a c c a c a a t a c t t a g a a a t a a c t

3900

t a c t t g t t t t t t c t g a c a t g g t a a g c t c t t g a c c a c c g t t c c c t a t c a t t
a t g a a c a a a a a a g a c t g t a c c a t t c g a g a a c t g g t g g c a a g g g a t a g t a a

a a g a a c a c g g t g a a t g t g t t c c a g t a t t t t g t a a a g a t g a g c c t g t a t a t
t t c t t g t g c c a c t t a c a c a a g g t c a t a a a a c a t t t c t a c t c g g a c a t a t a

4000

t a a g g t a q a t t g g c a t t a g g c c a t c t t t g g g a t g a a g g t a g g a c c a t t g a
a t t c c a t g t a a c c g t a a t c c g g t a g a a a c c c t a c t t c c a t c c t g g t a a c t

a g g g c t t c a t g t t c c g g g t t a g t g t t g t a g a t a a t c c a g t c a t a g c a a c a

t c c c g a a g t a c a a g g c c c a a t c a c a a c a t c t a t t a g g t c a g t a t c g t t g t

4100

a c g c t g g g c a t g g t g a t t a a a t a t a t c t t t t a a c a a c a a g c t a a t t g c t a

t g c g a c c c g t a c c a c t a a t t t a t a t a g a a a a t t g t t g t t c g a t t a a c g a t

a t g g a a g a c c t t t a g t a t a g g t a t t g a t a a a a c g g t t a a g c t g g g t g g g a

t a c c t t c t g g a a a t c a t a t c c a t a a c t a t t t t g c c a a t t c g a c c c a c c c t

4200

t g c a t c c g a g g t g a c a t g a t a t g a a g t t t t g a t t g t a t t t t g a g a t t g g c

a c g t a g g c t c c a c t g t a c t a t a c t t c a a a a c t a a c a t a a a a c t c t a a c c g

a a t g t t a c c t g c c a a a t c t c t t c t t g g a t t c a t a t t g t g g a g a a c c a c g a

t t a c a a t g g a c g g t t t a g a g a a g a a c c t a a g t a t a a c a c c t c t t g g t g c t

4300

a a a c g g t g t a g c c a g t a c a c t t g g g a a a t t t g t c a t g g a g t t t a g a a g g a

t t t g c c a c a t c g g t c a t g t g a a c c c t t t a a a c a g t a c c t c a a a t c t t c c t

a a g g c a t g g a a a a a c t t g g a a a c g c c t t t g t g a c t t c c c a a a t t t t c c a t

t t c c g t a c c t t t t t g a a c c t t t g c g g a a a c a c t g a a g g g t t t a a a a g g t a

4400

a c a c t c a t c c a t t a t t a t g g c a a t t g g a c c g c g a g c a g c g g c t t g a g c a a

t g t g a g t a g g t a a t a a t a c c g t t a a c c t g g c g c t c g t c g c c g a a c t c g t t

a a a t g t t t t c t g g a t c a g a a a c a t c a t a g t t g t g g t c t a g a g t t a g g t c a

t t t a c a a a a g a c c t a g t c t t t g t a g t a t c a a c a c c a g a t c t c a a t c c a g t

4500

t c g t a g g a c a a c t t a a c a a a t t t a g g a c a c a g c g t t c c a g a t t g t g g a a t

a g c a t c c t g t t g a a t t g t t t a a a t c c t g t g t c g c a a g g t c t a a c a c c t t a

a a t a g t t c c c t c t g g t c c t g g g a c a t a a t t t c c c t c a c a a a t t t g c a t t t
t t a t c a a g g g a g a c c a g g a c c c t g t a t t a a a g g g a g t g t t t a a a c g t a a a

4600

c c c a a g a t t t a a t t t c a g a t g g g g g a a t c a t g t c c a c t t g c g g a a c a a t a
g g g t t c t a a a t t a a a g t c t a c c c c c t t a g t a c a g g t g a a c g c c t t g t t a t

a a a a a a a c a g t t t c t g g a g c a g g t g t a a c c a g c t g g g c a g a a a g c a a a t t
t t t t t t t g t c a a a g a c c t c g t c c a c a t t g g t c g a c c c g t c t t t c g t t t a a

4700

a c g c a a c a a c t g a g a c t t c c c a c a g c c a g t g g g t c c a t a a a t t a c c c c a a
t g c g t t g t t g a c t c t g a a g g g t g t c g g t c a c c c a g g t a t t t a a t g g g g t t

t t a c a g g t t g c a a g t g a t a g t t t a a c g a g g t g c a g c t g c c g t c t t c g t g g
a a t g t c c a a c g t t c a c t a t c a a a t t g c t c c a c g t c g a c g g c a g a a g c a c c

4800

a g a a g c g g a g c c a c t t c a t t c a t c a t t t g t c g g a c g c g g a t g t t t t g c t t
t c t t c g c c t c g g t g a a g t a a g t a g t a a a c a g c c t g c g c c t a c a a a a c g a a

g g c c a g t t c c c c t a a c a g a c g c t c t c c g c c t a a g g a a a g t a a c t c t t g t a
c c g g t c a a g g g g a t t g t c t g c g a g a g g c g g a t t c c t t t c a t t g a g a a c a t

4900

a a g a t t t g a a a t t t t t a a g t g g c t t t a g g c c a t c g g c c a t a g g c a t g t g g
t t c t a a a c t t t a a a a a t t c a c c g a a a t c c g g t a g c c g g t a t c c g t a c a c c

t c c a g g g t t t g c t t c a g c a g t t g c a a g c g a t c c c a t a g c t c a g t t a t a t t
a g g t c c c a a a c g a a g t c g t c a a c g t t c g c t a g g g t a t c g a g t c a a t a t a a

5000

t t c t a t g c c a t c t c g a t c c a g c a a a c t t c c t c g t t g c g g g g g t t t g g c t g
a a g a t a c g g t a g a g c t a g g t c g t t t g a a g g a g c a a c g c c c c c a a a c c g a c

```
g c t g t t g c t g t a a g g a a c g a g g c g g t g a g c a t c c a a a t g g a c g a g g g t t t
c g a c a a c g a c a t t c c t t g c t c c g c c a c t c g t a g g t t t a c c t g c t c c c a a a

                                                                                                  5100
t g t c c t t c c a g g g a c g t a a t g t g c g c g t c a g g g t t g t t t c g g t c a c g g t g
a c a g g a a g g t c c c t g c a t t a c a c g c g c a g t c c c a a c a a a g c c a g t g c c a c

a a t g g a t g c g c t c c t g g t t g a g c g c t g g c c a g t g t g c g c t t t a a a c t g a g
t t a c c t a c g c g a g g a c c a a c t c g c g a c c g g t c a c a c g c g a a a t t t g a c t c

                                                                                                  5200
g c g g c t g g t g c t g a a g c g c g t g t c t t c t c c c t g t g c t t c g g c a a g g t a g c
c g c c g a c c a c g a c t t c g c g c a c a g a a g a g g g a c a c g a a g c c g t t c c a t c g

a t t t t a a c a t a a g a t c a t a a g a c a a a g c c t c t g t a g c g t g g c c t t t a g c c
t a a a a t t g t a t t c t a g t a t t c t g t t t c g g a g a c a t c g c a c c g g a a a t c g g

                                                                                                  5300
c g t a t t t t t c c t t t g g a g g t g c t c c c g c a g t g a g g a c a c t g a a g g c a t t t
g c a t a a a a a g g a a a c c t c c a c g a g g g c g t c a c t c c t g t g a c t t c c g t a a a

a a g g g c g t a c a g t t t t g g a g c c a a a a a a a c a g a t t c t g g a g a a t a a g c a t
t t c c c g c a t g t c a a a a c c t c g g t t t t t t t g t c t a a g a c c t c t t a t t c g t a

                                                                                                  5400
c t g c g c c a c a a t a a c t a c a a a c a g t t t c a c a t t c a a c t g a c c a g g t c a g c
g a c g c g g t g t t a t t g a t g t t t g t c a a a g t g t a a g t t g a c t g g t c c a g t c g

t c a g g a c a t g a t g g a t c a a a a a c a a g t t t c c c t c c g t a c t t t t t g a t g c g
a g t c c t g t a c t a c c t a g t t t t t g t t c a a a g g g a g g c a t g a a a a a c t a c g c

                                                                                                  5500
t t t c t t a c c t t g c g a c t c c a t a a g g c g g c g t c c t t t c t c t g t g a c a a a a a
a a a g a a t g g a a c g c t g a g g t a t t c c g c c g c a g g a a a g a g a c a c t g t t t t t
```

```
g a c t g t c a g t g t c t c c g t a t a c a g a t t t a a g g g g t c t a t c c t t c a g t g g t
c t g a c a g t c a c a g a g g c a t a t g t c t a a a t t c c c c a g a t a g g a a g t c a c c a
                                                                                            5600
a t t c c g c g g t c c t c c t c g t a c a g g a a t t c t g a c c a c t c t ·g a c a c a a a a g c
t a a g g c g c c a g g a g g a g c a t g t c c t t a a g a c t g g t g a g a c t g t g t t t t c g

t c t a g t c c a a g c a a g t a c a a a g g a a g c c a c a t g g g a a g g g t a c c g a t c g t
a g a t c a g g t t c g t t c a t g t t t c c t t c g g t g t a c c c t t c c c a t g g c t a g c a
                                                                                            5700
t g t t a a t t a a a g g g t t a g a a c t t t c t a a g g t g t g t a a a c a c a t g t c t c c t
a c a a t t a a t t t c c c a a t c t t g a a a g a t t c c a c a c a t t t g t g t a c a g a g g a

t c t t c a g c g t c c a t g a a t g t g a t t g g t t t g t a g g t g t a a g t c a c g t g t t c
a g a a g t c g c a g g t a c t t a c a c t a a c c a a a c a t c c a c a t t c a g t g c a c a a g
                                                                                            5800
a c a a t t t t c t g g t g g t g g g c t a t a a a a a g g g g c g g g t c c t t g g t c t t c a t
t g t t a a a a g a c c a c c a c c c g a t a t t t t t c c c c g c c c a g g a a c c a g a a g t a

c g c t t t c t t c t g c t t c g c t g t t t a c g a g c g c c a a c t g g t t g g g t g a g t a c
g c g a a a g a a g a c g a a g c g a c a a a t g c t c g c g g t t g a c c a a c c c a c t c a t g
                                                                                            5900
a c g c g c t c a a a g g c a g g c a t t a c c t c t g t a c t c a a c g t g t c a g t t t c t a t
t g c g c g a g t t t c c g t c c g t a a t g g a g a c a t g a g t t g c a c a g t c a a a g a t a

a a a c g a t g a g g a t t t g a t g t t t a a t c g c c c c g c t g c a a t t t c t t t c a t t a
t t t g c t a c t c c t a a a c t a c a a a t t a g c g g g g c g a c g t t a a a g a a a g t a a t
                                                                                            6000
g g c t t t c t t c c a t t t g a t c a g a a a a a a c t a t t t t t t t g t t a t c t a g t t t g
c c g a a a g a a g g t a a a c t a g t c t t t t t t g a t a a a a a a a c a a t a g a t c a a a c
```

g t a g c a a a a g a t c c g t a c a a g g c a t t g g a a a g c a g c t t g g c t a t a g a t c t
c a t c g t t t t c t a g g c a t g t t c c g t a a c c t t t c g t c g a a c c g a t a t c t a g a

6100
t a g g g t t t g a t t t t t g t c c c t a t c g g c c c g t t c t t t t g c g g c a a t a t t g a
a t c c c a a a c t a a a a a c a g g g a t a g c c g g g c a a g a a a a c g c c g t t a t a a c t

g t t g c a c a t a t t c g c g t g c c a g g c a t t t c c a g g t g g g g a a a a t g g t g g t g
c a a c g t g t a t a a g c g c a c g g t c c g t a a a g g t c c a c c c c t t t t a c c a c c a c

6200
c g c t c g t c a g a t a g c a a g c g t a a g c g c c a c c c g c g a t t a t g c a g t g t a a c
g c g a g c a g t c t a t c g t t c g c a t t c g c g g t g g g c g c t a a t a c g t c a c a t t g

c a g a t c t a c g c t g g t a a c t a c t t c a c c g c g c a a g c t t t c a t t g g t c c a g g
g t c t a g a t g c g a c c a t t g a t g a a g t g g c g c g t t c g a a a g t a a c c a g g t c c

6300
c t a a a c g a c c g c c t t t t c t a g a a c a a a a a g g a g g a a g a a c a t c c a a c t g a
g a t t t g c t g g c g g a a a a g a t c t t g t t t t t c c t c c t t c t t g t a g g t t g a c t

t t t t c a t c t g g g g g g t c g g c a t c t a t a g t a a a a a t g c c a g g a c a a a g a t t
a a a a g t a g a c c c c c c a g c c g t a g a t a t c a t t t t t a c g g t c c t g t t t c t a a

6400
t t t g t c a a a a t a a t c a a t t t t g c a a g t g t a a t t t t c c a g c g c c a c c t g c c
a a a c a g t t t t a t t a g t t a a a a c g t t c a c a t t a a a a g g t c g c g g t g g a c g g

a t t g c c g c a c g g c c a a t g c c c g c t c a t a g g g g t t a a g g g g a g g a c c c c a a
t a a c g g c g t g c c g g t t a c g g g c g a g t a t c c c c a a t t c c c c t c c t g g g g t t

6500
g g c a t g g g g t g t g t g a g g g c c g a t g c a t a c a t g c c g c a a a t a t c a t a t a c
c c g t a c c c c a c a c a c t c c c g g c t a c g t a t g t a c g g c g t t t a t a g t a t a t g

a t a t a t g g g c t c t t t t a g t a c t c c t a t g t a a g t a g g a t a g c a c c t g c c g c
t a t a t a c c c g a g a a a a t c a t g a g g a t a c a t t c a t c c t a t c g t g g a c g g c g

6600

c a c g a a t g c t g g c g c g a a c g t a g t c a t a t a g c t c a t g t g a a g g c g c c a g g
g t g c t t a c g a c c g c g c t t g c a t c a g t a t a t c g a g t a c a c t t c c g c g g t c c

a t g t t g g g c c c a a g a t g t g t g c g c t g t g g t t t t t c g g c g c g g t a c a a a a t
t a c a a c c c g g g t t c t a c a c a c g c g a c a c c a a a a a g c c g c g c c a t g t t t t a

6700

t t g t c t g a a a a t t g c a t g a g a g t t a g a g g a a a t g g t a g g a c g c t g a a a c a
a a c a g a c t t t t a a c g t a c t c t c a a t c t c c t t t a c c a t c c t g c g a c t t t g t

c a t t a a a a t g t g c c g c g t c a a g a c c c a c t g c g t c a g t a a c a a a c t g g g c g
g t a a t t t t a c a c g g c g c a g t t c t g g g t g a c g c a g t c a t t g t t t g a c c c g c

6800

t a t g a g c t a c g c a g t t t t t c t a c c a a t g a g g c a g t c a c a a g t a c a t c c a g
a t a c t c g a t g c g t c a a a a a g a t g g t t a c t c c g t c a g t g t t c a t g t a g g t c

g g c a c a a t a g t t t a a t g t t t c c c c g a t a a g a t t g t a a t t t t t t t c t c c t t
c c g t g t t a t c a a a t t a c a a a g g g g c t a t t c t a a c a t t a a a a a a a g a g g a a

6900

t t t t t t t c c a t a g t t c t t g a t t t a g g a g g t a t t c c t c c t t a t c c t t c c a g
a a a a a a a g g t a t c a a g a a c t a a a t c c t c c a t a a g g a g g a a t a g g a a g g t c

t a c t c c t c c a g g g g a a a c c c a t t t g c a t c t g c a c g g t a a g a a c c a a g c a t
a t g a g g a g g t c c c c t t t g g g t a a a c g t a g a c g t g c c a t t c t t g g t t c g t a

7000

a t a a a a c t g a t t t a c c g c c t t g t a c g g a c a a c a t c c t t t t t c t a c a g g c a
t a t t t t g a c t a a a t g g c g g a a c a t g c c t g t t g t a g g a a a a a g a t g t c c g t

```
g g g c a t a c g c t t g t g c a g c c t t t c t t a a a g a t g t a t g a g t a a g a g c a a a g
c c c g t a t g c g a a c a c g t c g g a a a g a a t t t c t a c a t a c t c a t t c t c g t t t c
                                                                                               7100
g t a t c t c t g a c c a t t a c t t t t a a a t a c t g g t a t t t a a a a t c t t g g t c g t c
c a t a g a g a c t g g t a a t g a a a a t t t a t g a c c a t a a a t t t t a g a a c c a g c a g

a c a c c c t c c g t g t t c c c a c a g t a g g a a g t t a g t t c g c t t t t t g t a g t g g g
t g t g g g a g g c a c a a g g g t g t c a t c c t t c a a t c a a g c g a a a a a c a t c a c c c
                                                                                               7200
g a t t g g g a a g g g c a a a a g t a a t a t c a t t a a a t a a t a t t t t g c c a g c t c t t
c t a a c c c t t c c c g t t t t c a t t a t a g t a a t t t a t t a t a a a a c g g t c g a g a a

g g a a t a a a a t t t c t a g a a a t t t t a a a g g g t c c a g g g a c g t c c a a g c g g t t
c c t t a t t t t a a a g a t c t t t a a a a t t t c c c a g g t c c c t g c a g g t t c g c c a a
                                                                                               7300
a t t g a t t a c c t g a g c g g c a a g a a c a a t t t c a t c a a a t c c a t t a a t a t t g t
t a a c t a a t g g a c t c g c c g t t c t t g t t a a a g t a g t t t a g g t a a t t a t a a c a

g t c c t a c t a t a t a c a a c t c t a c a a a t c t t g g c t c a c c c t t a a t t g c a g g g
c a g g a t g a t a t a t g t t g a g a t g t t t a g a a c c g a g t g g g a a t t a a c g t c c c
                                                                                               7400
g c t c t t t t a a g a t c t t c g t a g g a a a g a t c t t c a a g c g c g a c t a g t c c g t t
c g a g a a a a t t c t a g a a g c a t c c t t t c t a g a a g t t c g c g c t g a t c a g g c a a

t t c t t c t t g a g c c c a t t g a g a c a a g t g t g g a t t t t t t t g t a a a a a a g t c a
a a g a a g a a c t c g g g t a a c t c t g t t c a c a c c t a a a a a a a c a t t t t t t c a g t
                                                                                               7500
t c c a a a g a t c a g t a g c t a a g g a g g t t t g t a a g c g g t t t c t a t a g g t a c g a
a g g t t t c t a g t c a t c g a t t c c t c c a a a c a t t c g c c a a a g a t a t c c a t g c t
```

a a c t g t t g a c c g a c c t t c a t t t t t t c t g g g g t t a a g c a g t a g a a a g t a g t
t t g a c a a c t g g c t g g a a g t a a a a a a g a c c c c a a t t c g t c a t c t t t c a t c a

7600
a g a g t c t t t t t c c c a t t g g t c c c a t c c a a g t t c t a a t g c a a g t t g t a a g g
t c t c a g a a a a a g g g t a a c c a g g g t a g g t t c a a g a t t a c g t t c a a c a t t c c

c a t g t t t g a c a a g a t t g t c a t c c c c a g a c a g t t t c a t c a c c a g c a t a a a t
g t a c a a a c t g t t c t a a c a g t a g g g g t c t g t c a a a g t a g t g g t c g t a t t t a

7700
g g g a c a a g t t g c t t t c c a a a t g c c c c c a t c c a g g t g t a g g t t t c t a c a t c
c c c t g t t c a a c g a a a g g t t t a c g g g g g t a g g t c c a c a t c c a a a g a t g t a g

a t a g g t a a t a a a a a g g c g c t c a g t g c g a g g a t g c g a a c c g a t t g g g a a a a
t a t c c a t t a t t t t t c c g c g a g t c a c g c t c c t a c g c t t g g c t a a c c c t t t t

7800
a g t g g a t c t c c t g c c a/c c a g t t g g a a g a a t g g c t g t t g a t g t g a t g a a a g
t c a c c t a g a g g a c g g t g g t c a a c c t t c t t a c c g a c a a c t a c a c t a c t t t c

t a g a a a t c t c g t c g g c g g a c a g a g c a t t c a t g c t g a t g t t t g t a a a a g c g
a t c t t t a g a g c a g c c g c c t g t c t c g t a a g t a c g a c t a c a a a c a t t t t c g c

7900
t g c g c a g t g t t c g c a t c g t t g c a c g g g c t g t a t c t g t t g a a t g a g g t g t a
a c g c g t c a c a a g c g t a g c a a c g t g c c c g a c a t a g a c a a c t t a c t c c a c a t

c c t g g c g g c c t c g c a c c a g a a a g c a g a t g g g a a a a t c a a t a c c a c t t g g c
g g a c c g c c g g a g c g t g g t c t t t c g t c t a c c c t t t t a g t t a t g g t g a a c c g

8000
a g c t g c c g t t c g t c c t c t t c c t c t t c t g c t g c a t t g c c a c t a c c g t t t g g
t c g a c g g c a a g c a g g a g a a g g a g a a g a c g a c g t a a c g g t g a t g g c a a a c c

a t c c t c g a a a g c g a g a a c g g a g a g g g t g a c g g t g c c c c t c g a c c t g c a t g
t a g g a g c t t t c g c t c t t g c c t c t c c c a c t g c c a c g g g g a g c t g g a c g t a c

8100
t c c a g a t t t c a g c a c g a g a g g g g c g g a a a c g g g a a a t c a g g g c g t a c a g c
a g g t c t a a a g t c g t g c t c t c c c c g c c t t t g c c c t t t a g t c c c g c a t g t c g

c t g g a g c t g t c c a t g g t a t c a g t c a g a g a g a a a a g c a t g t c c g c g g g g a c
g a c c t c g a c a g g t a c c a t a g t c a g t c t c t c t t t t c g t a c a g g c g c c c c t g

8200
a g c g c g c a a g t t g a c t t c g c a c a g g c g g g t a a g a g c a g g c t g g a g g t g c a
t c g c g c g t t c a a c t g a a g c g t g t c c g c c c a t t c t c g t c c g a c c t c c a c g t

g g t a a t a c t t a a t t t c t a g a g g c g t g c c g t t g g c a g a g t c t a t t g c g t g a
c c a t t a t g a a t t a a a g a t c t c c g c a c g g c a a c c g t c t c a g a t a a c g c a c t

8300
a g t a t t c c a t g a g c c c g g g g a c t a a c c a c g g t t c c a c g g t g c a c t t t t c c
t c a t a a g g t a c t c g g g c c c c t g a t t g g t g c c a a g g t g c c a c g t g a a a a g g

a a t g c g c c t g c t t a a a a t c g g c g g c g c g g a c g a g c t c c c g g a g g a a g c g g
t t a c g c g g a c g a a t t t t a g c c g c c g c g c c t g c t c g a g g g c c t c c t t c g c c

8400
c g g t t c g g g t c c t g c g g g a a g c g g g g g a a g c g g t a t g t c g g c c t g a c g c t
g c c a a g c c c a g g a c g c c c t t c g c c c c c t t c g c c a t a c a g c c g g a c t g c g a

c t g g c a g g g g a a g g t g t t g a g c c c g a a g t t g a c t g g c a t g g g c g a c t a c c
g a c c g t c c c c t t c c a c a a c t c g g g c t t c a a c t g a c c g t a c c c g c t g a t g g

8500
c g g c g a t t g a t a t c t t g a a t c t g t c g g c g t t g t g t a a a c a c t a c c g g c c c
g c c g c t a a c t a t a g a a c t t a g a c a g c c g c a a c a c a t t t g t g a t g g c c g g g

```
t g t t g t t t t g a a c c t g a a a g a a a g t t c a a c a g a a t c a a t c t c a g t g t c a t
a c a a c a a a a c t t g g a c t t t c t t t c a a g t t g t c t t a g t t a g a g t c a c a g t a

                                                                                              8600
t t a c t g c a g c c t g t c t t a a a a t c t c c t g a a c g t c g c c t g a g t t a t c t t g g
a a t g a c g t c g g a c a g a a t t t t a g a g g a c t t g c a g c g g a c t c a a t a g a a c c

t a g g c a a t t t c t g c c a t t a a t t g a t c a a t t t c t t c c t c c t g g a g g t c t c c
a t c c g t t a a a g a c g g t a a t t a a c t a g t t a a a g a a g g a g g a c c t c c a g a g g

                                                                                              8700
a t g t c c c g c a c g t t c a a t a g t g g c t g c a a g g t c a t t a g a t a t c c g a c t c a
t a c a g g g c g t g c a a g t t a t c a c c g a c g t t c c a g t a a t c t a t a g g c t g a g t

t a a g c t g t g a a a a t g c g t t t a g t c c a a t t t c g t t c c a g a c t c g g c t g t a t
a t t c g a c a c t t t t a c g c a a a t c a g g t t a a a g c a a g g t c t g a g c c g a c a t a

                                                                                              8800
a c t a c c c c t c c t t c g c t g t c c c g a g c g c g c a t a a c c a c t t g c g c c a a g t t
t g a t g g g g a g g a a g c g a c a g g g c t c g c g c g t a t t g g t g a a c g c g g t t c a a

g a g t t c c a c g a g c c g t g c g a a c a c g c c g t a g t t g c g c a a g c g c t g a a a c a
c t c a a g g t g c t c g g c a c g c t t g t g c g g c a t c a a c g c g t t c g c g a c t t t g t

                                                                                              8900
g g t a g t t t a a g g t g g t g g c a a c g t g t t c t g a g a c g a a g a a a t a c a g a a t c
c c a t c a a a t t c c a c c a c c g t t g c a c a a g a c t c t g c t t c t t t a t g t c t t a g

c a c c g a c g a a g c g t c a g c t c g t t g a t g t c a c c t a a g g c t t c a a g a c g t t c
g t g g c t g c t t c g c a g t c g a g c a a c t a c a g t g g a t t c c g a a g t t c t g c a a g

                                                                                              9000
c a t g g c t t c g t a a a a g t c t a c t g c a a a a t t g a a a a a c t g g g a g t t g c g a g
g t a c c g a a g c a t t t t c a g a t g a c g t t t t a a c t t t t t g a c c c t c a a c g c t c
```

c t g c c a c c g t c a a t t c t t c t t c c a a c a g a c g a a t a a g c t c g g c c a c c g t c
g a c g g t g g c a g t t a a g a a g a a g g t t g t c t g c t t a t t c g a g c c g g t g g c a g

9100
t c g c g c a c t t c t t g c t g a a a t g c g c c c g g a a c t a t t t c t t g t t c t t c c t c
a g c g c g t g a a g a a c g a c t t t a c g c g g g c c t t g a t a a a g a a c a a g a a g g a g

t t c t a c c t c c a t t a t t t c t t c c t c g a c c a c a g g t g g t g g g g g t t g t c t t c
a a g a t g g a g g t a a t a a a g a a g g a g c t g g t g t c c a c c a c c c c c a a c a g a a g

9200
t t c g a c g c c g g c g a a c g g g c a g c c t g t c t a c a a a t c t t t c a a t c a t t t c g
a a g c t g c g g c c g c t t g c c c g t c g g a c a g a t g t t t a g a a a g t t a g t a a a g c

c c g c g a c g g c g g c g c a t a g t t t c g g t t a c t g c t c g a c c g t t t t c a c g t g g
g g c g c t g c c g c c g c g t a t c a a a g c c a a t g a c g a g c t g g c a a a a g t g c a c c

9300
t c g t a a c t c a a a a a c t c c a c c t c t a a g t t c t g t t t c a t g t a a a a t g g g a a
a g c a t t g a g t t t t t g a g g t g g a g a t t c a a g a c a a a g t a c a t t t t a c c c t t

a t g a g g c g t t g c g a g g g g c g t t a g g t a g g g a t a c a g c g c t g a t t a t g c a t
t a c t c c g c a a c g c t c c c c g c a a t c c a t c c c t a t g t c g c g a c t a a t a c g t a

9400
t t t a t t a t t t g c t g c g t a g g a a c t c c g c g c a a g g a g c t a a g c g t c t g c a t
a a a t a a t a a a c g a c g c a t c c t t g a g g c g c g t t c c t c g a t t c g c a g a c g t a

a t c c a c c g g g t c g g a g a a c c t t t c a a g a a a g g c a t c t a g c c a g t c a c a g t
t a g g t g g c c c a g c c t c t t g g a a a g t t c t t t c c g t a g a t c g g t c a g t g t c a

9500
c a c a a g g t a g g c t a a g t t t t g t t t c t t c t a a a g t a c c a g g a a g c t g a g c a
g t g t t c c a t c c g a t t c a a a a c a a a g a a g a t t t c a t g g t c c t t c g a c t c g t

```
a t g c t a c t a a t a a t g t a a t t g a a g t a a g c t g t t t t a a g c c c a c g a a t g g t
t a c g a t g a t t a t t a c a t t a a c t t c a t t c g a c a a a a t t c g g g t g c t t a c c a
                                                                                              9600
t t t a a g a a g c a c c a c a t c t t t g g g t c c g g c t t g t t g a a t t c g c a g g c g g t
a a a t t c t t c g t g g t g t a g a a a c c c a g g c c g a a c a a c t t a a g c g t c c g c c a

c t g c c a t t c c c c a c a c g t c a c t t t g a c a t c g t c c a a g a t c t t t g t a g t a g
g a c g g t a a g g g g t g t g c a g t g a a a c t g t a g c a g g t t c t a g a a a c a t c a t c
                                                                                              9700
t c t t g c a t t a a c c t t t c c a c c t c t a c c t c g c g g t t t c c g c g a t c a g c c a t
a g a a c g t a a t t g g a a a g g t g g a g a t g g a g c g c c a a a g g c g c t a g t c g g t a

g t g c g t g c t t c c g t a g c c t t g c a g c g g t t g t a a t a a a g c t a a a t c t g c c a
c a c g c a c g a a g g c a t c g g a a c g t c g c c a a c a t t a t t t c g a t t t a g a c g g t
                                                                                              9800
c t a c c c g t t c c g c a a g c a c t g c c t g t t g a a t t t g g g t a a g g g t g g t t g c a
g a t g g g c a a g g c g t t c g t g a c g g a c a a c t t a a a c c c a t t c c c a c c a a c g t

a a g t c a t c c a c a t c t a c a a a g c g g t g a t a a g c t c c t g c a t t a a t g g t g t a
t t c a g t a g g t g t a g a t g t t t c g c c a c t a t t c g a g g a c g t a a t t a c c a c a t
                                                                                              9900
g c t g c a g t t t g t c a t t a c t g a c c a a t t a a c a g t t t g c g t g c c t g g c t g t a
c g a c g t c a a a c a g t a a t g a c t g g t t a a t t g t c a a a c g c a c g g a c c g a c a t

c a g t t t c t g t g t a t c g c a a g c g t g a g t a a g c c c g a g a g t c a a a a a c a t a g
g t c a a a g a c a c a t a g c g t t c g c a c t c a t t c g g g c t c t c a g t t t t t g t a t c
                                                                                             10000
t c a t t g c a g g t g c g c a c t a g g t a t t g a t a g c c c a c a a g g a a a t g a g g a g g
a g t a a c g t c c a c g c g t g a t c c a t a a c t a t c g g g t g t t c c t t t a c t c c t c c
```

a g g t t c g c g a t a c a a c g g c c a g c c a a g c g t a g c c g c a g c a c c t g g a g c g a
t c c a a g c g c t a t g t t g c c g g t c g g t t c g c a t c g g c g t c g t g g a c c t c g c t

10100
g a t c t t c c a a c a t g a g g c g g t g g t a t t c a t a t a t g t a t c t g g a c a t c c a t
c t a g a a g g t t g t a c t c c g c c a c c a t a a g t a t a t a c a t a g a c c t g t a g g t a

g t g a t g c c g g c a g c g g t a g t t g t t g c t c g c a t a a a t t c g c g g g c t c g g t t
c a c t a c g g c c g t c g c c a t c a a c a a c g a g c g t a t t t a a g c g c c c g a g c c a a

10200
c c a a a t a t t g c g c a g g g g t a a a a a g c g t t c a a t a g t t g c c a c g c t t t g a c
g g t t t a t a a c g c g t c c c c a t t t t t c g c a a g t t a t c a a c g g t g c g a a a c t g

c g g t c a g g c g t g c g c a g t c t t g a a t g c t c t g g a c a t g g a a a a a a t g a a a g
g c c a g t c c g c a c g c g t c a g a a c t t a c g a g a c c t g t a c c t t t t t t a c t t t c

10300
t t g g t a a g c g a c t c c c t t c c g t g g t t t g g t g g a a a a g t c a c a a g g g t a c c
a a c c a t t c g c t g a g g g a a g g c a c c a a a c c a c c t t t t c a g t g t t c c c a t g g

a t a g c g a g g a a c c c c g g t t c g a a a c c g g c a g g a t c c g c t a t g a g c a c a a g
t a t c g c t c c t t g g g g c c a a g c t t t g g c c g t c c t a g g c g a t a c t c g t g t t c

10400
t g a g g c g c t t g c g c g t t g a a c c c g g c c a a g g a c c c c c a g a c a c g g a g a g g
a c t c c g c g a a c g c g c a a c t t g g g c c g g t t c c t g g g g g t c t g t g c c t c t c c

a g t c t t t t t t t a t t t a t t t t t t c t t a g a t g c a t c c t g t c c t g c g a c a a a t
t c a g a a a a a a a t a a a t a a a a a a g a a t c t a c g t a g g a c a g g a c g c t g t t t a

10500
g c g a c c t c a g c c c a g g g c a a c c a c g g c c t c a g c a g c g g t g g c g c t t t c g g
c g c t g g a g t c g g g t c c c g t t g g t g c c g g a g t c g t c g c c a c c g c g a a a g c c

g c t c t g g c g a a c a g g a a g a g c c t c a a t g t c c t a c a t t g g a g t t g g a a g a a
c g a g a c c g c t t g t c c t t c t c g g a g t t a c a g g a t g t a a c c t c a a c c t t c t t

10600

g g a g a a g g c a t a g c c c g a t t g g g c g c c c a c t c t c c t g a g c g t c a c c c a a g
c c t c t t c c g t a t c g g g c t a a c c c g c g g g t g a g a g g a c t c g c a g t g g g t t c

g g t g c a g c t c g c c c g g g a c a g t c g c g t g g c a t t t g t g c c t c g t c a g a a c a
c c a c g t c g a g c g g g c c c t g t c a g c g c a c c g t a a a c a c g g a g c a g t c t t g t

10700

t g t t t c g c g a c a a c a g c g g g g a g g a a g c t g a g g a a a t g c g a g a c t g c a g g
a c a a a g c g c t g t t g t c g c c c c t c c t t c g a c t c c t t t a c g c t c t g a c g t c c

t t t a g g g c c g g t c g c g a g c t g c g c c g c g g a t t t a a t c g c g a g c g a c t g c t
a a a t c c c g g c c a g c g c t c g a c g c g g c g c c t a a a t t a g c g c t c g c t g a c g a

10800

g c g t g a g g a g g a c t t t g a g c c a g a t g a a c a t t c g g g g a t t a g t t c t g c a c
c g c a c t c c t c c t g a a a c t c g g t c t a c t t g t a a g c c c c t a a t c a a g a c g t g

g g g c c c a t g t a t c a g c a g c c a a c t t a g t a a c a g c a t a t g a a c a a a c g g t t
c c c g g g t a c a t a g t c g t c g g t t g a a t c a t t g t c g t a t a c t t g t t t g c c a a

10900

a c a g a g g a a c g t a a c t t t c a a a a a a g c t t t a a t a a c c a t g t g c g c a c a c t
t g t c t c c t t g c a t t g a a a g t t t t t t c g a a a t t a t t g g t a c a c g c g t g t g a

a a t a g c g c g a g a a g a a g t a g c c a t t g g t t t a a t g c a t c t t t g g g a c t t t g
t t a t c g c g c t c t t c t t c a t c g g t a a c c a a a t t a c g t a g a a a c c c t g a a a c

11000

t a g a a g c t t a t g t a c a t a a t c c a g c a a g t a a a c c c c t a a c t g c c c a g c t g
a t c t t c g a a t a c a t g t a t t a g g t c g t t c a t t t g g g g a t t g a c g g g t c g a c

```
t t c t t a a t a g t t c a a c a t a g t a g a g a c a a t g a a a c t t t t a g g g a t g c a a t
a a g a a t t a t c a a g t t g t a t c a t c t c t g t t a c t t t g a a a a t c c c t a c g t t a

                                                                                         11100
g c t t a a c a t a g c t g a a c c c a g g g t c g g t g g t t a c t c g a t t t a a t t a a c a
c g a a t t g t a t c g a c t t g g g t c c c a g c c a c c a a t g a g c t a a a t t a a t t g t

t t c t g c a g a g c a t t g t g g t t c a g g a a c g c a g t c t t a g t t t g g c a g a c a a g
a a g a c g t c t c g t a a c a c c a a g t c c t t g c g t c a g a a t c a a a c c g t c t g t t c

                                                                                         11200
g t g g c c g c c a t t a a t t a c t c c a t g t t a a g t t t g g g a a a g t t t t a t g c t c g
c a c c g g c g g t a a t t a a t g a g g t a c a a t t c a a a c c c t t t c a a a a t a c g a g c

t a a a a t c t a c a a a a g t c c g t a t g t t c c c a t t g a c a a g g a a g t g a a g a t a g
a t t t t a g a t g t t t t c a g g c a t a c a a g g g t a a c t g t t c c t t c a c t t c t a t c

                                                                                         11300
a c a g c t t t t a t a t g c g c a t g g c t t t a a a g g t a c t a a c a t t a a g c g a c g a t
t g t c g a a a a t a t a c g c g t a c c g a a a t t t c c a t g a t t g t a a t t c g c t g c t a

c t t g g a g t g t a c c g c a a t g a c c g a a t c c a c a a a g c a g t a a g c g c c a g t c g
g a a c c t c a c a t g g c g t t a c t g g c t t a g g t g t t t c g t c a t t c g c g g t c a g c

                                                                                         11400
c c g c a g a g a g c t a a g c g a c a a a g a g c t t a t g c a t a g c t t a c a a a g g g c g c
g g c g t c t c t c g a t t c g c t g t t t c t c g a a t a c g t a t c g a a t g t t t c c c g c g

t g a c g g g a g c a g g a a c a g a g g a c g a g t c g t t c t t t g a t a t g g g c g c a g a c
a c t g c c c t c g t c c t t g t c t c c t g c t c a g c a a g a a a c t a t a c c c g c g t c t g

                                                                                         11500
c t a c g g t g g c a g c c a a g c g c t c g c g c t t t g g a g g c a g c t g g a g t g g c g t c
g a t g c c a c c g t c g g t t c g c g a g c g c g a a a c c t c c g t c g a c c t c a c c g c a g
```

```
t g c t g a c g t c a c t g g c g a t g a c g a t g a c g a a g a c c a g t a c g a g g a c t g a t
a c g a c t g c a g t g a c c g c t a c t g c t a c t g c t t c t g g t c a t g c t c c t g a c t a
                                                                                               11600
c g g c c g t a c c t t t t g t t a g a t g c a g c g a c c g g c g a t c a t c g c g g a g a g g g
g c c g g c a t g g a a a a c a a t c t a c g t c g c t g g c c g c t a g t a g c g c c t c t c c c

c t c c t a a c c t g g a t c c c g c g g t t t t g g c g g c c a t g c a a a g c c a g c c t t c t
g a g g a t t g g a c c t a g g g c g c c a a a a c c g c c g g t a c g t t t c g g t c g g a a g a
                                                                                               11700
g g c g t t a c a g c t t c a g a t g a c t g g a c a g c g g c c a t g g a t c g t a t t a t g g c
c c g c a a t g t c g a a g t c t a c t g a c c t g t c g c c g g t a c c t a g c a t a a t a c c g

t t t a a c g g c g c g c a g t c c t g a t g c t t t t c g c c a g c a g c c c c a a g c t a a c c
a a a t t g c c g c g c g t c a g g a c t a c g a a a a g c g g t c g t c g g g g t t c g a t t g g
                                                                                               11800
g c t t t t c g g c c a t t t t g g a a g c a g t a g t g c c g t c t c g t a c t a a c c c t a c t
c g a a a a g c c g g t a a a a c c t t c g t c a t c a c g g c a g a g c a t g a t t g g g a t g a

c a c g a g a a a g t g t t a a c c a t t g t a a a t g c t t t g t t g g a t a g c a a a g c c a t
g t g c t c t t t c a c a a t t g g t a a c a t t t a c g a a a c a a c c t a t c g t t t c g g t a
                                                                                               11900
c c g c a a a g a t g a g g c t g g t t t a a t a t a c a a c g c t t t g c t t g a g c g c g t g g
g g c g t t t c t a c t c c g a c c a a a t t a t a t g t t g c g a a a c g a a c t c g c g c a c c

c a c g c t a t a a c a g t a c c a a t g t g c a g g c t a a c t t a g a c c g g a t g g g t a c a
g t g c g a t a t t g t c a t g g t t a c a c g t c c g a t t g a a t c t g g c c t a c c c a t g t
                                                                                               12000
g a t g t a a a g g a g g c g c t g g c t c a a c g a g a g c g c t t t c a t c g c g a t g g t a a
c t a c a t t t c c t c c g c g a c c g a g t t g c t c t c g c g a a a g t a g c g c t a c c a t t
```

t c t t g g t t c g c t a g t a g c a t t a a a c g c t t t t t t g a g t a c t c a g c c g g c t a
a g a a c c a a g c g a t c a t c g t a a t t t g c g a a a a a a c t c a t g a g t c g g c c g a t

12100

a t g t t c c g c g t g g t c a g g a a g a t t a t a c a a a c t t c a t c a g c g c c t t g c g a
t a c a a g g c g c a c c a g t c c t t c t a a t a t g t t t g a a g t a g t c g c g g a a c g c t

c t a a t g g t t a c t g a a g t g c c t c a a a g t g a a g t g t a t c a g t c t g g a c c c g a
g a t t a c c a a t g a c t t c a c g g a g t t t c a c t t c a c a t a g t c a g a c c t g g g c t

12200

t t a c t t t t t t c a a a c g t c c a g g c a g g g t t t g c a a a c c g t a a a c t t a a c t c
a a t g a a a a a a g t t t g c a g g t c c g t c c c a a a c g t t t g g c a t t t g a a t t g a g

a g g c t t t t a a a a t t t g c a a g g t t t g t g g g g g g t t c g t g c t c c a g t a g g c
t c c g a a a a t t t t a a a c g t t c c a a a c a c c c c c c a a g c a c g a g g t c a t c c g

12300

g a t c g t t c a a c t t t g t c c a g t t t a c t a a c a c c a a a c t c g c g c c t a t t a c t
c t a g c a a g t t g a a a c a g g t c a a a t g a t t g t g g t t t g a g c g c g g a t a a t g a

g t t g c t a a t t g c c c c c t t t a c c a a c a c c a a c a g t t t a a g t c g a g a t t c a t
c a a c g a t t a a c g g g g g a a a t g g t t g t g g t t g t c a a a t t c a g c t c t a a g t a

12400

a c c t g g g t c a c t t a g t t a c t t t g t a c c g c g a a g c c a t t g g t c a a g c g c a g
t g g a c c c a g t g a a t c a a t g a a a c a t g g c g c t t c g g t a a c c a g t t c g c g t c

g t a g a c g a a c a a a c t t a t c a a g a a a t a a c c a g t g t t a g t c g c g c a c t g g g
c a t c t g c t t g t t t g a a t a g t t c t t t a t t g g t c a c a a t c a g c g c g t g a c c c

12500

c c a g g a g g a c a c t g g c a g t t t a g a g g c c a c a c t t a a c t t t t t a c t a a c t a
g g t c c t c c t g t g a c c g t c a a a t c t c c g g t g t g a a t t g a a a a a t g a t t g a t

```
a c c g t c g c c a g c a a g t g c c t c c t c a g t a c a c t t t a a a t g c g g a a g a a g a a
t g g c a g c g g t c g t t c a c g g a g g a g t c a t g t g a a a t t t a c g c c t t c t t c t t
                                                                                            12600
c g c a t a t t g c g c t a t g t a c a g c a a t c t g t a a g t t t g t a t c t t a t g c g t g a
g c g t a t a a c g c g a t a c a t g t c g t t a g a c a t t c a a a c a t a g a a t a c g c a c t

g g g t g c c a c c c c c a g t g c c g c c t t a g a c a t g a c a g c g c g c a a t a t g g a g c
c c c a c g g t g g g g g t c a c g g c g g a a t c t g t a c t g t c g c g c g t t a t a c c t c g
                                                                                            12700
c g t c c t t c t a c g c t t c c a a t c g a g c t t t c a t t a a t c g c t t g a t g g a t t a c
g c a g g a a g a t g c g a a g g t t a g c t c g a a a g t a a t t a g c g a a c t a c c t a a t g

c t t c a c c g c g c t g c g g c c a t g a a c g g g g a a t a c t t t a c a a a t g c a a t t c t
g a a g t g g c g c g a c g c c g g t a c t t g c c c c t t a t g a a a t g t t t a c g t t a a g a
                                                                                            12800
a a a t c c g c a t t g g t t g c c c c c t c c t g g a t t t t a c a c t g g t g a a t t t g a t t
t t t a g g c g t a a c c a a c g g g g g a g g a c c t a a a a t g t g a c c a c t t a a a c t a a

t g c c g g a a g g a a a t g a t g g c t t t t t g t g g g a t g a t g t t a c g g a c a g t c t g
a c g g c c t t c c t t t a c t a c c g a a a a a c a c c c t a c t a c a a t g c c t g t c a g a c
                                                                                            12900
t t t a g t c c t g c a g t t a t t g g a c a c c a t g g t a a a a a g g a a g c a g g t g a t g a
a a a t c a g g a c g t c a a t a a c c t g t g g t a c c a t t t t t c c t t c g t c c a c t a c t

a g g t c c c t t g c t t g a c t c t c g g g c g a g t t c t c c a t t c c c c a g t t t a a c t a
t c c a g g g a a c g a a c t g a g a g c c c g c t c a a g a g g t a a g g g g t c a a a t t g a t
                                                                                            13000
g t t t a c c c g c c a g t g t t a a c a g c g g t c g t a c c a c c a g a c c c c g a c t a a c a
c a a a t g g g c g g t c a c a a t t g t c g c c a g c a t g g t g g t c t g g g g c t g a t t g t
```

```
g g t g a a a g t g a a t a c t t a a a t g a c c c c a t c t t g t t t c c a g t g c g c g a c a a
c c a c t t t c a c t t a t g a a t t t a c t g g g g t a g a a c a a a g g t c a c g c g c t g t t

                                                                                             13100
a a a t t t t c c c a a c a a t g g c a t a g a a a g t t t g g t a g a t a a a a t g t c t c g c t
t t t a a a a g g g t t g t t a c c g t a t c t t t c a a a c c a t c t a t t t t a c a g a g c g a

g g a a a a c a t a t g c a c a a g a g c g g c g a g a a t g g g a g g a a a g a c a g c c a a g a
c c t t t t g t a t a c g t g t t c t c g c c g c t c t t a c c c t c c t t t c t g t c g g t t c t

                                                                                             13200
c c a g t t c g c c c t c c t a g g c a a c g t t g g c a g c g a c g c a a a a a a g g g g c a c a
g g t c a a g c g g g a g g a t c c g t t g c a a c c g t c g c t g c g t t t t t t c c c c g t g t

t g c g g g g g a t g a a g g a a g c g a t g a c t c a g c t g a c g a c a g t a g t g t a t t a g
a c g c c c c c t a c t t c c t t c g c t a c t g a g t c g a c t g c t g t c a t c a c a t a a t c

                                                                                             13300
a t t t a g g a g g g t c a g g a a a c c c a t t t g c t c a t t t g c g c c c a c a g g g t t g c
t a a a t c c t c c c a g t c c t t t g g g t a a a c g a g t a a a c g c g g g t g t c c c a a c g

a t a g g g t c a t t g t a t t a a a t t g a a t a a a a g c a t a c t t a c c a a a g c c a t g g
t a t c c c a g t a a c a t a a t t t a a c t t a t t t t c g t a t g a a t g g t t t c g g t a c c

                                                                                             13400
c g a c c a g t g t t c g t c t t a t t t t c c t t c t t c c g t t a g c t g t g a a a t g a g g c
g c t g g t c a c a a g c a g a a t a a a a g g a a g a a g g c a a t c g a c a c t t t a c t c c g

g c g c g g t g g a a c t g c a g a c a g t g g c t t t t c c t g a g a c a c c a c c t c c c t c t
c g c g c c a c c t t g a c g t c t g t c a c c g a a a a g g a c t c t g t g g t g g a g g g a g a

                                                                                             13500
t a c g a a a c c g t g a t g g c a g c g g c g c c a c c c t a c g t g c c t c c c c g c t a t t t
a t g c t t t g g c a c t a c c g t c g c c g c g g t g g g a t g c a c g g a g g g g c g a t a a a
```

g g g t c c t a c g g a g g g a a g a a a c a g t a t c c g t t a c t c g g a a t t g t c a c c g t
c c c a g g a t g c c t c c c t t c t t t g t c a t a g g c a a t g a g c c t t a a c a g t g g c a

13600

t g t a c g a t a c c a c t c g a g t g t a c t t g g t g g a c a a c a a g t c t t c t g a c a t t
a c a t g c t a t g g t g a g c t c a c a t g a a c c a c c t g t t g t t c a g a a g a c t g t a a

g c t t c a t t g a a t t a c c a g a a t g a t c a c a g c a a c t t t t t a a c c a c t g t a g t
c g a a g t a a c t t a a t g g t c t t a c t a g t g t c g t t g a a a a a t t g g t g a c a t c a

13700

g c a a a a t a a t g a c t a t t c c c c t a t a g a g g c t g g c a c g c a a a c t a t t a a c t
c g t t t t a t t a c t g a t a a g g g g a t a t c t c c g a c c g t g c g t t t g a t a a t t g a

t t g a t g a a a g g t c t a g a t g g g g t g g a g a t t t a a a a a c c a t c t t a c a t a c c
a a c t a c t t t c c a g a t c t a c c c c a c c t c t a a a t t t t t g g t a g a a t g t a t g g

13800

a a c a t g c c a a a c g t g a a c g a t t t t a t g t t t a c c a c c a a a t t t a a g g c c a g
t t g t a c g g t t t g c a c t t g c t a a a a t a c a a a t g g t g g t t t a a a t t c c g g t c

g g t a a t g g t g g c t a g g a a a a a c a a a c a a c g a a g g c c a a a c c a t t t t a g a a t
c c a t t a c c a c c g a t c c t t t t t g t t t g t t g c t t c c g g t t t g g t a a a a t c t t a

13900

a t g a g t g g g c a g a a t t t g t g c t a c c c g a g g g t a a c t a t t c g g a a a c c a t g
t a c t c a c c c g t c t t a a a c a c g a t g g g c t c c c a t t g a t a a g c c t t t g g t a c

a c t a t t g a c t t a a t g a a c a a t g c t a t t a t t g a g c a t t a t t t g c g a g t a g g
t g a t a a c t g a a t t a c t t g t t a c g a t a a t a a c t c g t a a t a a a c g c t c a t c c

14000

a a g a c a g c a t g g a g t g c t g g a a a g t g a c a t t g g a g t t a a g t t t g a c a c c a
t t c t g t c g t a c c t c a c g a c c t t t c a c t g t a a c c t c a a t t c a a a c t g t g g t

g a a a c t t t c g t c t g g g t t g g g a c c c c g a a a c c c a a t t a g t a a c t c c g g g a
c t t t g a a a g c a g a c c c a a c c c t g g g g c t t t g g g t t a a t c a t t g a g g c c c t

14100
g t g t a c a c t a a t g a g g c t t t t c a t c c a g a t a t a g t a c t g c t t c c a g g t t g
c a c a t g t g a t t a c t c c g a a a a g t a g g t c t a t a t c a t g a c g a a g g t c c a a c

c g g g g t t g a t t t a c a g a g a g c a g a t t a a g c a a c a t a c t a g g t a t a a g a a
g c c c c a a c t a a a a t g t c t c t c g t c t a a t t c g t t g t a t g a t c c a t a t t c t t

14200
a g a g g c a g c c g t t t c a g g a a g g a t t t g t g a t t a t g t a t g a a c a c t t a g a g
t c t c c g t c g g c a a a g t c c t t c c t a a a c a c t a a t a c a t a c t t g t g a a t c t c

g g a g g c a a t a t t c c a g c t c t t t t g g a t g t a a a a a a a t a c g a a a a c a g t c t
c c t c c g t t a t a a g g t c g a g a a a a c c t a c a t t t t t t a t g c t t t t g t c a g a

14300
g c a g g a t c a a a a c a c t g t a a g a g g a g a c a a c t t t a t t g c c t t a a a t a a g g
c g t c c t a g t t t t g t g a c a t t c t c c t c t g t t g a a a t a a c g g a a t t t a t t c c

c t g c t a g g a t t g a a c c g g t t g a a a c a g a c c c c a a a g g a c g c a g t t a c a a c
g a c g a t c c t a a c t t g g c c a a c t t t g t c t g g g g t t t c c t g c g t c a a t g t t g

14400
t t g c t t c c a g a c a a a a a a a a t a c t a a a t a t c g c a g c t g g t a t t t g g c a t a
a a c g a a g g t c t g t t t t t t t t a t g a t t t a t a g c g t c g a c c a t a a a c c g t a t

c a a c t a c g g a g a c c c a g a a a a a g g a g t t c g g t c a t g g a c t c t a c t a a c a a
g t t g a t g c c t c t g g g t c t t t t t c c t c a a g c c a g t a c c t g a g a t g a t t g t t

14500
c t c c a g a t g t a a c a g g c g g c t c c g a a c a g g t g t a c t g g t c c c t a c c c g a t
g a g g t c t a c a t t g t c c g c c g a g g c t t g t c c a c a t g a c c a g g g a t g g g c t a

a t g a t g c a a g a t c c g g t g a c t t t t c g c t c c t c g c g t c a a g t t a g c a a c t a
t a c t a c g t t c t a g g c c a c t g a a a a g c g a g g a g c g c a g t t c a a t c g t t g a t

14600

t c c t g t a g t t g c a g c a g a a t t a c t g c c a g t t c a t g c t a a a a g c t t c t a c a
a g g a c a t c a a c g t c g t c t t a a t g a c g g t c a a g t a c g a t t t t c g a a g a t g t

a c g a g c a a g c c g t c t a c t c a c a g c t t a t t c g c c a g t c a a c c g c g c t t a c g
t g c t c g t t c g g c a g a t g a g t g t c g a a t a a g c g g t c a g t t g g c g c g a a t g c

14700

c g c g t g t t t a a t c g c t t t c c c g a g a a c c a g a t a c t g g t g c g t c c a c c a g c
g c g c a c a a a t t a g c g a a a g g g c t c t t g g t c t a t g a c c a c g c a g g t g g t c g

c g c t a c c a t c a c t a c c g t c a g t g a a a a c g t t c c c g c c c t t a c a g a t c a c g
g c g a t g g t a g t g a t g g c a g t c a c t t t t g c a a g g g c g g g a a t g t c t a g t g c

14800

g g a c c c t g c c g c t g c g t a g c a g t a t c a g t g g a g t t c a g c g a g t c a c c a t c
c c t g g g a c g g c g a c g c a t c g t c a t a g t c a c c t c a a g t c g c t c a g t g g t a g

a c t g a c g c c c g c c g c c g g a c c t g t c c c t a c g t t t a c a a a g c a c t g g g c a t
t g a c t g c g g g c g g c g g c c t g g a c a g g g a t g c a a a t g t t t c g t g a c c c g t a

14900

a g t t t c t c c a c g a g t g c t t t c t a g t c g c a c t t t t t a a a a a a g t g t g g t a a
t c a a a g a g g t g c t c a c g a a a g a t c a g c g t g a a a a a t t t t t t c a c a c c a t t

c a t g t c c a t t t t g g t t t c g c c a a g t a a c a a c a c g g g c t g g g g a c t g g g t g
g t a c a g g t a a a a c c a a a g c g g t t c a t t g t t g t g c c c g a c c c c t g a c c c a c

15000

c c g c c c g c a t g t a t g g a g g a g c t a a a a c a a g g t c t a g c c a a c a t c c a g t g
g g c g g g c g t a c a t a c c t c c t c g a t t t t g t t c c a g a t c g g t t g t a g g t c a c

```
c g c g t a c g c g g a c a t t a c c g a g c t c c a t g g g g c g c g c a t a c c c g a g g a c g
g c g c a t g c g c c t g t a a t g g c t c g a g g t a c c c c g c g c g t a t g g g c t c c t g c
                                                                                            15100
c a c t g g t c g c a c c a c t g t a g a c g a t g t t a t t g a c t c g g t a g t g g c c g a t g
g t g a c c a g c g t g g t g a c a t c t g c t a c a a t a a c t g a g c c a t c a c c g g c t a c

c t c g c a a g t a c c g c g c g c c c g c t g a a a c a g c a g g g t c t a c t g t t g a t g c a
g a g c g t t c a t g g c g c g c g g g c g a c t t t g t c g t c c c a g a t g a c a a c t a c g t
                                                                                            15200
g t a a t t g a t g a g g t a g t g g c a a a c g c g c g g g c t t a t g c a a g g c g c c g c a g
c a t t a a c t a c t c c a t c a c c g t t t g c g c g c c c g a a t a c g t t c c g c g g c g t c

a c g g c t g c g t c g c c g g c g t a g a c c a a c c a c c g c c a t g c g c g c g g c c a g a g
t g c c g a c g c a g c g g c c g c a t c t g g t t g g t g g c g g t a c g c g c g c c g g t c t c
                                                                                            15300
c g t t g g t t c g a c g g g c c a g g c g c a t t g g g c g g c g a g c t a t g a t g c g g g c a
g c a a c c a a g c t g c c c g g t c c g c g t a a c c c g c c g c t c g a t a c t a c g c c c g t

g c c a g g c g g g c t g c a a c g c c t g c c g g t c g a g c g c g g a g a c g g g c c g c a g c
c g g t c c g c c c g a c g t t g c g g a c g g c c a g c t c g c g c c t c t g c c c g g c g t c g
                                                                                            15400
t g c g g c c g c a a c a g c t a t t g c a a a c c t a g c t g c t c c g c g a c g a g g a a a t g
a c g c c g g c g t t g t c g a t a a c g t t t g g a t c g a c g a g g c g c t g c t c c t t t a c

t a t a c t g g g t g c g c g a c t c a g t g a c c g g g a c g c g t g t g c c a g t t c g t a c g
a t a t g a c c c a c g c g c t g a g t c a c t g g c c c t g c g c a c a c g g t c a a g c a t g c
                                                                                            15500
c g t c c a c c t c a c c c t t a g a a g a c a a a g a g t g a c t c a a t g t c t g t t a t g t a
g c a g g t g g a g t g g g a a t c t t c t g t t t c t c a c t g a g t t a c a g a c a a t a c a t
```

t g c c c a g c a t g a c c a a a c g c a a g t t c a a a g a a g a g c t g c t g c a g g c c t t a
a c g g g t c g t a c t g g t t t g c g t t c a a g t t t c t t c t c g a c g a c g t c c g g a a t

15600

g c g c c t g a a a t a t a t g g c c c a t c g g a t a a c c t t a c c a a g c g c g a t a t c a a
c g c g g a c t t t a t a t a c c g g g t a g c c t a t t g g a a t g g t t c g c g c t a t a g t t

g c a t g t t a a a a a a c g g g a a a a a a a a g a g g a a g a a g t c g c c g c g g c g t c a g
c g t a c a a t t t t t t g c c c t t t t t t t t c t c c t t c t t c a g c g g c g c c g c a g t c

15700

c a g a c g g c g t c g a g t t t g t g c g c t c a t t t g c g c c c a g a c g t a g g g t a c a g
g t c t g c c g c a g c t c a a a c a c g c g a g t a a a c g c g g g t c t g c a t c c c a t g t c

t g g a a g g g a c g g c a a g t a a a a c g c a t t t t g c g a c c g g g c a c c a c a g t g g t
a c c t t c c c t g c c g t t c a t t t t g c g t a a a a c g c t g g c c c g t g g t g t c a c c a

15800

t t t t t c t c c c g g a g a g c g a a c g a t t a t g c g t c c c c t a a a g c g c g a g t a c g
a a a a a g a g g g c c t c t c g c t t g c t a a t a c g c a g g g g a t t t c g c g c t c a t g c

a c g a a g t g t a c g c a g a c g a t g a c a t t t t g g a g c a a g c g g c a c a a c a g a c t
t g c t t c a c a t g c g t c t g c t a c t g t a a a a c c t c g t t c g c c g t g t t g t c t g a

15900

g g g g a a t t t g c a t a t g g a a a a a a a g g g c g t t a c g g a g a c a a a a t t g c t a t
c c c c t t a a a c g t a t a c c t t t t t t t c c c g c a a t g c c t c t g t t t t a a c g a t a

t c c t t t g g a c g a g g g a a a t c c a a c a c c c a g t t t a a a g g c t g t c a c t t t g c
a g g a a a c c t g c t c c c t t t a g g t t g t g g g t c a a a t t t c c g a c a g t g a a a c g

16000

a a c a a g t g t t g c c c g t c c t t g g g c c t t c g g a a g a a a a g c g t g g a a t t a a a
t t g t t c a c a a c g g g c a g g a a c c c g g a a g c c t t c t t t t c g c a c c t t a a t t t

a g g g a a g c c a t g g a t g a a t t g c a g c c t a c a a t g c a a c t g a t g g t g c c t a a
t c c c t t c g g t a c c t a c t t a a c g t c g g a t g t t a c g t t g a c t a c c a c g g a t t

16100
g c g g c a a a a g t t a g a g g a c g t a c t a g a g c a c a t g a a g g t g g a t c c t a g c g
c g c c g t t t t c a a t c t c c t g c a t g a t c t c g t g t a c t t c c a c c t a g g a t c g c

t a c a g c c a g a t g t a a a a g t a c g t c c g a t a a a a a a g g t a g c t c c a g g a t t g
a t g t c g g t c t a c a t t t t c a t g c a g g c t a t t t t t t c c a t c g a g g t c c t a a c

16200
g g a g t t c a a a c a g t g g a c a t t c a a a t t c c t g t g c a a a c t g c a t t g g g t g a
c c t c a a g t t t g t c a c c t g t a a g t t t a a g g a c a c g t t t g a c g t a a c c c a c t

a a c t a t g g a a a t c c a a a c t t c g c c a a t a a a a a c a a c g g t g a a c g c a a g c g
t t g a t a c c t t t a g g t t t g a a g c g g t t a t t t t t g t t g c c a c t t g c g t t c g c

16300
t g c a a a c a g a c c c t t g g t a c c c g c c a g t g c t t t c a a c a a a a a a a a a g c g t
a c g t t t g t c t g g g a a c c a t g g g c g g t c a c g a a a g t t g t t t t t t t t t c g c a

c a c t a c a g a c a a a c a a g t t c g c t t t t g c c a g a c t a c g t t t t a c a t c c t t c
g t g a t g t c t g t t t g t t c a a g c g a a a a c g g t c t g a t g c a a a a t g t a g g a a g

16400
c a t t g t g c c c a c g c c t g g g t a c c g t g g g a c a a c t t t t c a g c g c c g a g c c a
g t a a c a c g g g t g c g g a c c c a t g g c a c c c t g t t g a a a a g t c g c g g c t c g g t

c a g c c c c t a g c c g t a g a c g a g g t c c a t c a c g c c g t a g a c g t c g a c g c a a a
g t c g g g g a t c g g c a t c t g c t c c a g g t a g t g c g g c a t c t g c a g c t g c g t t t

16500
g c c a c t t t a g c c c c a g c g g c a g t a c g t c g c g t t g t a c a a a g g g g g c g c a c
c g g t g a a a t c g g g g t c g c c g t c a t g c a g c g c a a c a t g t t t c c c c c g c g t g

```
a c t a a t a c t t c c a t c c g t g c g t t a c c a c c c t a g c a t t c t c t a a c a a g c t g
t g a t t a t g a a g g t a g g c a c g c a a t g g t g g g a t c g t a a g a g a t t g t t c g a c
                                                                                              16600
c g c t g c c g t t t t t t c a g a t g g c t c t t a c t t g c c g a a t g c g c a t a c c c a t t
g c g a c g g c a a a a a a g t c t a c c g a g a a t g a a c g g c t t a c g c g t a t g g g t a a

c c a g g a t a c a g a g g a c g a c c c c g c c g g a g g a a a g g g c t g a c c g g g a a c g g
g g t c c t a t g t c t c c t g c t g g g g c g g c c t c c t t t c c c g a c t g g c c c t t g c c
                                                                                              16700
t c g a t t t c g g c g g c g t a g t a t g c g c a g a c g c a t g a a g g g t g g g g t g c t g c
a g c t a a a g c c g c c g c a t c a t a c g c g t c t g c g t a c t t c c c a c c c c a c g a c g

c c t t c c t a a t t c c a c t t a t t g c t g c g g c c a t t g g a g c c g t t c c c g g a a t t
g g a a g g a t t a a g g t g a a t a a c g a c g c c g g t a a c c t c g g c a a g g g c c t t a a
                                                                                              16800
g c c t c a g t a g c c t t g c a g g c t t c t c g a a a a a a t t a a a a t a a a a t a a a a c t
c g g a g t c a t c g g a a c g t c c g a a g a g c t t t t t t a a t t t t a t t t t a t t t t g a

t c c a a c t t a t t a c t g g t a c t a t g a c t g t t t t a t g c a g a c t a a a t g g a a g a
a g g t t g a a t a a t g a c c a t g a t a c t g a c a a a a t a c g t c t g a t t t a c c t t c t
                                                                                              16900
c a t c a a t t t t t c g t c g c t g g c c c c g c g a c a c g g c a c g c g g c c g t a c a t g g
g t a g t t a a a a a g c a g c g a c c g g g g c g c t g t g c c g t g c g c c g g c a t g t a c c

g c a c c t g g a a c g a g a t c g g c a c g a g c c a g c t g a a c g g g g g c g c c t t c a a t
c g t g g a c c t t g c t c t a g c c g t g c t c g g t c g a c t t g c c c c c g c g g a a g t t a
                                                                                              17000
t g g a a c a g t a t c t g g a g c g g t c t t a a a a a t t t t g g t t c c a c g a t t a a g a c
a c c t t g t c a t a g a c c t c g c c a g a a t t t t t a a a a c c a a g g t g c t a a t t c t g
```

a t a t g g c a c c a a g g c g t g g a a c a g c c a a a c c g g c c a g a t g c t a a g g g a c a
t a t a c c g t g g t t c c g c a c c t t g t c g g t t t g g c c g g t c t a c g a t t c c c t g t

17100

a g t t a a a a g a c c a a a a t t t t c a a c a g a a a g t t g t a g a t g g t c t g g c t t c g
t c a a t t t t c t g g t t t t a a a a g t t g t c t t t c a a c a t c t a c c a g a c c g a a g c

g g a a t t a a t g g a g t t g t a g a c a t a g c c a a t c a g g c t g t a c a g a a a a a a a t
c c t t a a t t a c c t c a a c a t c t g t a t c g g t t a g t c c g a c a t g t c t t t t t t t a

17200

t g c c a a c c g t t t a g a g c c g c g g c c c g a c g a g g t a a t g g t a g a g g a a a a g c
a c g g t t g g c a a a t c t c g g c g c c g g g c t g c t c c a t t a c c a t c t c c t t t t c g

t g c c a c c t c t a g a a a c t g t g c c c g g a t c c g t t c c a a c c a a a g g a g a a a a g
a c g g t g g a g a t c t t t g a c a c g g g c c t a g g c a a g g t t g g t t t c c t c t t t t c

17300

c g g c c a c g g c c g g a t g c a g a g g a a a c c t t a g t a a c g c a c a c a a c a g a a c c
g c c g g t g c c g g c c t a c g t c t c c t t t g g a a t c a t t g c g t g t g t t g t c t t g g

g c c g t c c t a t g a g g a a g c a a t a a a a c a a g g a g c c g c t c t g t c a c c t a c c a
c g g c a g g a t a c t c c t t c g t t a t t t t g t t c c t c g g c g a g a c a g t g g a t g g t

17400

c c t a t c c c a t g a c c a a g c c t a t t t t a c c c a t g g c t a c t a g a g t g t a t g g a
g g a t a g g g t a c t g g t t c g g a t a a a a t g g g t a c c g a t g a t c t c a c a t a c c t

a a a a a c g a a a a t g t g c c t a t g a c c c t t g a g c t g c c t c c t t t g c c a g a a c c
t t t t t g c t t t t a c a c g g a t a c t g g g a a c t c g a c g g a g g a a a c g g t c t t g g

17500

c a c t a t c g c g g a t c c c g t a g g t t c c g t t c c t g t t g c a t c t g t t c c a g t t g
g t g a t a g c g c c t a g g g c a t c c a a g g c a a g g a c a a c g t a g a c a a g g t c a a c

c a t c g a c a g t g a g c c g t c c a g c a g t g c g g c c t g t t g c c g t g g c t a g c t t g
g t a g c t g t c a c t c g g c a g g t c g t c a c g c c g g a c a a c g g c a c c g a t c g a a c

17600

c g a a a c c c a c g a t c c a g t a a t t g g c a a a g t a c c c t a a a c a g t a t t g t g g g
g c t t t g g g t g c t a g g t c a t t a a c c g t t t c a t g g g a t t t g t c a t a a c a c c c

a c t g g g a g t a a a g t c t c t c a a a c g c c g a c g c t g c t a c t a a c a t t a a a a g a
t g a c c c t c a t t t c a g a g a g t t t g c g g c t g c g a c g a t g a t t g t a a t t t t c t

17700

c g a g t g t t a a t t c c c a t c t g t g t a t a c g c c t c c t a t g t t a g c g c c a g a g g
g c t c a c a a t t a a g g g t a g a c a c a t a t g c g g a g g a t a c a a t c g c g g t c t c c

a c c a a c g c g t g a a t c g c a g t c a c c a c c a g c g c t t t c a a g a t g g c c a c t c c
t g g t t g c g c a c t t a g c g t c a g t g g t g g t c g c g a a a g t t c t a c c g g t g a g g

17800

c t c g a t g a t g c c g c a g t g g t c t t a c a t g c a c a t c g c c g g t c a g g a t g c c t
g a g c t a c t a c g g c g t c a c c a g a a t g t a c g t g t a g c g g c c a g t c c t a c g g a

c g g a g t a c c t g a g t c c c g g t c t g g t g c a a t t c g c c c g c g c c a c g g a c a c c
g c c t c a t g g a c t c a g g g c c a g a c c a c g t t a a g c g g g c g c g g t g c c t g t g g

17900

t a c t t c a c c c t g g g a a a c a a g t t t a g a a a c c c c a c c g t g g c t c c c a c c c a
a t g a a g t g g g a c c c t t t g t t c a a a t c t t t g g g g t g g c a c c g a g g g t g g g t

t g a t g t t a c c a c c g a t c g c t c g c a g c g t c t g a c g c t g c g t t t t g t g c c c g
a c t a c a a t g g t g g c t a g c g a g c g t c g c a g a c t g c g a c g c a a a a c a c g g g c

18000

t g g a t c g g g a a g a t a c t a c c t a c t c c t a c a a g g c t c g c t t t a c g c t g g c t
a c c t a g c c c t t c t a t g a t g g a t g a g g a t g t t c c g a g c g a a a t g c g a c c g a

```
g t g g g t g a c a a c c g c g t g t t a g a c a t g g c t a g t t c t t a c t t t g a c a t t c g
c a c c c a c t g t t g g c g c a c a a t c t g t a c c g a t c a a g a a t g a a a c t g t a a g c

                                                                                           18100
a g g g g t a c t g g a t c g t g g t c c c a g t t t t a a g c c c t a t t c c g g a a c c g c c t
t c c c c a t g a c c t a g c a c c a g g g t c a a a a t t c g g g a t a a g g c c t t g g c g g a

a c a a t t c t t t g g c a c c a a a a g g c g c t c c t a a t g c t t c a c a a t g g t c a g a t
t g t t a a g a a a c c g t g g t t t t c c g c g a g g a t t a c g a a g t g t t a c c a g t c t a

                                                                                           18200
a a c g c t a a g c t t a a t a c c t t t g c t c a g g c g c c g t a t c t t a g c g a c a c t a t
t t g c g a t t c g a a t t a t g g a a a c g a g t c c g c g g c a t a g a a t c g c t g t g a t a

c a c c g c c g c c g a t g g t a t t a a a g t t g g a a c a g a c a c c g c c c a g g c a g g c g
g t g g c g g c g g c t a c c a t a a t t t c a a c c t t g t c t g t g g c g g g t c c g t c c g c

                                                                                           18300
c g g c g g t g t a t g c c a a c a a a a c t t a t c a g c c a g a g c c g c a a g t a g g a c c a
g c c g c c a c a t a c g g t t g t t t t g a a t a g t c g g t c t c g g c g t t c a t c c t g g t

a g t g a a t g g a a c a c c a g c a t t g a a a a c g t t a a a g c t g g c g g g a g g g c a t t
t c a c t t a c c t t g t g g t c g t a a c t t t t g c a a t t t c g a c c g c c c t c c c g t a a

                                                                                           18400
a a a g c a a a c c a c t g c a a t g c a g c c g t g c t a t g g c t c c t a c g c t c g t c c a a
t t t c g t t t g g t g a c g t t a c g t c g g c a c g a t a c c g a g g a t g c g a g c a g g t t

c c a a c g a a c a c g g a g g a c a a t c c a a g g a t g a c a a c a t t g a a c t t a a g t t c
g g t t g c t t g t g c c t c c t g t t a g g t t c c t a c t g t t g t a a c t t g a a t t c a a g

                                                                                           18500
t t t g a t t c a g c t a a c a a t g c a g c a a a c a c t g c t c a a g t t g t g t t c t a t a c
a a a c t a a g t c g a t t g t t a c g t c g t t t g t g a c g a g t t c a a c a c a a g a t a t g
```

c g a a g a c g t a a a c c t t g a a a t g c c a g a c a c g c a t c t t g t g t t t a a g c c t a
g c t t c t g c a t t t g g a a c t t t a c g g t c t g t g c g t a g a a c a c a a a t t c g g a t

18600

c t g t t a c c a a t g g a a c a a t t g c t t c t g a g t c g c t g t t g g g a c a g c a a g c a
g a c a a t g g t t a c c t t g t t a a c g a a g a c t c a g c g a c a a c c c t g t c g t t c g t

g c g c c a a a t a g a g c a a a c t a c a t t g c a t t c a g a g a t a a t t t t a t t g g c c t
c g c g g t t t a t c t c g t t t g a t g t a a c g t a a g t c t c t a t t a a a a t a a c c g g a

18700

g a t g t a t t a c a a c a g t a c a g g c a a c a t g g g t g t a t t g g c c g g g c a a g c t t
c t a c a t a a t g t t g t c a t g t c c g t t g t a c c c a c a t a a c c g g c c c g t t c g a a

c c c a a c t t a a c g c a g t a g t a g a c c t g c a a g a c a g a a a t a c a g a g c t g t c a
g g g t t g a a t t g c g t c a t c a t c t g g a c g t t c t g t c t t t a t g t c t c g a c a g t

18800

t a c c a g t t a a t g c t g g a t g c t t t g g g a g a c a g a a c a c g g t a c t t t t c c t t
a t g g t c a a t t a c g a c c t a c g a a a c c c t c t g t c t t g t g c c a t g a a a a g g a a

g t g g a a t t c c g c a g t g g a c a g t t a c g a c c c t g a c g t t c g c g t t a t t g a g a
c a c c t t a a g g c g t c a c c t g t c a a t g c t g g g a c t g c a a g c g c a a t a a c t c t

18900

a t c a c g g g g t a g a g g a t g a a c t a c c a a a t t a t t g c t t t c c t c t t a g c g c a
t a g t g c c c c a t c t c c t a c t t g a t g g t t t a a t a a c g a a a g g a g a a t c g c g t

g t a g g t g a a a t a a a a a a t t a c a a a g g c a t t a a g c c a g a t a a c g g a g g a g g
c a t c c a c t t t a t t t t t t a a t g t t t c c g t a a t t c g g t c t a t t g c c t c c t c c

19000

a g g t g g c t g g a c t g c c g a c a a c a c t g t c a g t g a a g c a a a c c a c a t a g g c a
t c c a c c g a c c t g a c g g c t g t t g t g a c a g t c a c t t c g t t t g g t g t a t c c g t

t t g g g a a t a t a g c c g c c a t g g a a a t t a a t t t g c a g g c t a a t t t g t g g a g a
a a c c c t t a t a t c g g c g g t a c c t t t a a t t a a a c g t c c g a t t a a a c a c c t c t

19100

a g c t t c t t g t a c t c a a a t g t g g g c t t a t a c c t a c c a g a c g a c t t a a a a t a
t c g a a g a a c a t g a g t t t a c a c c c g a a t a t g g a t g g t c t g c t g a a t t t t a t

c a c t c c a g g a a a c a t a a a a c t a c c t g a t a a c a a g a a c a c c t a c g a g t a c a
g t g a g g t c c t t t g t a t t t t g a t g g a c t a t t g t t c t t g t g g a t g c t c a t g t

19200

t g a a c g g g c g t g t g a c t g c c c c g g g g t t g g t g g a t a c c t a t g t c a a t a t c
a c t t g c c c g c a c a c t g a c g g g g c c c c a a c c a c c t a t g g a t a c a g t t a t a g

g g c g c t c g c t g g t c c c c a g a t g t g a t g g a t a a t g t a a a c c c t t t t a a c c a
c c g c g a g c g a c c a g g g g t c t a c a c t a c c t a t t a c a t t t g g g a a a a t t g g t

19300

c c a c c g a a a c g c a g g g t t g c g c t a c a g a t c c a t g t t g c t a g g c a a t g g g a
g g t g g c t t t g c g t c c c a a c g c g a t g t c t a g g t a c a a c g a t c c g t t a c c c t

g a t t t g t t c c t t t t c a c a t t c a g g t g c c g c a a a a a t t t t t t g c c a t c a g a
c t a a a c a a g g a a a a g t g t a a g t c c a c g g c g t t t t t a a a a a a c g g t a g t c t

19400

a a t t t g t t g c t g t t g c c c g g t t c c t a c a c t t a c g a a t g g a a c t t t a g a a a
t t a a a c a a c g a c a a c g g g c c a a g g a t g t g a a t g c t t a c c t t g a a a t c t t t

g g a t g t a a a c a t g a t t c t t c a g a g c a c a c t g g g a a a t g a t c t t c g g g t g g
c c t a c a t t t g t a c t a a g a a g t c t c g t g t g a c c c t t t a c t a g a a g c c c a c c

19500

a c g g a g c c a g c g t t c g c t t t g a c a a c a t t g c c c t g t a t g c t a a c t t t t t t
t g c c t c g g t c g c a a g c g a a a c t g t t g t a a c g g g a c a t a c g a t t g a a a a a a

```
c c c a t g g c a c a t a a c a c a g c t t c t a c t t t a g a a g c c a t g t t a a g a a a t g a
g g g t a c c g t g t a t t g t g t c g a a g a t g a a a t c t t c g g t a c a a t t c t t t a c t

                                                                                           19600
c a c c a a c g a c c a g t c t t t t a a c g a t t a t t t g t g t g c t g c a a a c a t g c t g t
g t g g t t g c t g g t c a g a a a a t t g c t a a t a a a c a c a c g a c g t t t g t a c g a c a

a t c c c a t c c c a g c t a a c g c c a c c a g c g t g c c c a t t t c a a t a c c t t c g c g a
t a g g g t a g g g t c g a t t g c g g t g g t c g c a c g g g t a a a g t t a t g g a a g c g c t

                                                                                           19700
a a t t g g g c g g c a t t t a g a g g c t g g a g c t t t a c t c g c c t a a a a a c t a a a g a
t t a a c c c g c c g t a a a t c t c c g a c c t c g a a a t g a g c g g a t t t t t g a t t t c t

a a c t c c t t c c c t g g g t t c a g g g t t t g a c c c c t a c t t t g t a t a c t c t g g a a
t t g a g g a a g g g a c c c a a g t c c c a a a c t g g g g a t g a a a c a t a t g a g a c c t t

                                                                                           19800
c c a t t c c c t a t t t a g a c g g c a c c t t t t a c c t a a a c c a c a c t t t t a a g a a g
g g t a a g g g a t a a a t c t g c c g t g g a a a a t g g a t t t g g t g t g a a a a t t c t t c

g t g t c a a t c a t g t t t g a c t c c t c c g t g a g t t g g c c t g g a a a t g a c c g t t t
c a c a g t t a g t a c a a a c t g a g g a g g c a c t c a a c c g g a c c t t t a c t g g c a a a

                                                                                           19900
g c t a a c c c c a a a t g a a t t t g a a a t a a a g c g t t c t g t g g a t g g g g a g g g a t
c g a t t g g g g t t t a c t t a a a c t t t a t t t c g c a a g a c a c c t a c c c c t c c c t a

a c a a t g t g g c c c a a t g c a a t a t g a c t a a g g a t t g g t t c c t a a t a c a a a t g
t g t t a c a c c g g g t t a c g t t a t a c t g a t t c c t a a c c a a g g a t t a t g t t t a c

                                                                                           20000
c t t a g t c a t t a c a a c a t t g g a t a c c a a g g t t t t t a c a t t c c a g a g a g c t a
g a a t c a g t a a t g t t g t a a c c t a t g g t t c c a a a a a t g t a a g g t c t c t c g a t
```

```
c a a g g a c c g c a t g t a t t c t t t c t t t a g a a a c t t t c a g c c c a t g a g t a g g c
g t t c c t g g c g t a c a t a a g a a a g a a a t c t t t g a a a g t c g g g t a c t c a t c c g

                                                                                          20100
a a g t t g t g g a t a c c a c a g a a t a t a a g a a c t a c a a a a a a g t a a c c g t a g a g
t t c a a c a c c t a t g g t g t c t t a t a t t c t t g a t g t t t t t t c a t t g g c a t c t c

t t t c a a c a t a a c a a c t c a g g a t t c g t g g g a t a c c t g g g c c c c a c t a t g c g
a a a g t t g t a t t g t t g a g t c c t a a g c a c c c t a t g g a c c c g g g g t g a t a c g c

                                                                                          20200
g g a g g g a c a a g c t t a c c c c g c c a a c t a t c c c t a c c c t c t t a t a g g c c a a a
c c t c c c t g t t c g a a t g g g g c g g t t g a t a g g g a t g g g a g a a t a t c c g g t t t

c a g c t g t g g a a a g c a t c a c a c a g a a a a a g t t t c t a t g c g a t c g t g t t a t g
g t c g a c a c c t t t c g t a g t g t g t c t t t t t c a a a g a t a c g c t a g c a c a a t a c

                                                                                          20300
t g g c g c a t c c c a t t t c t a g t a a c t t c a t g t c t a t g g g g g c g c t a a c g g a a
a c c g c g t a g g g t a a a g a t c a t t g a a g t a c a g a t a c c c c c g c g a t t g c c t t

t c t t g g g c a a a a t a t g c t g t a c g c a a a c t c a g c c c a t g c t c t a g a c a t g a
a g a a c c c g t t t t a t a c g a c a t g c g t t t g a g t c g g g t a c g a g a t c t g t a c t

                                                                                          20400
c a t t t g a g g t g g a t c c a a t g g a t g a g c c t a c c c t t c t t t a t g t t t t a t t t
g t a a a c t c c a c c t a g g t t a c c t a c t c g g a t g g g a a g a a a t a c a a a a t a a a

g a a g t t t t c g a c g t g g t a c g c a t t c a c c a g c c a c a c c g c g g c g t c a t t g a
c t t c a a a a g c t g c a c c a t g c g t a a g t g g t c g g t g t g g c g c c g c a g t a a c t

                                                                                          20500
a g c g g t c t a c c t g c g c a c g c c c t t c t c g g c g g g t a a c g c t a c c a c c t a a g
t c g c c a g a t g g a c g c g t g c g g g a a g a g c c g c c c a t t g c g a t g g t g g a t t c
```

a a g g c a c c c t c c c a g a c t g c t g t a a t g g g t t c a a g c g a a c a g g a g c t g a c
t t c c g t g g g a g g g t c t g a c g a c a t t a c c c a a g t t c g c t t g t c c t c g a c t g

20600

g g c c a t t g t t c g a g a t c t a g g c t g t g g a c c c t a t t t t t t g g g a a c c t t t g
c c g g t a a c a a g c t c t a g a t c c g a c a c c t g g g a t a a a a a a c c c t t g g a a a c

a c a a a c g t t t t c c g g g t t t t g t g t c t c g c g a c c g c t t a t c a t g t g c t a t t
t g t t t g c a a a a g g c c c a a a a c a c a g a g c g c t g g c g a a t a g t a c a c g a t a a

20700

g t t a a c a c t g c c g g t c g c g a a a c t g g g g g c g t a c a c t g g c t g g c t t t t g g
c a a t t g t g a c g g c c a g c g c t t t g a c c c c c g c a t g t g a c c g a c c g a a a a c c

a t g g a a c c c c a a a t c g c a c a c t t g c t a t t t a t t c g a t c c a t t t g g a t t t t
t a c c t t g g g g t t t a g c g t g t g a a c g a t a a a t a a g c t a g g t a a a c c t a a a a

20800

c t g a t c a a c g a c t a a a a c a a a t c t a t c a g t t t g a g t a c g a a a g t c t g t t g
g a c t a g t t g c t g a t t t t g t t t a g a t a g t c a a a c t c a t g c t t t c a g a c a a c

c g c c g t a g t g c g c t a g c g g c c a c t a a a g a c c g a t g c g t t a c c c t a g a a a a
g c g g c a t c a c g c g a t c g c c g g t g a t t t c t g g c t a c g c a a t g g g a t c t t t t

20900

g t c a a c c c a a a c t g t a c a a g g a c c g t t t t c t g c a g c g t g c g g c c t g t t t t
c a g t t g g g t t t g a c a t g t t c c t g g c a a a a g a c g t c g c a c g c c g g a c a a a a

g t t g t a t g t t c t t a c a c g c t t t t a c t c a c t g g c c t g a c c a t c c a a t g g a t
c a a c a t a c a a g a a t g t g c g a a a a t g a g t g a c c g g a c t g g t a g g t t a c c t a

21000

a a a a a t c c c a c t a t g g a c c t a c t t a c t g g g g t g c c t a a t t g t a t g c t a c a
t t t t t a g g g t g a t a c c t g g a t g a a t g a c c c c a c g g a t t a a c a t a c g a t g t

a a g t c c t c a g g t a g t g g g c a c a t t g c a a c g c a a t c a g a a t g a a t t g t a t a
t t c a g g a g t c c a t c a c c c g t g t a a c g t t g c g t t a g t c t t a c t t a a c a t a t

21100

a a t t c t t a a a c a a t c t g t c c c c t t a c t t t c g t c a c a a c c g c g a g c g c a t a
t t a a g a a t t t g t t a g a c a g g g g a a t g a a a g c a g t g t t g g c g c t c g c g t a t

g a a a a a g c t a c a t c t t t t a c t a a a a t g c a a a a t g g a c t c a a a t a a a c g t g
c t t t t t c g a t g t a g a a a a t g a t t t t a c g t t t t a c c t g a g t t t a t t t g c a c

21200

t a c a c a a t g c a t t a a t a a t a a a a c c a t t t t a t t a g c t c a t t g g a g t a c a a
a t g t g t t a c g t a a t t a t t a t t t t g g t a a a a t a a t c g a g t a a c c t c a t g t t

g c t t g a c t g t t t t a t t a a a a a t c a a a t g g c t c t t c g c g a c a g t c g c c g t g
c g a a c t g a c a a a a t a a t t t t t a g t t t a c c g a g a a g c g c t g t c a g c g g c a c

21300

g t t g g t g g g c a g g g a t a t g t t t c t g t a c t g c a a a c g c t g a t g c c a c t t g a
c a a c c a c c c g t c c c t a t a c a a a g a c a t g a c g t t t g c g a c t a c g g t g a a c t

a t t c t g g a a t a a c a a g c c t a g g g g g g g a g c c g t c a a a a t t t t c t c c c c a c
t a a g a c c t t a t t g t t c g g a t c c c c c c c t c g g c a g t t t t a a a a g a g g g g t g

21400

a g c t g g c g c a c a a g t t g c a g g g c g c c c a t a a c a t c a g g a g c a g a a a t c t t
t c g a c c g c g t g t t c a a c g t c c c g c g g g t a t t g t a g t c c t c g t c t t t a g a a

g a a g t c g c a a t t a g g g c c a g c a t t g c c g c g c g c a t t g c g a t a a a c t g g a t
c t t c a g c g t t a a t c c c g g t c g t a a c g g c g c g c g t a a c g c t a t t t g a c c t a

21500

t t g c g c a c t g a a a a a c c a a c a a a c a c g g a t a c t t a a t a c t g g c t a a c g c t
a a c g c g t g a c t t t t t g g t t g t t t g t g c c t a t g a a t t a t g a c c g a t t g c g a

```
c c a g g g t c g g t t a c t t c g t t g a t a t c a a t g t t a t c c a c a t t g c t g a g g t t
g g t c c c a g c c a a t g a a g c a a c t a t a g t t a c a a t a g g t g t a a c g a c t c c a a

                                                                                               21600
a a a a g g a g t g a t t t t a c a c a g t t g a c g c c c c a t c c g t g g c a g g c c a t c t t
t t t t c c t c a c t a a a a t g t g t c a a c t g c g g g g t a g g c a c c g t c c g g t a g a a

g c t t g t t t a a a c a t t c g c a g c g c a c t g g c a t a a g g a g a c g t t t t t g c c c a
c g a a c a a a t t t g t a a g c g t c g c g t g a c c g t a t t c c t c t g c a a a a a c g g g t

                                                                                               21700
t g t c g c a t g t g a g g g t a g t c g g c c a g c a t a a a a g c t t c a a t t t g c c t a a a
a c a g c g t a c a c t c c c a t c a g c c g g t c g t a t t t t c g a a g t t a a a c g g a t t t

a g c t a t t t g a g c c t t c a t t c c t t c a g a a t a a a a c a a g c c g c a g g a c t t t c
t c g a t a a a c t c g g a a g t a a g g a a g t c t t a t t t t g t t c g g c g t c c t g a a a g

                                                                                               21800
c g g a g a a a g a a t t a t t c c c g c a g c c a a c a t c a t g a a a a c a g c a g c g g g c a
g c c t c t t t c t t a a t a a g g g c g t c g g t t g t a g t a c t t t t g t c g t c g c c c g t

t c g t c g t t t t t a a t t t g a a c t a c a t t a c g c c c c c a g c g g t t t t g c g c c a c
a g c a g c a a a a a t t a a a c t t g a t g t a a t g c g g g g g t c g c c a a a a c g c g g t g

                                                                                               21900
c t t g g c t t t c g a g g g g t t c t c t t t c a a c g c t c g t t g c c c a c t t t c g c t g g
g a a c c g a a a g c t c c c c a a g a g a a a g t t g c g a g c a a c g g g t g a a a g c g a c c

t t a c a t c c a t t t c c a c c a a a t g c t c t t t g c g c a c c a t c t c c a t t c c a t g c
a a t g t a g g t a a a g g t g g t t t a c g a g a a a c g c g t g g t a g a g g t a a g g t a c g

                                                                                               22000
a g g c a t c t a a g c t c c c c t t c g c g c t c g g t a c a c t t a t g c t c c c a c a c g c a
t c c g t a g a t t c g a g g g g a a g c g c g a g c c a t g t g a a t a c g a g g g t g t g c g t
```

g c a a c c g g t g g g t t c c c a g g a a t t c t g t t g g a c a c c g g c a t a a g c t t g c a
c g t t g g c c a c c c a a g g g t c c t t a a g a c a a c c t g t g g c c g t a t t c g a a c g t

22100
t a t a t c c t t g c a a a a a g c g t c c c a t g a g c t c c t g a a a g g t t t t t t g g g a t
a t a t a g g a a c g t t t t t c g c a g g g t a c t c g a g g a c t t t c c a a a a a a c c c t a

g a a a a a g t c a g c t g c a a a c c g c g c t t t t c t t c g t t g a g c c a t g t t g t g c a
c t t t t t c a g t c g a c g t t t g g c g c g a a a a g a a g c a a c t c g g t a c a a c a c g t

22200
t a t t t t c t t g t a c a c g c t g c c c t g a t c c g g c a a a a a a c g a a a g g t g g c g c
a t a a a a g a a c a t g t g c g a c g g g a c t a g g c c g t t t t t t g c t t t c c a c c g c g

g c t c g t c g t g a t c c a c a t g g t a c t t t t c c a t t a g c a t a g c c a t g g c t t c c
c g a g c a g c a c t a g g t g t a c c a t g a a a a g g t a a t c g t a t c g g t a c c g a a g g

22300
a t g c c t t t t t c c c a a g c t g a a a c t a g g g g c t g g c t t g c c g g a t t g c g a a c
t a c g g a a a a a g g g t t c g a c t t t g a t c c c c g a c c g a a c g g c c t a a c g c t t g

a a c a a c a a c a t t c t t t t c a t t t t c g t c g c t g t t t t g a g c g g a a g c c t t c a
t t g t t g t t g t a a g a a a a g t a a a a g c a g c g a c a a a a c t c g c c t t c g g a a g t

22400
a a a c g t g t a c c t g c c t g g t t t c c a t t t t t t g a a a a g a c t g a g a a c c g t c t
t t t g c a c a t g g a c g g a c c a a a g g t a a a a a a c t t t t c t g a c t c t t g g c a g a

g c a t g a t g c a t a a t g c g g a c g g g c g g c a t g c t g a a a c c c a t t a c t c c t a a
c g t a c t a c g t a t t a c g c c t g c c c g c c g t a c g a c t t t g g g t a a t g a g g a t t

22500
a a c t g c t c t t g g t g g t t c t g c c t c t t c t t c t t c t g c a c t c t c t g g g g a a a
t t g a c g a g a a c c a c c a a g a c g g a g a a g a a g a a g a c g t g a g a g a c c c c t t t

```
g a g g t a t c g c a g c c a t a g a t t t c t t g a c t t t t t t c t t t g g a g g t a a a g g c
c t c c a t a g c g t c g g t a t c t a a a g a a c t g a a a a a a g a a a c c t c c a t t t c c g

                                                                                            22600
a c a g c t t c c a g t t c t t c t t c g c t t t c g g a a t c c a g a a a g t a t c t g c c c a t
t g t c g a a g g t c a a g a a g a a g c g a a a g c c t t a g g t c t t t c a t a g a c g g g t a

t t t t g g c g g c g g c g g c t g a g c g c t g c g g t c t g g g g t g c g c t c c c t c t g t g
a a a a c c g c c g c c g c c g a c t c g c g a c g c c a g a c c c c a c g c g a g g g a g a c a c

                                                                                            22700
a g t g c t g a t t g c t g g c c a t t a t t t a a t c c t a g g c a a a g a a a c a c a t g a t g
t c a c g a c t a a c g a c c g g t a a t a a a t t a g g a t c c g t t t c t t t g t g t a c t a c

g a t c t g g a g c c a c a g g a a a g c t t a a c c g c c c c c a c c g c t c c c g c c a t t g g
c t a g a c c t c g g t g t c c t t t c g a a t t g g c g g g g g t g g c g a g g g c g g t a a c c

                                                                                            22800
c g c t a c g g c t g t c a t g g a g a a g g a c a a a a g t c t a c t c a t a c c c c a a g a c g
g c g a t g c c g a c a g t a c c t c t t c c t g t t t t c a g a t g a g t a t g g g g t t c t g c

c a c c g g t t g a g c a g a a c t t g g g c t a c g a g a c t c c c c c c g a g g a a t t t g a a
g t g g c c a a c t c g t c t t g a a c c c g a t g c t c t g a g g g g g g c t c c t t a a a c t t

                                                                                            22900
g g c t t t c t t c a a a t c c a a a a g c a a c c a a a t g a g c a a a a c g c t g g g c t c g a
c c g a a a g a a g t t t a g g t t t t c g t t g g t t t a c t c g t t t t g c g a c c c g a g c t

g g a c c a t g a c t a c c t a a a c g a g g g a g a t g t c c t g t t t a a a c a t c t a c a g c
c c t g g t a c t g a t g g a t t t g c t c c c t c t a c a g g a c a a a t t t g t a g a t g t c g

                                                                                            23000
g a c a a a g c a c t a t c g t t c g c g a c g c c a t a t c t g a t c g c t c t t c a a t a c c a
c t g t t t c g t g a t a g c a a g c g c t g c g g t a t a g a c t a g c g a g a a g t t a t g g t
```

```
g t t t c a a t t g c a g a a c t a t c t t g c a t c t a c g a a c g c a a c c t g t t c t c c c c
c a a a g t t a a c g t c t t g a t a g a a c g t a g a t g c t t g c g t t g g a c a a g a g g g g
                                                                                          23100
a c g t g t g c c c c t a a a c g g c a a g c c a a c g g c a c a t g c g a g c c a a a t c c t c
t g c a c a c g g g g g a t t t g c c g t t c g g t t g c c g t g t a c g c t c g g t t t a g g a g

g c c t t a a c t t c t a c c c a g t t t t t g c a g t g c c a g a a g c a c t g g c a a c a t a c
c g g a a t t g a a g a t g g g t c a a a a a c g t c a c g g t c t t c g t g a c c g t t g t a t g
                                                                                          23200
c a t a t t t t c t t t a a a a a t c a c a a a a t a c c c c t a t c c t g t c g a g c t a a c c g
g t a t a a a a g a a a t t t t t a g t g t t t t a t g g g g a t a g g a c a g c t c g a t t g g c

c a g c c g c g c a g a t g a g c t t c t t g c t t t a a g g g c t g g c g c t t c c a t a c c t g
g t c g g c g c g t c t a c t c g a a g a a c g a a a t t c c c g a c c g c g a a g g t a t g g a c
                                                                                          23300
g g a t t g t g t c c t t g g a a g a g g t g c c t a a a a t t t t t g a a g g t t t a g g t c g g
c c t a a c a c a g g a a c c t t c t c c a c g g a t t t t a a a a a c t t c c a a a t c c a g c c

g a t g a a a a a c g a g c a g c a a a t g c c c t g c a a a a a g a a a a t g a a c a a a a t c a
c t a c t t t t t g c t c g t c g t t t a c g g g a c g t t t t t c t t t t a c t t g t t t t a g t
                                                                                          23400
c c a t g g g a a t a g t g c t c t a a t a g a a c t g g a a g g t g a c a a t g c c c g c c t g g
g g t a c c c t t a t c a c g a g a t t a t c t t g a c c t t c c a c t g t t a c g g g c g g a c c

c a g t t t t a a a g c g c a a t a t t g a g g t t a c t c a c t t t g c c t a c c c g g c a g t a
g t c a a a a t t t c g c g t t a t a a c t c c a a t g a g t g a a a c g g a t g g g c c g t c a t
                                                                                          23500
a a t c t t c c g c c a a a g g t a a t g a g c g c a g t g a t g a a t c a g c t a c t a a t t a a
t t a g a a g g c g g t t t c c a t t a c t c g c g t c a c t a c t t a g t c g a t g a t t a a t t
```

g c g a g c c c a a c c c a t t g a c a a a g a t g c a a a c t t g c a a g a c c c g g a g g c a a
c g c t c g g g t t g g g t a a c t g t t t c t a c g t t t g a a c g t t c t g g g c c t c c g t t

23600
c a g a t g a t g g a a a g c c g g t t g t a a g c g a c g a g c a a t t a a c t a a g t g g t t g
g t c t a c t a c c t t t c g g c c a a c a t t c g c t g c t c g t t a a t t g a t t c a c c a a c

g g a a c a g a c a a t t c c a a c g a a c t a c a a c a g c g g c g t a a a c t c a t g a t g g c
c c t t g t c t g t t a a g g t t g c t t g a t g t t g t c g c c g c a t t t g a g t a c t a c c g

23700
c g c c g t a c t t g t a a c t g t g g a a c t c g a g t g c a t g c a t c g t t t t t t c t c c g
g c g g c a t g a a c a t t g a c a c c t t g a g c t c a c g t a c g t a g c a a a a a a g a g g c

a c a t c a c c a c a t t g c g c a a a a t t g a g g a a t g t c t t c a c t a c a c t t t c c g c
t g t a g t g g t g t a a c g c g t t t t a a c t c c t t a c a g a a g t g a t g t g a a a g g c g

23800
c a t g g c t a c g t g c g c c a a g c c t g t a a a a t t t c t a a t g t g g a g c t g a g c a a
g t a c c g a t g c a c g c g g t t c g g a c a t t t t a a a g a t t a c a c c t c g a c t c g t t

t c t a g t t t c t t a c a t g g g c a t c t t g c a t g a a a a c c g a t t g g g a c a g a a c g
a g a t c a a a g a a t g t a c c c g t a g a a c g t a c t t t t g g c t a a c c c t g t c t t g c

23900
t g c t a c a c t c a a c a c t a c g c g a t g a a g c a c g c a g a g a t t a c g t g c g a g a c
a c g a t g t g a g t t g t g a t g c g c t a c t t c g t g c g t c t c t a a t g c a c g c t c t g

t g c a t t t a c c t t t t c c t g t t a c a t a c c t g g c a a a c t g g g a t g g g t g t t t g
a c g t a a a t g g a a a a g g a c a a t g t a t g g a c c g t t t g a c c c t a c c c a c a a a c

24000
g c a g c a a t g c t t g g a a g a a a a a a a c c t t c g a g a a c t a a a c a a a c t g t t a g
c g t c g t t a c g a a c c t t c t t t t t t t g g a a g c t c t t g a t t t g t t t g a c a a t c

```
a c a g a g c a c t a a a a t c c c t a t g g a c c g g t t t t g a c g a a c g g a c a g t a g c t
t g t c t c g t g a t t t t a g g g a t a c c t g g c c a a a a c t g c t t g c c t g t c a t c g a

                                                                                      24100
g c a g a g c t a g c t g a c a t a a t t t t c c c a g a a a g g t t a a t g a t a a c c t t g c a
c g t c t c g a t c g a c t g t a t t a a a a g g g t c t t t c c a a t t a c t a t t g g a a c g t

a a a c g g c t t g c c t g a c t t t a t g a g t c a a a g t a t g c t g c a c a a t t a t c g c t
t t t g c c g a a c g g a c t g a a a t a c t c a g t t t c a t a c g a c g t g t t a a t a g c g a

                                                                                      24200
c t t t t a t a t t a g a g c g t t c t g g g a t g c t t c c t a g c a t g t g t t g t g c a c t t
g a a a a t a t a a t c t c g c a a g a c c c t a c g a a g g a t c g t a c a c a a c a c g t g a a

c c t t c a g a t t t t g t g c c t a t a t a t t t a g a g a g t g c c c c c t c c c c t g t g
g g a a g t c t a a a a c a c g g a t a t a t a a a t c t c t c a c g g g g g a g g g g a c a c

                                                                                      24300
g a g c c a c t g c t a c t t a c t a c g a c t t g c t a a c t a c c t a g c t t a c c a c t c a g
c t c g g t g a c g a t g a a t g a t g c t g a a c g a t t g a t g g a t c g a a t g g t g a g t c

a c c t t a t g a c a g a t t c a a g c g g c g a a g g c c t a a t g g a g t g t c a c t g c c g c
t g g a a t a c t g t c t a a g t t c g c c g c t t c c g g a t t a c c t c a c a g t g a c g g c g

                                                                                      24400
t g c a a t c t t t g c a c c c c c c a c c g t t c t t t g g t t t g c a a t a c t g a a c t a t t
a c g t t a g a a a c g t g g g g g g t g g c a a g a a a c c a a a c g t t a t g a c t t g a t a a

a a g t g a a a g t c a a g t c a t t g g t a c c t t c g a a a t g c a g g g a c c g c a g t c t g
t t c a c t t t c a g t t c a g t a a c c a t g g a a g c t t t a c g t c c c t g g c g t c a g a c

                                                                                      24500
a c a g c a a t t t c a c g a c g a a c c t a a g a c t t a c c c c t g g g c t t t g g a c t t c t
t g t c g t t a a a g t g c t g c t t g g a t t c t g a a t g g g g a c c c c g a a a c c t g a a g a
```

g c c t a c c t g c g c a a a t t t g a a c c c c a a g a t t a c c a c g c c c a c a g t a t c a a
c g g a t g g a c g c g t t t a a a c t t g g g g t t c t a a t g g t g c g g g t g t c a t a g t t

24600

t t t t t a c g a a g a c c a a t c c a a a c c c c c a a a a g c g c c a c t a a c g g c t t g c g
a a a a a t g c t t c t g g t t a g g t t t g g g g g t t t t c g c g g t g a t t g c c g a a c g c

t c a t t a c g c a g g g a a a a a t t c t a g c c c a a t t g c a t g c t a t t a a g c a a g c g
a g t a a t g c g t c c c t t t t t a a g a t c g g g t t a a c g t a c g a t a a t t c g t t c g c

24700

c g c g a a g a g t t t t t a c t t a a a a a a g g a c a c g g a g t g t a c c t t g a t c c c c a
g c g c t t c t c a a a a a t g a a t t t t t t c c t g t g c c t c a c a t g g a a c t a g g g g t

a a c c g g c g a g g a a c t a a a c c t t c c a t c a c c t t t g t g t g c t a c t g c g t c t c
t t g g c c g c t c c t t g a t t t g g a a g g t a g t g g a a a c a c a c g a t g a c g c a g a g

24800

c c c a t t c g c a g c a t g t c c c c g a a a g c c g c a a a a c a g g c t a t t g c g c a g c a
g g g t a a g c g t c g t a c a g g g g c t t t c g g c g t t t t g t c c g a t a a c g c g t c g t

a c g c t c a a a g a a a c a g c a g c a a c g g c a g g a a a t c t g g g a g g a a g a a t c t t
t g c g a g t t t c t t t g t c g t c g t t g c c g t c c t t t a g a c c c t c c t t c t t a g a a

24900

g g g a g a g t c a g g c a g a g g a c g a g g t c g a g g a c t t g g a a g a a t g g g a g g a g
c c c t c t c a g t c c g t c t c c t g c t c c a g c t c c t g a a c c t t c t t a c c c t c c t c

g a g g a g g c g g a c a g c c t a g a c g a g g a t c c a g a g g a g g a g g a g g a a g g t t c
c t c c t c c g c c t g t c g g a t c t g c t c c t a g g t c t c c t c c t c c t c c t t c c a a g

25000

c a a g g a c g g a g c g a c c g c c g c c a a a c c g t c g c t t t c a a c c a a g c c c t c t c
g t t c c t g c c t c g c t g g c g g c g g t t t g g c a g c g a a a g t t g g t t c g g g a g a g

```
c a a t g a a a c c c g c t g t g a g c a a a t c t c a g a a a g c c a a c c g t a g a t g g g a c
g t t a c t t t g g g c g a c a c t c g t t t a g a g t c t t t c g g t t g g c a t c t a c c c t g
                                                                                             25100
a c c a t t g a a a c c a g c g c c g c a a a c t t g g g t a a g a a t c g c a a g c a g g c g c g
t g g t a a c t t t g g t c g c g g c g t t t g a a c c c a t t c t t a g c g t t c g t c c g c g c

t c g g g g c t a c t g c t c a t g g c g g g c t c a c c a a a g t a a t a t t g t a g c c t g c t
a g c c c c g a t g a c g a g t a c c g c c c g a g t g g t t t c a t t a t a a c a t c g g a c g a
                                                                                             25200
t t c a g c a c t g c g g g g g g a a t a t c t c a t t t g c a a g g c g g t a t t t g c t a t a c
a a g t c g t g a c g c c c c c c t t a t a g a g t a a a c g t t c c g c c a t a a a c g a t a t g

c a t g a t g g a g t g g c g a t t c c a a g g a a t g t c c t c c a t t a c t a c c g t c a t c t
g t a c t a c c t c a c c g c t a a g g t t c c t t a c a g g a g g t a a t g a t g g c a g t a g a
                                                                                             25300
c t a c a g c c c c t t t g a a g a g c t c g a c a a g g a a c c g a c c t g c a a c a g c c a a g
g a t g t c g g g g a a a c t t c t c g a g c t g t t c c t t g g c t g g a c g t t g t c g g t t c

c g g c c c a c t a g a a t c g g c a a c a g c a g c a a c a a g g a a a g t c c t g a g g c g c g
g c c g g g t g a t c t t a g c c g t t g t c g t c g t t g t t c c t t t c a g g a c t c c g c g c
                                                                                             25400
c g a g t t a a g a a a a c g c a t t t t t c c c a c t t t a t a t g c t a t t t t t c a g c a g a
g c t c a a t t c t t t t g c g t a a a a a g g g t g a a a t a t a c g a t a a a a a g t c g t c t

g t c g a g g t c a a g a a c a c g a a c t g a a a a t a a a a a a c c g t t c c c t g c g t t c a
c a g c t c c a g t t c t t g t g c t t g a c t t t t a t t t t t t g g c a a g g g a c g c a a g t
                                                                                             25500
c t t a c c c g c a g c t g t c t c t a c c t c a a a a g c g a a g a t c a g t t g c a a c g c a c
g a a t g g g c g t c g a c a g a g a t g g a g t t t t c g c t t c t a g t c a a c g t t g c g t g
```

```
cttgcaggacgcagaagctctgttcaataaatactgctccctctcgctta
gaacgtcctgcgtcttcgagacaagttatttatgacgagggagagcgaat

                                                  25600
aagagtaaaaaaagcccgcgcgcggactttcaacaggcgggaaaagtgac
ttctcatttttttcgggcgcgcgcctgaaagttgtccgcccttttcactg

gtcacaacaagatgagtaaagatattcccacgccttacatgtggagcttt
cagtgttgttctactcatttctataagggtgcggaatgtacacctcgaaa

                                                  25700
caaccccaaatgggactggcggccggcgcggctcaagactattctagcaa
gttggggtttaccctgaccgccggccgcgccgagttctgataagatcgtt

aatgaattggttaagcgccggaccccacatgatttccagggtgaatgggg
ttacttaaccaattcgcggcctggggtgtactaaaggtcccacttacccc

                                                  25800
tacgagcccggcgtaaccaaatactgctagaacaagccgctctcaccgct
atgctcgggccgcattggtttatgacgatcttgttcggcgagagtggcga

acaccacgtaatcaacttaaccctccctcttggccagctgccctgatata
tgtggtgcattagttgaattgggagggagaaccggtcgacgggactatat

                                                  25900
tcaggaaaatccccctcctaccactgtacttttgcctcgcgacgcccagg
agtccttttagggggaggatggtgacatgaaaacggagcgctgcgggtcc

ccgaagtccatatgactaacgctggggcacagcttgcgggcggtgcacgt
ggcttcaggtatactgattgcgaccccgtgtcgaacgcccgccacgtgca

                                                  26000
cacagtttcaggtataaaggtcgcactgagccctatccgtctccagctat
gtgtcaaagtccatatttccagcgtgactcgggataggcagaggtcgata
```

```
a a a a a g a g t a c t c a t c a g a g g g a a a g g t a t t c a g c t g a a c g a c g a a g t c a
t t t t t c t c a t g a g t a g t c t c c c t t t c c a t a a g t c g a c t t g c t g c t t c a g t
                                                                                               26100
c a t c g c c a t t g g g a g t c a g a c c c g a c g g a g t g t t t c a g c t c g g a g g g t c c
g t a g c g g t a a c c c t c a g t c t g g g c t g c c t c a c a a a g t c g a g c c t c c c a g g

g g a c g t t c c t c c t t t a c c g c t c g t c a a g c c t a c c t g a c a c t a c a g a g c t c
c c t g c a a g g a g g a a a t g g c g a g c a g t t c g g a t g g a c t g t g a t g t c t c g a g
                                                                                               26200
a t c c t c a g c t c c g a g a t c t g g t g g t a t t g g a a c t c t c c a a t t t g t g g a g g
t a g g a g t c g a g g c t c t a g a c c a c c a t a a c c t t g a g a g g t t a a a c a c c t c c

a a t t t a c t c c a t c t g t t t a c t t c a a t c c t t t t t c g g g c t c g c c t g g a c a c
t t a a a t g a g g t a g a c a a a t g a a g t t a g g a a a a a g c c c g a g c g g a c c t g t g
                                                                                               26300
t a t c c t g a c g c c t t c a t a c c c a a c t t t g a c g c a g t g a g t g a a t c t g t g g a
a t a g g a c t g c g g a a g t a t g g g t t g a a a c t g c g t c a c t c a c t t a g a c a c c t

t g g c t a t g a t t a a t g t c t a a t g g a g c g g c t g a c a g a g c g c g g c t g c g a c a
a c c g a t a c t a a t t a c a g a t t a c c t c g c c g a c t g t c t c g c g c c g a c g c t g t
                                                                                               26400
t t t a g a c c a c t g t c g c c a a c c t c a c t g c t t t g c t c g a g a c a t c t g t g t c t
a a a t c t g g t g a c a g c g g t t g g a g t g a c g a a a c g a g c t c t g t a g a c a c a g a

t t a c c t a c t t t g a g c t t c c a g a g g a g c a c c c c c a g g g g c c a g c t c a c g g t
a a t g g a t g a a a c t c g a a g g t c t c c t c g t g g g g g t c c c c g g t c g a g t g c c a
                                                                                               26500
g t c a g a a t a a c a g t t g a a a a a g g a a t t g a t a c a c a c c t c a t t a a a t t t t t
c a g t c t t a t t g t c a a c t t t t t c c t t a a c t a t g t g t g g a g t a a t t t a a a a a
```

c a c c a a a c g c c c g c t a t t g g t g g a a a a a g a t c a a g g a a a t a c t a t a t t a a
g t g g t t t g c g g g c g a t a a c c a c c t t t t t c t a g t t c c t t t a t g a t a t a a t t

26600
c t t t a t a t t g c a t t t g t c c t g t t c c c g g a t t a c a t g a a g a t t t c t g c t g t
g a a a t a t a a c g t a a a c a g g a c a a g g g c c t a a t g t a c t t c t a a a g a c g a c a

c a t t t g t g t g c t g a a t t t a a t c a t c t g t a g t g g c g c t g t a c c g c c t g a a g
g t a a a c a c a c g a c t t a a a t t a g t a g a c a t c a c c g c g a c a t g g c g g a c t t c

26700
a a g a a c c t a a c t g t c a t c c g c a t t t a a g c a a c a t t a a a a t c a a c c t t t c g
t t c t t g g a t t g a c a g t a g g c g t a a a t t c g t t g t a a t t t t a g t t g g a a a g c

a t c c c t c a t a t c a c t c t t c g c t g c a g t t t t t t t t c c a c a c a t c t c a c c t g
t a g g g a g t a t a g t g a g a a g c g a c g t c a a a a a a a a g g t g t g t a g a g t g g a c

26800
g a c c t t t a a c g g a a a a c a c g t t a c c a a t a c a g a t a t a a a g t t t a a a c t a c
c t g g a a a t t g c c t t t t g t g c a a t g g t t a t g t c t a t a t t t c a a a t t t g a t g

a c a a a g a a a a c a t c a c t c t a t t t c a a c c t a t t a a c c t g g g a t a c t a c c g c
t g t t t c t t t t g t a g t g a g a t a a a g t t g g a t a a t t g g a c c c t a t g a t g g c g

26900
t g c t c a g c t c c a c c c t g t a c g c a a g c a t t t t t t g t t g c t c c a g t t a t t g a
a c g a g t c g a g g t g g g a c a t g c g t t c g t a a a a a a c a a c g a g g t c a a t a a c t

c a a a c g c c c t g c t c c g a c a a c a g c t g c t g t c a c t g a g c a c a t c a c c g a g g
g t t t g c g g g a c g a g g c t g t t g t c g a c g a c a g t g a c t c g t g t a g t g g c t c c

27000
c a g t t t c t c c t t c t a a a g g t a c a g a g g a a a t t g t g t a c t t t t c a a a c t t t
g t c a a a g a g g a a g a t t t c c a t g t c t c c t t t a a c a c a t g a a a a g t t t g a a a

```
a c a a a c c a c t t a g t t t t a a a t t g t t c c t g t t c t a a c t c c t t a a t t t c a t g
t g t t t g g t g a a t c a a a a t t t a a c a a g g a c a a g a t t g a g g a a t t a a a g t a c
                                                                                          27100
g t t t g c t a a c a g c t c t c t g t g c a a a a c t t t c t a c c a a g g a a a a c t t t t g t
c a a a c g a t t g t c g a g a g a c a c g t t t t g a a a g a t g g t t c c t t t t g a a a a c a

a t t c t g c t a a a c t c a c a t t g t g t a a c c a g a g c a c c c c t t c c c a c c t t a c t
t a a g a c g a t t t g a g t g t a a c a c a t t g g t c t c g t g g g g a a g g g t g g a a t g a
                                                                                          27200
c t a t t g c c a c c t t t t g t t g c c g g t c g t t a c t t t t g c a t a g g a g c t g c a c g
g a t a a c g g t g g a a a a c a a c g g c c a g c a a t g a a a a c g t a t c c t c g a c g t g c

t a c t a g c c c c t g t c a a c a g c a t t g g a a t t t a a c t t a c t g t c c c c c a c c a g
a t g a t c g g g g a c a g t t g t c g t a a c c t t a a a t t g a a t g a c a g g g g g t g g t c
                                                                                          27300
t g t c g c c c t t t g t g a t c a a t a c t g a a t a t t t a g a c[t a t a a t]c c c t t g c t t
a c a g c g g g a a a c a c t a g t t a t g a c t t a t a a a t c t g a t a t t a g g g a a c g a a

g c t t a c g g c g g t c t c g c a g c t c t t a t t t t a t t c c t g a t t t c t a a c t t g t t
c g a a t g c c g c c a g a g c g t c g a g a a t a a a a t a a g g a c t a a a g a t t g a a c a a
                                                                                          27400
t c t a g t g c a a c a t t t g t a t t c a t a c t a a c a a t g c t t t c c a t t t t t c t t t t
a g a t c a c g t t g t a a a c a t a a g t a t g a t t g t t a c g a a a g g t a a a a a g a a a a

a t t t c t c t t t t c t t t a c c t t c t g g c t t g t a t g c t c a a a c a g c c g a a a g a c
t a a a g a g a a a a g a a a t g g a a g a c c g a a c a t a c g a g t t t g t c g g c t t t c t g
                                                                                          27500
c a c t a a a a g t c g t g g t g g a a g c t g g c c a t a a t g t a a c c c t t c c c c a c c t t
g t g a t t t t c a g c a c c a c c t t c g a c c g g t a t t a c a t t g g g a a g g g g t g g a a
```

t c t g g t t c a c a c c a a a c t g g c c a t g t t a c t t g g c t a g t a g a g a c a t c a g a
a g a c c a a g t g t g g t t t g a c c g g t a c a a t g a a c c g a t c a t c t c t g t a g t c t

27600
t t a t g g t t c a g c t t c t c c a g a c a a c t t c a t t t t c a g t g g a c a a a a a c t a t
a a t a c c a a g t c g a a g a g g t c t g t t g a a g t a a a a g t c a c c t g t t t t t g a t a

g c c a g t t t a c t g a c a g a a c c a t g g t g t g g c c t t a t t a c a a t t t a c a t t t t
c g g t c a a a t g a c t g t c t t g g t a c c a c a c c g g a a t a a t g t t a a a t g t a a a a

27700
a a c t g t g a a a a t t a t g a c c t t a a t c t g t t t t g g c t t a a g g t g g a a a a t t c
t t g a c a c t t t t a a t a c t g g a a t t a g a c a a a a c c g a a t t c c a c c t t t t a a g

g g c t a t t t a c a a c g t t a a a a a t a c a g t c a a t g c t t c t g a a a c a a a t a t t t
c c g a t a a a t g t t g c a a t t t t t a t g t c a g t t a c g a a g a c t t t g t t t a t a a a

27800
a c t a t g a t t t a a g a g t a g t a c a a a t t t t t c c a c c t a a a t g c a t c a t t a c t
t g a t a c t a a a t t c t c a t c a t g t t t a a a a a g g t g g a t t t a c g t a g t a a t g a

t c a a a g t a c c t t a c a a a c g a t t a t t g t c a c a t t a c a a t t a a c t g c a c t a a
a g t t t c a t g g a a t g t t t g c t a a t a a c a g t g t a a t g t t a a t t g a c g t g a t t

27900
c t c t g a t t a c c c c a a t a a g g t t g t g t t t a a t a a t g t c a g t c g a t g g t a c t
g a g a c t a a t g g g g t t a t t c c a a c a c a a a t t a t t a c a g t c a g c t a c c a t g a

a c g g a t a c g g t a a g g g c a g c c c a a c c c t t c c c a a c t a c t t t a t a a c t a a c
t g c c t a t g c c a t t c c c g t c g g g t t g g g a a g g g t t g a t g a a a t a t t g a t t g

28000
t t t a a t g t t t c a g g t a t t a c t a a a a g c t t t a a t c a c a c t t a c c c t t t t a a
a a a t t a c a a a g t c c a t a a t g a t t t t c g a a a t t a g t g t g a a t g g g a a a a t t

```
t g a g c t c t g t g a t t a t c c c a c a t c c c a a t c t c a a c a c a g t t t a a c a c a t a
a c t c g a g a c a c t a a t a g g g t g t a g g g t t a g a g t t g t g t c a a a t t g t g t a t
                                                                                            28100
c a g t a a g c a c a g t a a t c t t t t t a g g a a t a a t t g g c t t c a g c a t t t g a t t t
g t c a t t c g t g t c a t t a g a a a a a t c c t t a t t a a c c g a a g t c g t a a a a c t a a

a t t a t a g c a g c c t t t a t t t a t c t g t g c t g g c a t a g a a a a t c t t t g t g t g t
t a a t a t c g t c g g a a a t a a a t a g a c a c g a c c g t a t c t t t t a g a a a c a c a c a
                                                                                            28200
t t c t a a a a c a g a a c c t c t t a t g c c g a t t c c t t a c t a g t t t t c t t t t t t c t
a a g a t t t t g t c t t g g a g a a t a c g g c t a a g g a a t g a t c a a a a g a a a a a a g a

t a c a g t a t g g t g a c g g t t c t t c t c a t c t t t t t a t g c c t g c c a g t c a t t t t
a t g t c a t a c c a c t g c c a a g a a g a g t a g a a a a a t a c g g a c g g t c a g t a a a a
                                                                                            28300
t t c t t c t t c g a c t t t t g c c g c a g t c a g t g a c c t t g a t c c c g a g t g t t t a g
a a g a a g a a g c t g a a a a c g g c g t c a g t c a c t g g a a c t a g g g c t c a c a a a t c

c c c c c t t t g c g g t g t a c c t g a t t t t c a c a t t t g t g a c t g c t a c c t g c g t c
g g g g g a a a c g c c a c a t g g a c t a a a a g t g t a a a c a c t g a c g a t g g a c g c a g
                                                                                            28400
t g c a g t a t t a t t a c t c t g c t a a t c a c c t c g c t c c a a t t t t t t g a t t a c t a
a c g t c a t a a t a a t g a g a c g a t t a g t g g a g c g a g g t t a a a a a a c t a a t g a t

c t a c g t g a g a a t t g t t t a c c g c a g a c a c c a c c c c c g t t a c c a a a a c c c t c
g a t g c a c t c t t a a c a a a t g g c g t c t g t g g t g g g g g c a a t g g t t t t g g g a g
                                                                                            28500
a a a t t g c g g c t c t t t t g c a g c t c c a a c c a t g a a a a c a g c a t t a g t t c t t t
t t t a a c g c c g a g a a a a c g t c g a g g t t g g t a c t t t t g t c g t a a t c a a g a a a
```

t c t t t a t g t t a a t c c c a g t t t g g g c t a g t t c t t g t c a a c t a c a t a a a c c a
a g a a a t a c a a t t a g g g t c a a a c c c g a t c a a g a a c a g t t g a t g t a t t t g g t

28600

t g g a a t t t t t t a g a t t g t t a t a c t a a a g a a a c a a a c t a c a t a g g c t g g g t
a c c t t a a a a a a t c t a a c a a t a t g a t t t c t t t g t t t g a t g t a t c c g a c c c a

t t a t g g a a t t a t g t c t g g c t t a g t a t t t g t c t c c t c t g t a g t t t c t t t a c
a a t a c c t t a a t a c a g a c c g a a t c a t a a a c a g a g g a g a c a t c a a a g a a a t g

28700

a a c t g t a t g c g c g c c t t a a t t t t a g t t g g a a t a a g t a t a c t g a t g a t c t t
t t g a c a t a c g c g c g g a a t t a a a a t c a a c c t t a t t c a t a t g a c t a c t a g a a

c c c g a a t a t c c a a a c c c c c a g g a t g a t t t a c c c c t a a a t a t t g t a t t t c c
g g g c t t a t a g g t t t g g g g g t c c t a c t a a a t g g g g a t t t a t a a c a t a a a g g

28800

a g a g c c c c c g c g t c c t c c t t c t g t t g t t a g c t a t t t t a a g t t c a c c g g t g
t c t c g g g g g c g c a g g a g g a a g a c a a c a a t c g a t a a a a t t c a a g t g g c c a c

a a g a t g a t t g a a c c t g a t c t a g a a a t t g a t g g a a g a a t c a c c g a a c a g a g
t t c t a c t a a c t t g g a c t a g a t c t t t a a c t a c c t t c t t a g t g g c t t g t c t c

28900

g c t c c t c a c t g a t c g c g c t a g g c g a c g c c a a c a g g a t c a a a a a a a t a a a g
c g a g g a g t g a c t a g c g c g a t c c g c t g c g g t t g t c c t a g t t t t t t t a t t t c

a g t t a a t t g a t t t a c a a a c c g t g c a t c a g t g t a a a a a a g g a c t t t t t t g c
t c a a t t a a c t a a a t g t t t g g c a c g t a g t c a c a t t t t t t c c t g a a a a a a c g

29000

c t g g t a a a a c a a g c t a c c c t t c g c t a t g a a t c t t t a c c a g g c a a a g a a c a
g a c c a t t t t g t t c g a t g g g a a g c g a t a c t t a g a a a t g g t c c g t t t c t t g t

t c a a c t g t g c t a c a c g c t g c c c a c t c a g c g a c a a a c c t t t t a c t g c a a t g g

a g t t g a c a c g a t g t g c g a c g g g t g a g t c g c t g t t t g g a a a a t g a c g t t a c c

29100

t g g g c t c g g t a c c t a t t a a a g t g t c c c a a c a a g c a g g a g a a c a a g a a g g c

a c c c g a g c c a t g g a t a a t t t c a c a g g g t t g t t c g t c c t c t t g t t c t t c c g

t c t a t t c g g t g c c t a t g t g a t a a c c c t g a a t g t t t g t a c a c t t t a a t a a a

a g a t a a g c c a c g g a t a c a c t a t t g g g a c t t a c a a a c a t g t g a a a t t a t t t

29200

a a c a c t g t g c g g t t t a a g a a a t c t t t t a c c a a t g a a t t a a a t a a a t t a c t

t t g t g a c a c g c c a a a t t c t t t a g a a a a t g g t t a c t t a a t t t a t t t a a t g a

t a c c g g a a a t c t g a a a a t a c a t c a t g g t c t c c g t g t a c t c t t a t a a a a t t

a t g g c c t t t a g a c t t t t a t g t a g t a c c a g a g g c a c a t g a g a a t a t t t t a a

29300

t c c c t c t t c c c a a c t g t c a a a c c t g a c a g a c t t g c a a a c a g c a a a c t t t c

a g g g a g a a g g g t t g a c a g t t t g g a c t g t c t g a a c g t t t g t c g t t t g a a a g

t c c a a a t c t t a a a t g g a a g g t c a g a t t c t t c t t c c c a a t c c c t a c c c a c c

a g g t t t a g a a t t t a c c t t c c a g t c t a a g a a g a a g g g t t a g g g a t g g g t g g

29400

a t c t t c a t c t t t t c t a g a t g a a g c g c a g c a g a a c c c a g t a t g c t g a a g a a

t a g a a g t a g a a a a g a t c t a c t t c g c g t c g t c t t g g g t c a t a c g a c t t c t t

a c a g a a g a a a a t g a t g a c t t c a a c c c c g t t t a c c c t t t t g a c c c a t t t g a

t g t c t t c t t t t a c t a c t g a a g t t g g g g c a a a t g g g a a a a c t g g g t a a a c t

29500

c a c a t c a g a c g t a c c c t t t g t t a c a c c c c c t t t t a c t t c t t c c a a t g g t c

g t g t a g t c t g c a t g g g a a a c a a t g t g g g g g a a a a t g a a g a a g g t t a c c a g

```
t t c a a g a a a a a c c a c c a g g t g t a t t a g c a c t t a a t t a c a a a g a c c c c a t t
a a g t t c t t t t t g g t g g t c c a c a t a a t c g t g a a t t a a t g t t t c t g g g g t a a

                                                                                          29600
g t a a c t g a a a a t g g a a c c c t t a c a c t c a a g c t a g g g g a c g g a a t a a a a c t
c a t t g a c t t t t a c c t t g g g a a t g t g a g t t c g a t c c c c t g c c t t a t t t t g a

t a a t g c c c a a g g t c a a c t t a c a g c t a g t a a t a a t a t c a a t g t t t t g g a g c
a t t a c g g g t t c c a g t t g a a t g t c g a t c a t t a t t a t a g t t a c a a a a c c t c g

                                                                                          29700
c c c t t a c c a a c a c c t c a c a a g g t c t t a a a c t t t c t t g g a g c g c c c c c c t a
g g g a a t g g t t g t g g a g t g t t c c a g a a t t t g a a a g a a c c t c g c g g g g g g a t

g c a g t a a a g g c t a g t g c c c t c a c a c t t a a c a c a a g a g c g c c c t t a a c c a c
c g t c a t t t c c g a t c a c g g g a g t g t g a a t t g t g t t c t c g c g g g a a t t g g t g

                                                                                          29800
a a c g g a t g a a a g c t t a g c c t t a a t a a c c g c c c c t c c c a t t a c a g t a g a g t
t t g c c t a c t t t c g a a t c g g a a t t a t t g g c g g g g a g g g t a a t g t c a t c t c a

c t t c g c g t t t g g g c t t g g c c a c c a t a g c c c c t c t a a g c t t a g a t g g a g g t
g a a g c g c a a a c c c g a a c c g g t g g t a t c g g g g a g a t t c g a a t c t a c c t c c a

                                                                                          29900
g g a a a c c t a g g t t t a a a t c t t t c t g c t c c c c t g g a c g t t a g t a a c a a c a a
c c t t t g g a t c c a a a t t t a g a a a g a c g a g g g g a c c t g c a a t c a t t g t t g t t

t t t g c a t c t c a c c a c t g a a a c t c c c t t a g t t g t a a a t t c t a g c g g t g c c c
a a a c g t a g a g t g g t g a c t t t g a g g g a a t c a a c a t t t a a g a t c g c c a c g g g

                                                                                          30000
t a t c t g t t g c t a c t g c a g a c c c c a t a a g t g t t c g c a a c a a c g c t c t t a c c
a t a g a c a a c g a t g a c g t c t g g g g t a t t c a c a a g c g t t g t t g c g a g a a t g g
```

```
ctacctacggcagatccgttaatggtgagctccgatgggttgggaataag
gatggatgccgtctaggcaattaccactcgaggctacccaacccttattc
                                             30100
tgtcactagtcccattacagtaataaacggttccttagccttgtctacaa
acagtgatcagggtaatgtcattatttgccaaggaatcggaacagatgtt

ctgctcccctcaacagcacaggatccactttaagtctgtctgttgccaat
gacgaggggagttgtcgtgtcctaggtgaaattcagacagacaacggtta
                                             30200
cctctgactatttcacaagacacattgactgtttccactggtaacggtct
ggagactgataaagtgttctgtgtaactgacaaaggtgaccattgccaga

tcaagtgtcggggtctcaattagtaacaagaataggggatggtttaacat
agttcacagccccagagttaatcattgttcttatcccctaccaaattgta
                                             30300
tcgataatggggtcatgaaagtaaacgttgccgggggaatgagaacttct
agctattaccccagtactttcatttgcaacggcccccttactcttgaaga

ggcggtagaataattttagatgttaattatccctttgatgcgagcaataa
ccgccatcttattaaaatctacaattaatagggaaactacgctcgttatt
                                             30400
cctgtccttaagacggggattgggactaatttataaccaatctacaaact
ggacaggaattctgcccctaaccctgattaaatattggttagatgtttga

ggaacttaacaactgatattagtaccgaaaaaggtttaatgtttagtggc
ccttgaattgttgactataatcatggctttttccaaattacaaatcaccg
                                             30500
aatcaaatagctcttaatgcaggtcaggggcttacatttaataatggcca
ttagtttatcgagaattacgtccagtccccgaatgtaaattattaccggt
```

```
a c t t a g g g t t a a g t t g g g a g c t g g a c t t a t t t t t g a t t c a a a c a a t a a c a
t g a a t c c c a a t t c a a c c c t c g a c c t g a a t a a a a a c t a a g t t t g t t a t t g t

                                                                                         30600
t t g c c t t a g g c a g c a g c a g c a a c a c t c c a t a c g a c c c t c t g a c a c t g t g g
a a c g g a a t c c g t c g t c g t c g t t g t g a g g t a t g c t g g g a g a c t g t g a c a c c

a c a a c t c c t g a c c c a c c a c c a a a c t g c a g c c t c a t a c a a g a g c t a g a t g c
t g t t g a g g a c t g g g t g g t g g t t t g a c g t c g g a g t a t g t t c t c g a t c t a c g

                                                                                         30700
a a a a c t c a c c c t g t g c t t a a c a a a a a a c g g a t c t a t t g t t a a t g g c a t t g
t t t t g a g t g g g a c a c g a a t t g t t t t t t g c c t a g a t a a c a a t t a c c g t a a c

t a a g t t t a g t g g g t g t t a a g g g t a a t c t c c t a a a t a t c c a a a g t a c t a c t
a t t c a a a t c a c c c a c a a t t c c c a t t a g a g g a t t t a t a g g t t t c a t g a t g a

                                                                                         30800
a c c a c t g t a g g a g t g c a t t t a g t g t t t g a t g a a c a g g g a a g a t t a a t c a c
t g g t g a c a t c c t c a c g t a a a t c a c a a a c t a c t t g t c c c t t c t a a t t a g t g

a t c a a c c c c t a c t g c c c t g g t t c c c c a a g c t t c g t g g g g a t a t a g a c a a g
t a g t t g g g g a t g a c g g g a c c a a g g g g t t c g a a g c a c c c c t a t a t c t g t t c

                                                                                         30900
g c c a a t c a g t g t c t a c c a a t a c t g t t a c c a a t g g t c t a g g t t t t a t g c c t
c g g t t a g t c a c a g a t g g t t a t g a c a a t g g t t a c c a g a t c c a a a a t a c g g a

a a t g t g a g t g c t t a c c c t a g a c c a a a t g c c a g t g a g g c t a a a a g c c a a a t
t t a c a c t c a c g a a t g g g a t c t g g t t t a c g g t c a c t c c g a t t t t c g g t t t a

                                                                                         31000
g g t a a g t c t c a c g t a c t t a c a g g g a g a t a c a t c t a a a c c t a t a a c a a t g a
c c a t t c a g a g t g c a t g a a t g t c c c t c t a t g t a g a t t t g g a t a t t g t t a c t
```

a a g t t g c a t t t a a t g g c a t t a c g t c g c t a a a t g g a t a c t c t t t a a c a t t c
t t c a a c g t a a a t t a c c g t a a t g c a g c g a t t t a c c t a t g a g a a a t t g t a a g

31100

a t g t g g t c a g g t c t a t c a a a c t a t a t a a a t c a g c c t t t c t c t a c a c c a t c
t a c a c c a g t c c a g a t a g t t t g a t a t a t t t a g t c g g a a a g a g a t g t g g t a g

c t g c t c c t t t t c t t a c a t t a c c c a a g a a t a a a a a c a c a c a c a a a a c a c a a
g a c g a g g a a a a g a a t g t a a t g g g t t c t t a t t t t t g t g t g t g t t t t g t g t t

31200

a t t g c g t a c t t a t t g t t t a t t t t t t t t t t t t t t t a c a c t a t a c g c g t g g t
t a a c g c a t g a a t a a c a a a t a a a a a a a a a a a a a a a t g t g a t a t g c g c a c c a

t a a a c t g c c t c c t t c c c a t t t t a c c t t g t a t a c c t c c c t t t c c c c c t t t g
a t t t g a c g g a g g a a g g g t a a a a t g g a a c a t a t g g a g g g a a a g g g g g a a a c

31300

t a g c t g a a a a c a a c t g c a c t t g a a t a t t t c g a c t t a g g t t t t t t g g c g t t
a t c g a c t t t t g t t g a c g t g a a c t t a t a a a g c t g a a t c c a a a a a a c c g c a a

a g c g t c c a c a c a g t t t c t t t a c g g g c a a a g c g a g g g t c g g t g a t g g a a a c
t c g c a g g t g t g t c a a a g a a a t g c c c g t t t c g c t c c c a g c c a c t a c c t t t g

31400

g a a t c c c t c g c c c g c a c a g t c a c t c a a g c g g c a t t c c c c a t c c a a a a c c a
c t t a g g g a g c g g g c g t g t c a g t g a g t t c g c c g t a a g g g g t a g g t t t t g g t

g g t c c a t g a t t t t a t c c t a c a a a a a g t a a c a a c a g t c a g t g t c c a t c a g c
c c a g g t a c t a a a a t a g g a t g t t t t t c a t t g t t g t c a g t c a c a g g t a g t c g

31500

c g c c c a a g g a t t c t c t c g t t g a t t a t a a t c t c c a a a t a a a a t t g c t c g a t
g c g g g t t c c t a a g a g a g c a a c t a a t a t t a g a g g t t t a t t t t a a c g a g c t a

```
g a t g c a t a a t t a a a c c c t t t a g c a g t t g c t g a c g a t a a c g t t c a t g c c g a
c t a c g t a t t a a t t t g g g a a a t c g t c a a c g a c t g c t a t t g c a a g t a c g g c t

                                                                                          31600
c t a t g t t t t a g a g g g c g a a c a g t g t t t t c a g c a a t t a c t t g a a c a a c t t t
g a t a c a a a a t c t c c c g c t t g t c a c a a a a g t c g t t a a t g a a c t t g t t g a a a

t a a c a t t a g c a g t c t g g t a c g a c g a g c g c a a c a g c g c a t g c g t a t c t c a c
a t t g t a a t c g t c a g a c c a t g c t g c t c g c g t t g t c g c g t a c g c a t a g a g t g

                                                                                          31700
t t a a g t c t t t a c a a t a a t c a c a a c a c a g c a c t a a c a t g t t a t t t a a a a t t
a a t t c a g a a a t g t t a t t a g t g t t g t g t c g t g a t t g t a c a a t a a a t t t t a a

c c a t a a t t a a a g g c g c t c c a t c c a a a a c t a a c t t t t t c t a a c g c t a a c c a
g g t a t t a a t t t c c g c g a g g t a g g t t t t g a t t g a a a a a g a t t g c g a t t g g t

                                                                                          31800
g g c a t g g c c a t c a t a c a t a a t t t t a a a g t a a a t t a a a t g g c g a c c t c t a a
c c g t a c c g g t a g t a t g t a t t a a a a t t t c a t t t a a t t t a c c g c t g g a g a t t

c a a a g g t g c t t c c c a c a t a c a t c a c c t c t t t a g g c a t t a a a t g g t t a a c a
g t t t c c a c g a a g g g t g t a t g t a g t g g a g a a a t c c g t a a t t t a c c a a t t g t

                                                                                          31900
a c c t c c c g a t a c c a a a a a c a c c t t t t t g t t a a t t a a g g c g c c a t a t a c g g c
t g g a g g g c t a t g g t t t t t g t g g a a a a a c a a t t a a t t c c g c g g t a t a t g c c g

c a t t t t g a a c c a g c g t c c c a a a a g c a t c c c a g c t g a c a t a c a c t g t a g t g
g t a a a a c t t g g t c g c a g g g t t t t c g t a g g g t c g a c t g t a t g t g a c a t c a c

                                                                                          32000
a a c c c g g a c g c t g g c a a t g a c a a t g a a t a a g c c a c c g c t c a t g a c c a t g t
t t g g g c c t g c g a c c g t t a c t g t t a c t t a t t c g g t g g c g a g t a c t g g t a c a
```

```
a a t a a t t g a g t a a c t t c a a c a t t t a t a g t g g c a c a a c a c a t a c a t a c a c t
t t a t t a a c t c a t t g a a g t t g t a a a t a t c a c c g t g t t g t g t a t g t a t g t g a

                                                                                                  32100
c a t g t a t t t t t t c a a a a t a a a c a t c t c a t a a t c a g t t a g a a t c a t a t c c c
g t a c a t a a a a a a g t t t t a t t t g t a g a g t a t t a g t c a a t c t t a g t a t a g g g

a c g g t a t t g g c c a t t c c t g c a g c a c t g t a a a a c c t a c a c a t g a a g g a a t g
t g c c a t a a c c g g t a a g g a c g t c g t g a c a t t t t g g a t g t g t a c t t c c t t a c

                                                                                                  32200
c c t c t t a c c t c a c t t a c a t t a t g t a a a g t c a g a c t a t t a c a c t c a g g c c a
g g a g a a t g g a g t g a a t g t a a t a c a t t t c a g t c t g a t a a t g t g a g t c c g g t

t a a a g a a t t t t c c g a a g t a c t c a a c g t a g c t t t t g a c t g t t c c t c a c a g g
a t t t c t t a a a a g g c t t c a t g a g t t g c a t c g a a a a c t g a c a a g g a g t g t c c

                                                                                                  32300
g c g g t a g t t g g t a c t t g t t g t a t g g t g c c a a t c t g t a g c g a t a c c g t c t g
c g c c a t c a a c c a t g a a c a a c a t a c c a c g g t t a g a c a t c g c t a t g g c a g a c

t c g c g c t g c a t c g t a a a c a a c a g a c t t g c g a g c g t c t t c g t a c t t a a a a a
a g c g c g a c g t a g c a t t t g t t g t c t g a a c g c t c g c a g a a g c a t g a a t t t t t

                                                                                                  32400
a a c a a a a c c a c g t a c g a c c a c t g g t t a t c g c a c c t c g t c c t t t t t g t t t g
t t g t t t t g g t g c a t g c t g g t g a c c a a t a g c g t g g a g c a g g a a a a a c a a a c

c a g c g t t g g c g t t c c g t c a a a a a a g c a a a g t a c a a c c a c t c t c g c a g g c t
g t c g c a a c c g c a a g g c a g t t t t t t c g t t t c a t g t t g g t g a g a g c g t c c g a

                                                                                                  32500
t g c t a a a a t g t a t t c a g c t t c a g g t g t t a t c t t c a a a t c a t g a t g t t t a a
a c g a t t t t a c a t a a g t c g a a g t c c a c a a t a g a a g t t t a g t a c t a c a a a t t
```

```
t a a a g c g c a g a g t a t c c a c a c a g g a t g c a t g g g c t a a a c c a a g c c a t g c t
a t t t c g c g t c t c a t a g g t g t g t c c t a c g t a c c c g a t t t g g t t c g g t a c g a

                                                                                           32600
a t g c a g g c a g c c g t g t c c c g a c t t a c a g g a g g a g g a g g a a t a c a a g g t a g
t a c g t c c g t c g g c a c a g g g c t g a a t g t c c t c c t c c t c c t t a t g t t c c a t c

a g g c a t a a a a a c t t a a t c a a g a c g g t c a g c a a g g a t t t g a a t g c g t a a a t
t c c g t a t t t t t g a a t t a g t t c t g c c a g t c g t t c c t a a a c t t a c g c a t t t a

                                                                                           32700
c t c g c a g g t g g c a g c g a t c g c c t c c g c t g t g c t g g t g a a a g a t c a c a g c c
g a g c g t c c a c c g t c g c t a g c g g a g g c g a c a c g a c c a c t t t c t a g t g t c g g

a g a t c a a a t t g t a a g c g a t t t t c c a a a t g t t c a a c a a c a g c t t c t a a a a g
t c t a g t t t a a c a t t c g c t a a a a g g t t t a c a a g t t g t t g t c g a a g a t t t t c

                                                                                           32800
a g c c a c a g c t c t g a t t t c g a t a a a c a a a a g c a a a g c a a a t g c a t t a t c a t
t c g g t g t c g a g a c t a a a g c t a t t t g t t t t c g t t t c g t t t a c g t a a t a g t a

g a a a c t c t t c t a t c a t c a a a c t g c c t g a c t g a a c c a t t c c c a g g t a a t t t
c t t t g a g a a g a t a g t a g t t t g a c g g a c t g a c t t g g t a a g g g t c c a t t a a a

                                                                                           32900
t c a t t c t t c c a c t g t t g t a t t a t t t g a a c a c a c t g a t t t t g c a g g t t t a a
a g t a a g a a g g t g a c a a c a t a a t a a a c t t g t g t g a c t a a a a c g t c c a a a t t

a c c g t g a a t a t t a a a a a g c t c t g t a a g g g c g c c c t c c a c c g c c a t c c g c a
t g g c a c t t a t a a t t t t t c g a g a c a t t c c c g c g g g a g g t g g c g g t a g g c g t

                                                                                           33000
g g c a g t a c t t c a t a t t t g c t g a a a a a a g t c t g g a t c t t c a a a c a c c t g c a
c c g t c a t g a a g t a t a a a c g a c t t t t t t c a g a c c t a g a a g t t t g t g g a c g t
```

```
g t a a a t t c a g t a g a t t t a c a t t a g g c t c c a c a c c t t g g t c t c g c a g c t g a
c a t t t a a g t c a t c t a a a t g t a a t c c g a g g t g t g g a a c c a g a g c g t c g a c t

                                                                                              33100
c a t c t t a a t g c c a g t t g t a t a a a a t c a t a c a a a t c a g a a g c c a g c a g c a a
g t a g a a t t a c g g t c a a c a t a t t t t a g t a t g t t t a g t c t t c g g t c g t c g t t

a g a a a g t t c a c c t c c a g g t a c a a g t t c c g g a g t t c c c a c a g a a c a t a c a a
t c t t t c a a g t g g a g g t c c a t g t t c a a g g c c t c a a g g g t g t c t t g t a t g t t
                                                                                              33200
c t t g c a c a a a t g g a c c c a t a t t a g t a a g c g t g g c g c c a a c g t a g a c a t c g
g a a c g t g t t t a c c t g g g t a t a a t c a t t c g c a c c g c g g t t g c a t c t g t a g c

c g c a t a g g a g g a g t t a a a t a a t g c a t t a c c a g c a g c c a a a a c t c a g g t a g
g c g t a t c c t c c t c a a t t t a t t a c g t a a t g g t c g t c g g t t t t g a g t c c a t c
                                                                                              33300
c a c g t c t t t a a g a a a c g t c a c c a c c t c a a a a t c t a a g c c a t g c a a a t a g t
g t g c a g a a a t t c t t t g c a g t g g t g g a g t t t t a g a t t c g g t a c g t t t a t c a

t c c g t a a a g a c t c c g g a a a c a a c a c g g a g t a a t g a a c a a g c g a c c t c t g a
a g g c a t t t c t g a g g c c t t t g t t g t g c c t c a t t a c t t g t t c g c t g g a g a c t
                                                                                              33400
a a c a t g c t t t a g g t t a g c c t g a a a a a t a a a a a t a t g t t a a a t t a a a g a t g
t t g t a c g a a a t c c a a t c g g a c t t t t t a t t t t t a t a c a a t t t a a t t t c t a c

c c t g g c a a a c g g g t g g a a a a a c a a c t c t a c t t a a a a g c a a g c g c g c g a c t
g g a c c g t t t g c c c a c c t t t t t g t t g a g a t g a a t t t t c g t t c g c g c g c t g a
                                                                                              33500
g g c t g c t t t g c g c g a a c a t c g c a a a a c a c g t c g g a a t g a t t a a a c a a c a a
c c g a c g a a a c g c g c t t g t a g c g t t t t g t g c a g c c t t a c t a a t t t g t t g t t
```

```
a a c a c t g a g c t c c a t t c t t g a g c c t g g a t a a a g c g t t t c a g c g c c a a c a a
t t g t g a c t c g a g g t a a g a a c t c g g a c c t a t t t c g c a a a g t c g c g g t t g t t

                                                                                          33600
a a a c c c c t c t g g c g t t c a t g t c g c a t a a t g a a a a c a a t g t t c c c a a a t a t
t t t g g g g a g a c c g c a a g t a c a g c g t a t t a c t t t t g t t a c a a g g g t t t a t a

c c a g g a g g a a t a t c a a c t g c t a t g t g c a a a t a t a a a a g c a c a a c t c c a t g
g g t c c t c c t t a t a g t t g a c g a t a c a c g t t t a t a t t t t c g t g t t g a g g t a c

                                                                                          33700
t g g a g g t a t a a c a a a a t t c g c a g g a g a a a a t a a c a c a t a a g c a t t a g a g t
a c c t c c a t a t t g t t t t a a g c g t c c t c t t t t a t t g t g t a t t c g t a a t c t c a

c g c c c t c t t g t t t a g g c a a c a t a g c c c c a g g t c c c g t a a a a t a c a c a t a a
g c g g g a g a a c a a a t c c g t t g t a t c g g g g t c c a g g g c a t t t t a t g t g t a t t

                                                                                          33800
a g a g t c t c a a a a g c a g c c a t a a t g c c t t a c c a g a a a a a c a g t a c a a a g c c
t c t c a g a g t t t t c g t c g g t a t t a c g g a a t g g t c t t t t t g t c a t g t t t c g g

a g g c a c a g c a g a c a c a a t c t g c c g c a a g t g c g c a c c t t t a a t a c t g a a a a
t c c g t g t c g t c t g t g t t a g a c g g c g t t c a c g c g t g g a a a t t a t g a c t t t t

                                                                                          33900
a t a g t g a c g t a a a t g g c c a a a g t t c g c c t a c a c a a c a c a a a a a a a a c c c c
t a t c a c t g c a t t t a c c g g t t t c a a g c g g a t g t g t t g t g t t t t t t t t g g g g

a a a a g c c c g c g a a a a a a a t c a c t t c c g c a t a t g a c t c g g c a t a a t a c g g t
t t t t c g g g c g c t t t t t t t a g t g a a g g c g t a t a c t g a g c c g t a t t a t g c c a

                                                                                          34000
g t t c t c a c g a c a c g t c a c a t c c g g c g c g c c c g g c t c c c a c g c c g c g c c c c
c a a g a g t g c t g t g c a g t g t a g g c c g c g c g g g c c g a g g g t g c g g c g c g g g g
```

a c t t c c t c a t c c g c c c a a a c t t a c a a g c a c g c c a a a g c c a c a c c t c c a c c
t g a a g g a g t a g g c g g g t t t g a a t g t t c g t g c g g t t t c g g t g t g g a g g t g g

34100
c a a t c a a a t t a c a c a c t a c g c c c a c t t c a t t t t a a t a t t g g c a c t a g t c c
g t t a g t t t a a t g t g t g a t g c g g g t g a a g t a a a a t t a t a a c c g t g a t c a g g

a g t a t a a g g t a t a t t a t t a g a t a g g
t c a t a t t c c a t a t a a t a a t c t a t c c

Fig. 10. Nucleotide sequence of Ad12 DNA

References

Akusjärvi G, Zabielski J, Perricaudet M, Pettersson U (1981) The sequence of the 3′ non-coding region of the hexon mRNA discloses a novel adenovirus gene. Nucleic Acids Res 9: 1–17

Barton GJ (1993) ALSCRIPT, a tool to format multiple sequence alignments. Protein Eng 6: 37–40

Bucher P, Trifonov EN (1986) Compilation and analysis of eukaryotic POL II promoter sequences. Nucleic Acids Res 14: 10009–10026

Burger H, Doerfler W (1974) Intracellular forms of adenovirus DNA. III. Integration of the DNA of adenovirus type 2 into host DNA in productively infected cells. J Virol 13: 975–992

Cai F, Bourbonnière M, Tang D, Hu S-L, Weber JM (l990a) Nucleotide and deduced amino acid sequence of the bovine adenovirus type 3 proteinase. Nucleic Acids Res 18: 5568

Cai F, Tang D, Hu S-L, Weber JM (l990b) Nucleotide and deduced amino acid sequence of the bovine adenovirus type 7 proteinase. Nucleic Acids Res 18: 5567

Chroboczek J, Jacrot B (1987) The sequence of adenovirus fiber: similarities and differences between serotypes 2 and 5. Virology 161: 549–554

Chroboczek J, Bieber F, Jacrot B (1992) The sequence of the genome of adenovirus type 5 and its comparison with the genome of adenovirus type 2. Virology 186: 280–285

Davison AJ, Telford EAR, Watson MS, McBride K, Mautner V (1993) The DNA sequence of adenovirus type 40. J Mol Biol 234: 1308–1316

Deuring R, Doerfler W (1983) Proof of recombination between viral and cellular genomes in human KB cells productively infected by adenovirus type 12: structure of the junction sites in a symmetric recombinant (SYREC). Gene 26: 283–289

Deuring R, Klotz G, Doerfler W (1981) An unusual symmetric recombinant between adenovirus type 12 DNA and human cell DNA. Proc Natl Acad Sci USA 78: 3142–3146

Devereux J, Haeberli P, Smithies O (1984) A comprehensive set of sequence analysis programs for the VAX. Nucleic Acids Res 12: 387–395

Doerfler W (1968) The fate of the DNA of adenovirus type 12 in baby hamster kidney cells. Proc Natl Acad Aci USA 60: 636–643

Doerfler W (1970) Integration of the deoxyribonucleic acid of adenovirus type 12 into the deoxyribonucleic acid of baby hamster kidney cells. J Virol 6: 652–666

Doerfler W (1981) DNA methylation a regulatory signal in eukaryotic gene expression. J Gen Virol 57: 1–20

Doerfler W (1983) DNA methylation—and gene activity. Annu Rev Biochem 52: 93–124

Doerfler W (1991) The abortive infection and malignant transformation by adenoviruses: integration of viral DNA and control of viral gene expression by specific patterns of DNA methylation. Adv Virus Res 39: 89–128

Doerfler W (1992) Transformation of cells by adenoviruses: less frequently discussed mechanisms. In: Doerfler W, Bohm P (eds) Malignant transformation by DNA viruses. Molecular mechanisms. Weinheim, pp 87–109

Doerfler W (1993) Adenoviral DNA integration and changes in DNA methylation patterns: a different view of insertional mutagenesis. Prog Nucleic Acid Res Mol Biol 46: 1–36

Doerfler W (1995) The insertion of foreign DNA into mammalian genomes and its consequences: a concept for oncogenesis. Adv Cancer Res 66: 313–344

Dragulev BP, Sira S, Abouhaidar MG, Campbell JB (1991) Sequence analysis of putative E3 and fiber genomic regions of two strains of canine adenovirus type 1. Virology 183: 298–305

Fechteler K, Tatzelt J, Huppertz S, Wilgenbus P, Doerfler W (1995) On the mechanism of adenovirus DNA integration: studies in a cell-free system. Curr Top Microbiol Immunol 199/II: 109–137

Gosh D (1992) TFD: The transcription factors database. Nucleic Acids Res 20: 2091–2093

Hérissé J, Rigolet M, de Dinechin SD, Galibert F (1981) Nucleotide sequence of adenovirus 2 DNA fragment encoding for the carboxylic region of the fiber protein and the entire E4 region. Nucleic Acids Res 9: 4023–4042

Higgins DG, Bleasby AJ, Fuchs R (1992) CLUSTAL V: improved software for multiple sequence alignment. Comput Appl Biosci 8: 189–191

Hong JS, Mullis KG, Engler JA (1988) Characterization of the early region 3 and fiber genes of Ad7. Virology 167: 545–553

Houde A, Weber JM (1987) Sequence of the protease of human subgroup E adenovirus type 4. Gene 54: 51–56

Houde A, Weber JM (1988a) Sequence of the human adenovirus type 3 protease. Nucleic Acids Res 16: 11374

Houde A, Weber JM (1988b) The primary structure of human adenovirus type 12 protease. Nucleic Acids Res 16: 7195

Kidd AH, Erasmus MJ (1989) Sequence characterization of the adenovirus 40 fiber gene. Virology 172: 134–144

Kristensen T, Lopez R, Prydz H (1992) An estimate of the sequencing error frequency in the DNA sequence databases. DNA Sequence. J DNA Sequencing Mapping 2: 343–346

Kruijer W, Van Schaik FMA, Sussenbach JS (1980) Nucleotide sequence analysis of a region of adenovirus 5 DNA encoding a hitherto unidentified gene. Nucleic Acids Res 8: 6033–6042

Lopez R (1992) Database contamination. Nature 355: 211

Mautner V, Steinthorsdottir V, Bailey V (1995) Euteric adenoviruses. Curr Top Microbiol Immunol 199/III: 229–282

Pearson WR, Lipman DJ (1988) Improved tools for biological sequence comparison. Proc Natl Acad Sci USA 85: 2444–2448

Petterson M, Schaffner W (1987) A purine-rich DNA sequence motif present in SV40 and lymphotropic papovavirus binds a lymphoid-specific factor and contributes to enhancer activity in lymphoid cells. Genes Dev 1: 962–972

Pieniazek NJ, Slemenda SB, Pieniazek D, Velarde J Jr, Luftig RB (1989) Sequence of human enteric adenovirus type 41 Tak fiber protein gene. Nucleic Acids Res 17: 9474

Raviprakash KS, Grunhaus A, el Kholy MA, Horwitz MS (1989) The mouse adenovirus type 1 contains an unusual E3 region. J Virol 63: 5455–5458

Roberts RJ, Akusjärvi G, Aleström P, Gelinas RE, Gingeras T, Sciaky D, Pettersson U (1986) A consensus sequence for the adenovirus-2 genome. Dev Mol Virol 8: 1–51

Schick J, Baczko K, Fanning E, Groneberg J, Burger H, Doerfler W (1976) Intracellular forms of adenovirus DNA: Integrated form of adenovirus DNA appears early in productive infection. Proc Natl Acad Sci USA 73: 1043–1047

Signäs C, Akusjärvi G, Pettersson U (1985) Adenovirus 3 fiber polypeptide gene: implications for the structure of the fiber protein. J Virol 53: 672–678

Sprengel J, Schmitz B, Heuss-Neitzel D, Zock C, Doerfler W (1994) Nucleotide sequence of human adenovirus type 12 DNA: a comparative analysis. J Virol 68: 379–389

Sutter D, Doerfler W (1980) Methylation of integrated adenovirus type 12 DNA sequences in transformed cells is inversely correlated with viral gene expression. Proc Natl Acad Sci USA 77: 253–256

Sutter D, Westphal M, Doerfler W (1978) Patterns of integration of viral DNA sequences in the genomes of adenovirus type 12-transformed hamster cells. Cell 14: 569–585

Vos HL, van der Lee FM, Reemst AMCB, van Loon AD, Sussenbach JS (1988) The genes encoding the DNA binding protein and the 23K protease of adenovirus types 40 and 41. Virology 163: 1–10

Note Added Proof: Restriction enzyme data base (REBASE) release version 412 indicates that – in extension of what has been stated in Sprengel et al. 1994 – the following restriction endonucleases do not cut in the Ad12 DNA sequence:

ClaI, FseI, RsrII, SfiI, SrfI, SnaBI, Ssc8387I, SwaI

(Richard J. Roberts, New England Biolabs, Inc. Nov. 30, 1994).

Subject Index

Current Topics in Microbiology and Immunology

Volumes published since 1989 (and still available)

Vol. 157: **Swanstrom, Roland; Vogt, Peter K. (Ed.):** Retroviruses. Strategies of Replication. 1990. 40 figs. XII, 260 pp. ISBN 3-540-51895-9

Vol. 158: **Muzyczka, Nicholas (Ed.):** Viral Expression Vectors. 1992. 20 figs. IX, 176 pp. ISBN 3-540-52431-2

Vol. 159: **Gray, David; Sprent, Jonathan (Ed.):** Immunological Memory. 1990. 38 figs. XII, 156 pp. ISBN 3-540-51921-1

Vol. 160: **Oldstone, Michael B. A.; Koprowski, Hilary (Eds.):** Retrovirus Infections of the Nervous System. 1990. 16 figs. XII, 176 pp. ISBN 3-540-51939-4

Vol. 161: **Racaniello, Vincent R. (Ed.):** Picornaviruses. 1990. 12 figs. X, 194 pp. ISBN 3-540-52429-0

Vol. 162: **Roy, Polly; Gorman, Barry M. (Eds.):** Bluetongue Viruses. 1990. 37 figs. X, 200 pp. ISBN 3-540-51922-X

Vol. 163: **Turner, Peter C.; Moyer, Richard W. (Eds.):** Poxviruses. 1990. 23 figs. X, 210 pp. ISBN 3-540-52430-4

Vol. 164: **Bækkeskov, Steinnun; Hansen, Bruno (Eds.):** Human Diabetes. 1990. 9 figs. X, 198 pp. ISBN 3-540-52652-8

Vol. 165: **Bothwell, Mark (Ed.):** Neuronal Growth Factors. 1991. 14 figs. IX, 173 pp. ISBN 3-540-52654-4

Vol. 166: **Potter, Michael; Melchers, Fritz (Eds.):** Mechanisms in B-Cell Neoplasia. 1990. 143 figs. XIX, 380 pp. ISBN 3-540-52886-5

Vol. 167: **Kaufmann, Stefan H. E. (Ed.):** Heat Shock Proteins and Immune Response. 1991. 18 figs. IX, 214 pp. ISBN 3-540-52857-1

Vol. 168: **Mason, William S.; Seeger, Christoph (Eds.):** Hepadnaviruses. Molecular Biology and Pathogenesis. 1991. 21 figs. X, 206 pp. ISBN 3-540-53060-6

Vol. 169: **Kolakofsky, Daniel (Ed.):** Bunyaviridae. 1991. 34 figs. X, 256 pp. ISBN 3-540-53061-4

Vol. 170: **Compans, Richard W. (Ed.):** Protein Traffic in Eukaryotic Cells. Selected Reviews. 1991. 14 figs. X, 186 pp. ISBN 3-540-53631-0

Vol. 171: **Kung, Hsing-Jien; Vogt, Peter K. (Eds.):** Retroviral Insertion and Oncogene Activation. 1991. 18 figs. X, 179 pp. ISBN 3-540-53857-7

Vol. 172: **Chesebro, Bruce W. (Ed.):** Transmissible Spongiform Encephalopathies. 1991. 48 figs. X, 288 pp. ISBN 3-540-53883-6

Vol. 173: **Pfeffer, Klaus; Heeg, Klaus; Wagner, Hermann; Riethmüller, Gert (Eds.):** Function and Specificity of γ/δ TCells. 1991. 41 figs. XII, 296 pp. ISBN 3-540-53781-3

Vol. 174: **Fleischer, Bernhard; Sjögren, Hans Olov (Eds.):** Superantigens. 1991. 13 figs. IX, 137 pp. ISBN 3-540-54205-1

Vol. 175: **Aktories, Klaus (Ed.):** ADP-Ribosylating Toxins. 1992. 23 figs. IX, 148 pp. ISBN 3-540-54598-0

Vol. 176: **Holland, John J. (Ed.):** Genetic Diversity of RNA Viruses. 1992. 34 figs. IX, 226 pp. ISBN 3-540-54652-9

Vol. 177: **Müller-Sieburg, Christa; Torok-Storb, Beverly; Visser, Jan; Storb, Rainer (Eds.):** Hematopoietic Stem Cells. 1992. 18 figs. XIII, 143 pp. ISBN 3-540-54531-X

Vol. 178: **Parker, Charles J. (Ed.):** Membrane Defenses Against Attack by Complement and Perforins. 1992. 26 figs. VIII, 188 pp. ISBN 3-540-54653-7

Vol. 179: **Rouse, Barry T. (Ed.):** Herpes Simplex Virus. 1992. 9 figs. X, 180 pp. ISBN 3-540-55066-6

Vol. 180: **Sansonetti, P. J. (Ed.):** Pathogenesis of Shigellosis. 1992. 15 figs. X, 143 pp. ISBN 3-540-55058-5

Vol. 181: **Russell, Stephen W.; Gordon, Siamon (Eds.):** Macrophage Biology and Activation. 1992. 42 figs. IX, 299 pp. ISBN 3-540-55293-6

Vol. 182: **Potter, Michael; Melchers, Fritz (Eds.):** Mechanisms in B-Cell Neoplasia. 1992. 188 figs. XX, 499 pp. ISBN 3-540-55658-3

Vol. 183: **Dimmock, Nigel J.:** Neutralization of Animal Viruses. 1993. 10 figs. VII, 149 pp. ISBN 3-540-56030-0

Vol. 184: **Dunon, Dominique; Mackay, Charles R.; Imhof, Beat A. (Eds.):** Adhesion in Leukocyte Homing and Differentiation. 1993. 37 figs. IX, 260 pp. ISBN 3-540-56756-9

Vol. 185: **Ramig, Robert F. (Ed.):** Rotaviruses. 1994. 37 figs. X, 380 pp. ISBN 3-540-56761-5

Vol. 186: **zur Hausen, Harald (Ed.):** Human Pathogenic Papillomaviruses. 1994. 37 figs. XIII, 274 pp. ISBN 3-540-57193-0

Vol. 187: **Rupprecht, Charles E.; Dietzschold, Bernhard; Koprowski, Hilary (Eds.):** Lyssaviruses. 1994. 50 figs. IX, 352 pp. ISBN 3-540-57194-9

Vol. 188: **Letvin, Norman L.; Desrosiers, Ronald C. (Eds.):** Simian Immunodeficiency Virus. 1994. 37 figs. X, 240 pp. ISBN 3-540-57274-0

Vol. 189: **Oldstone, Michael B. A. (Ed.):** Cytotoxic T-Lymphocytes in Human Viral and Malaria Infections. 1994. 37 figs. IX, 210 pp. ISBN 3-540-57259-7

Vol. 190: **Koprowski, Hilary; Lipkin, W. Ian (Eds.):** Borna Disease. 1995. 33 figs. IX, 134 pp. ISBN 3-540-57388-7

Vol. 191: **ter Meulen, Volker; Billeter, Martin A. (Eds.):** Measles Virus. 1995. 23 figs. IX, 196 pp. ISBN 3-540-57389-5

Vol. 192: **Dangl, Jeffrey L. (Ed.):** Bacterial Pathogenesis of Plants and Animals. 1994. 41 figs. IX, 343 pp. ISBN 3-540-57391-7

Vol. 193: **Chen, Irvin S. Y.; Koprowski, Hilary; Srinivasan, Alagarsamy; Vogt, Peter K. (Eds.):** Transacting Functions of Human Retroviruses. 1995. 49 figs. IX, 240 pp. ISBN 3-540-57901-X

Vol. 194: **Potter, Michael; Melchers, Fritz (Eds.):** Mechanisms in B-cell Neoplasia. 1995. 152 figs. XXV, 458 pp. ISBN 3-540-58447-1

Vol. 195: **Montecucco, Cesare (Ed.):** Clostridial Neurotoxins. 1995. 28 figs. XI, 278 pp. ISBN 3-540-58452-8

Vol. 196: **Koprowski, Hilary; Maeda, Hiroshi (Eds.):** The Role of Nitric Oxide in Physiology and Pathophysiology. 1995. 21 figs. IX, 90 pp. ISBN 3-540-58214-2

Vol. 197: **Meyer, Peter (Ed.):** Gene Silencing in Higher Plants and Related Phenomena in Other Eukaryotes. 1995. 17 figs. IX, 232 pp. ISBN 3-540-58236-3

Vol. 198: **Griffiths, Gillian M.; Tschopp, Jürg (Eds.):** Pathways for Cytolysis. 1995. 45 figs. IX, 224 pp. ISBN 3-540-58725-X

Vol. 199/I: **Doerfler, Walter; Böhm, Petra (Eds.):** The Molecular Repertoire of Adenoviruses I. 1995. Approx. 51 figs. Approx. XIII, 296 pp. ISBN 3-540-58828-0

Vol. 199/III: **Doerfler, Walter; Böhm, Petra (Eds.):** The Molecular Repertoire of Adenoviruses III. 1995. Approx. 50 figs. Approx. XIII, 324 pp. ISBN 3-540-58987-2

Springer-Verlag and the Environment

MIX
Papier aus verantwortungsvollen Quellen
Paper from responsible sources
FSC® C105338

If you have any concerns about our products,
you can contact us on
ProductSafety@springernature.com

In case Publisher is established outside the EU,
the EU authorized representative is:
Springer Nature Customer Service Center GmbH
Europaplatz 3, 69115 Heidelberg, Germany

Printed by Libri Plureos GmbH
in Hamburg, Germany